BIODIVERSITY AND CONSERVATION

Characterization and Utilization of Plants, Microbes, and Natural Resources for Sustainable Development and Ecosystem Management

Current Advances in Biodiversity, Conservation, and Environmental Sciences

BIODIVERSITY AND CONSERVATION

Characterization and Utilization of Plants, Microbes, and Natural Resources for Sustainable Development and Ecosystem Management

Edited by

Jeyabalan Sangeetha, PhD
Devarajan Thangadurai, PhD
Hong Ching Goh, PhD
Saher Islam, MPhil

AAP APPLE ACADEMIC PRESS

Apple Academic Press Inc.
3333 Mistwell Crescent
Oakville, ON L6L 0A2 Canada

Apple Academic Press Inc.
1265 Goldenrod Circle NE
Palm Bay, Florida 32905 USA

© 2019 by Apple Academic Press, Inc.

First issued in paperback 2021

Exclusive worldwide distribution by CRC Press, a member of Taylor & Francis Group

No claim to original U.S. Government works

ISBN 13: 978-1-77463-445-5 (pbk)
ISBN 13: 978-1-77188-748-9 (hbk)

Library and Archives Canada Cataloguing in Publication

Title: Biodiversity and conservation : characterization and utilization of plants, microbes, and natural resources for sustainable development and ecosystem management / edited by Jeyabalan Sangeetha, PhD, Devarajan Thangadurai, PhD, Hong Ching Goh, PhD, Saher Islam, MPhil

Other titles: Biodiversity and conservation (Oakville, Ont.)

Names: Sangeetha, Jeyabalan, editor. | Thangadurai, D., editor. | Goh, Hong Ching, editor. | Islam, Saher, editor.

Description: Series statement: Current advances in biodiversity, conservation, and environmental sciences | Includes bibliographical references and index.

Identifiers: Canadiana (print) 20190071648 | Canadiana (ebook) 20190071710 | ISBN 9781771887489 (hardcover) | ISBN 9780429425790 (PDF)

Subjects: LCSH: Biodiversity. | LCSH: Nature conservation.

Classification: LCC QH541.15.B56 B56 2019 | DDC 333.95/16—dc23

Library of Congress Cataloging-in-Publication Data

Names: Sangeetha, Jeyabalan, editor. | Thangadurai, D., editor. | Goh, Hong Ching, editor. | Islam, Saher, editor.

Title: Biodiversity and conservation : characterization and utilization of plants, microbes, and natural resources for sustainable development and ecosystem management / editors: Jeyabalan Sangeetha, Devarajan Thangadurai, Hong Ching Goh, Saher Islam.

Description: Waretown, NJ : Apple Academic Press, 2019. | Includes bibliographical references and index.

Identifiers: LCCN 2019006903 (print) | LCCN 2019008372 (ebook) | ISBN 9780429425790 (ebook) | ISBN 9781771887489 (hardcover : alk. paper)

Subjects: LCSH: Biodiversity. | Nature conservation.

Classification: LCC QH541.15.B56 (ebook) | LCC QH541.15.B56 B5668 2019 (print) | DDC 333.95/16--dc23

LC record available at hhttps://lccn.loc.gov/2019006903

Apple Academic Press also publishes its books in a variety of electronic formats. Some content that appears in print may not be available in electronic format. For information about Apple Academic Press products, visit our website at **www.appleacademicpress.com** and the CRC Press website at **www.crcpress.com**

ABOUT THE EDITORS

Jeyabalan Sangeetha, PhD
Assistant Professor, Central University of Kerala, Kasaragod, South India

Jeyabalan Sangeetha, PhD, is an Assistant Professor at the Central University of Kerala at Kasaragod, South India. She earned her BSc in Microbiology and PhD in Environmental Science from Bharathidasan University, Tiruchirappalli, Tamil Nadu, India. She holds an MSc in Environmental Science from Bharathiar University, Coimbatore, Tamil Nadu, India. She is the recipient of a Tamil Nadu Government Scholarship and a Rajiv Gandhi National Fellowship of the University Grants Commission, Government of India for her doctoral studies. She served as a Dr. D.S. Kothari Postdoctoral Fellow and a UGC Postdoctoral Fellow at Karnatak University, Dharwad, South India during 2012–2016 with funding from the University Grants Commission, Government of India, New Delhi. Her research interests are in environmental toxicology, environmental microbiology, and environmental biotechnology, and her scientific/community leadership has included serving as editor of an international journal, *Acta Biololgica Indica*.

Devarajan Thangadurai, PhD
Senior Assistant Professor, Karnatak University, Dharwad, South India

Devarajan Thangadurai, PhD, is a Senior Assistant Professor at Karnatak University in South India and President of the International Society for Applied Biotechnology and General Secretary for the Association for the Advancement of Biodiversity Science. In addition, Dr. Thangadurai is Editor-in-Chief of two international journals, *Biotechnology, Bioinformatics, and Bioengineering* and *Acta Biologica Indica*. He received his PhD in Botany from Sri Krishnadevaraya University in South India. During 2002–2004, he worked as a CSIR Senior Research Fellow with funding from the Ministry of Science and Technology, Government of India. He served as a Postdoctoral Fellow at the University of Madeira, Portugal; University of Delhi, India, and ICAR National Research Centre for Banana, India. He is the recipient of the Best Young Scientist Award with a Gold Medal from Acharya Nagarjuna University, India, and the VGST-SMYSR Young Scientist Award of the Government of Karnataka, Republic of India. He has authored and edited 19 books including *Genetic Resources and Biotechnology* (3 vols.); *Genes, Genomes, and Genomics* (2 vols.); *Mycorrhizal Biotechnology, Genomics and Proteomics, Industrial Biotechnology*; and *Environmental Biotechnology*, with publishers of national and international reputation. He has also visited Bangladesh, China, Egypt, Georgia, Indonesia, Italy, Malaysia, Maldives, Myanmar, Nepal, Oman, Portugal, Russia, Sri Lanka, Thailand, Ukraine, United Arab Emirates, and Vietnam for academic work, scientific meetings, and international collaborations.

Hong Ching Goh, PhD
Senior Lecturer, Department of Urban and Regional Planning,
Faculty of Built Environment, Universiti Malaya, Malaysia

Hong Ching Goh, PhD is a Senior Lecturer at the Department of Urban and Regional Planning, Faculty of Built Environment, Universiti Malaya, Malaysia. She holds a Doctor of Natural Science (Geography) from Rheinische Friedrich-Wilhelms-Universitaet Bonn, Germany. She received a bachelor's degree in Urban and Regional Planning and an MSc in Tourism Planning from Universiti Teknologi Malaysia. She worked for five years in the real estate sector before pursuing her doctoral study when she received a three-year scholarship from the German Academic Exchange Services (DAAD) in 2004 and was attached to the Centre for Development Research (ZEF) in Bonn as a Junior Researcher. She is a Corporate Member of the Malaysia Institute of Planners (MIP) and a Registered Member of the Board of Town Planners Malaysia. She is also an active member of the Forum for Urban Future in Southeast Asia, a network of Southeast Asian and German Experts. She was a Visiting Scholar and participated in the 2014/2015 MIT-UTM Sustainable Cities Program. She is also a member of the Global Young Academy, of which her membership serves a five-year term (2015–2019). At present, she is a core member of the Landuse Governance Unit under the Spatial-Environmental Governance for Sustainability Research Centre (UMSERGE). She has two main research interests: tourism and urban studies. For the former, she focuses on tourism governance, tourism planning, and impact management in natural heritage and protected areas. For the latter, she focuses on natural resource and environmental issues and governance, sustainable development of cities, urbanization, and the related risks.

Saher Islam, MPhil
Higher Education Commission (HEC) Scholar, Islamic Republic of Pakistan at the University of Veterinary and Animal Sciences (UVAS), Lahore, Pakistan

Saher Islam, MPhil, is a PhD candidate in Molecular Biology and Biotechnology and a Higher Education Commission (HEC) Scholar of the Islamic Republic of Pakistan at the University of Veterinary and Animal Sciences (UVAS), Lahore, where she received her MPhil in Molecular Biology and her BS in Biotechnology and Bioinformatics. She worked as a Research Associate in a project funded by Grand Challenges Canada at UVAS. She served as an internee at the Pakistan Council of Scientific and Industrial Research (PCSIR), Lahore. She has visited United Kingdom, Singapore, Germany, United Arab Emirates, Egypt, Italy, Russia, and Maldives for training, courses, and meetings. She is the recipient of the 2016 Boehringer Ingelheim Fonds Travel Grant and the 2016 Travel Grant from the Wildlife Conservation Society. She has research interests in molecular biology, bioinformatics, genetics, and conservation biology.

ABOUT THE SERIES

CURRENT ADVANCES IN BIODIVERSITY, CONSERVATION AND ENVIRONMENTAL SCIENCES

SERIES EDITORS:

Jeyabalan Sangeetha, PhD
Assistant Professor, Central University of Kerala, Kasaragod, South India

Devarajan Thangadurai, PhD
Assistant Professor, Karnatak University, Dharwad, South India

Nature is that entity composed of three main internally interactive components: biodiversity–human–environment. These are triangularly related aspects of life science, and the interdependency of these three reflects the crucial need of maintenance of equilibrium in their interaction.Both human and biodiversity are two interactive phases of environment that provide a necessary platform for interaction. Being part of biodiversity, human life is almost dependent on biodiversity and its products. This is the point where the human–biodiversity interaction fluctuates due to over-exploitation of biodiversity, where humans take more than the basic needs of human life. Over the past few centuries, this fluctuation in interaction has caused dramatic depletion of biodiversity and thus a drastic change in environment. This has now boomeranged on human life significantly. This variation in the interaction triangle has already travelled the long path of time. Until now, conservationists and life scientists are thinking of restoring the equilibrium in the interaction triangle by taking innovative steps in conserving biodiversity and protecting the environment.

Conservation Biology and Environmental Science are the hotspots of research activities over the past several decades. This has caused much-needed development in observing, identifying, gathering, storing, and analyzing information, social awareness, planning, and advancement in conservation tools and techniques. Conservation biologists have given more importance to building stronger networks of specialists and experts in biology, agriculture, food technology, medicine, mathematics, engineering, architecture, robotics, information technology, and other branches of science and technology to gain innovative ideas to successfully treat the threats to biodiversity and the environment.

Gathering and monitoring information about biodiversity, conservation, and environment is a crucial need of biologists. Surveys from biological survey agencies, establishment of bio-inventories, modern tools and technologies to monitor the biodiversity and collection of data, and strengthening the scientific networks to make awareness, to generate the data, and for the accumulation of both traditional and scientific knowledge about biodiversity conservation and environmental management have been achieved in the recent decades. Conservationists need technological innovations to resolve the threats to biodiversity and environment that are now within reach.

The Current Advances in Biodiversity, Conservation and Environmental Sciences book series publishes titles on a regular basis for a better understanding of biodiversity conservation and environmental management to achieve sustainable development goals for future generations.

Titles in the Series to Date:
- Biodiversity and Conservation: Characterization and Utilization of Plants, Microbes, and Natural Resources for Sustainable Development and Ecosystem Management
- Beneficial Microbes for Sustainable Agriculture and Environmental Management

CONTENTS

CONTRIBUTORS

Eitimad Hshim Abdel-Rahman Ahmed
Biology Department, Hail University, Hail, 2400, Kingdom of Saudi Arabia

Maia Akhalkatsi
Department of Plant Genetic Resources, Institute of Botany, Ilia State University, Tbilisi, Georgia

Wan Nur Syazana Wan Mohamad Ariffin
Center for Sustainable Urban Planning and Real Estate, Faculty of Built Environment, University of Malaya, 50603, Kuala Lumpur, Malaysia

Tapan Kumar Barik
Post Graduate Department of Zoology, Berhampur University, Berhampur 760007, Odisha, India

Shivanand S. Bhat
Government Arts and Science College, Karwar, Karnataka, 581301, India

Sze-Siong Chew
Centre for Sustainable Urban Planning and Real Estate, Faculty of Built Environment, University of Malaya, 50603 Kuala Lumpur, Malaysia

Fei Ern Ching
University of Malaya Spatial-Environmental Governance for Sustainability Research, University of Malaya, 50603 Kuala Lumpur, Malaysia

Muniswamy David
Department of Zoology, Karnatak University, Dharwad 580003, Karnataka, India

Hong Ching Goh
Water Engineering and Spatial-Governance (WESERGE) Research Centre, University of Malaya, 50603 Kuala Lumpur, Malaysia; Department of Urban and Regional Planning, Faculty of Built Environment, University of Malaya, 50603 Kuala Lumpur, Malaysia

Ravichandra Hospet
Department of Botany, Karnatak University, Dharwad 580003, Karnataka, India

Abhishek Mundaragi
Department of Botany, Karnatak University, Dharwad 580003, Karnataka, India

Abdelrahman Eltahir Ahmed Musa
Biology Department, Khartoum University, Khartoum, 11115, Sudan

Azizkhani Negin
Department of Biotechnology, Institute of Science, High Technology and Environmental Sciences, Graduate University of Advanced Technology, Kerman, 7631133131, Iran

Maršalkienė Nijolė
Institute of Environment and Ecology, Aleksandras Stulginskis University, Institute of Environment and Ecology, Studentu Str.11, LT–53361 Akademija, Kaunas, Lithuania

Wongsakorn Phongsopitanun
Department of Biochemistry and Microbiology, Faculty of Pharmaceutical Sciences, Chulalongkorn University, Bangkok 10330, Thailand

Purushotham Prathima
Department of Botany, Karnatak University, Dharwad 580003, Karnataka, India

L. Rajanna
Department of Botany, Bangalore University, Bengaluru, Karnataka, 560056, India

Mirzaei Saeid
Department of Biotechnology, Institute of Science, High Technology and Environmental Sciences, Graduate University of Advanced Technology, Kerman, 7631133131, Iran

Jeyabalan Sangeetha
Department of Environmental Science, Central University of Kerala, Periye, Kasaragod – 671316, Kerala, India

Abul Khayer Mohammad Golam Sarwar
Laboratory of Plant Systematics, Department of Crop Botany, Bangladesh Agricultural University, Mymensingh 2202, Bangladesh

Jaya Kishor Seth
Post-Graduate Department of Zoology, Berhampur University, Berhampur 760007, Odisha, India

Shafiquzzaman Siddiquee
Biotechnology Research Institute, University Malaysia Sabah, Jalan UMS, 88400 Kota Kinabalu, Sabah, Malaysia

Maxim Steffi Simmi
Department of Environmental Science, Central University of Kerala, Periye, Kasaragod – 671316, Kerala, India

Joko Sulistyo Soetikno
Faculty of Food Science and Nutrition, University Malaysia Sabah, Kota Kinabalu 88400, Sabah, Malaysia

Panneri Sreeshma
Department of Environmental Science, Central University of Kerala, Periye, Kasaragod 371316, Kerala, India

Somboon Tanasupawat
Department of Biochemistry and Microbiology, Faculty of Pharmaceutical Sciences, Chulalongkorn University, Bangkok 10330, Thailand

Devarajan Thangadurai
Department of Botany, Karnatak University, Dharwad 580003, Karnataka, India

Natalia Togonidze
Department of Plant Genetic Resources, Institute of Botany, Ilia State University, Tbilisi, Georgia

ACRONYMS AND ABBREVIATIONS

°C	degree celsius
ACC	1-aminocyclopropane-1-carboxylic acid
ACCD	1-aminocyclopropane-1-carboxylic acid deaminase
AFLP	amplified fragment length polymorphism
APCC	Asian Pacific Coconut Community
ASU	Aleksandras Stulginskis University
BARI	Bangladesh Agricultural Research Institute
BAU	Bangladesh Agricultural University
BFRI	Bangladesh Forest Research Institute
BGCI	Botanic Gardens Conservation International
BLAST	Basic Local Alignment Search Tool
BNF	biological nitrogen fixation
bp	base pairs
BRRI	Bangladesh Rice Research Institute
BTRI	Bangladesh Tea Research Institute
Ca	calcium
CAMPs	conservation assessment and management plans
CBA	cost-benefit analysis
CBD	Convention on Biological Diversity
CBM	Community Biodiversity Management
CBSG	Captive Breeding Specialist Group
CGIAR	Consultative Group on International Agricultural Research
CH_4	methane
CHD	coronary heart diseases
Cl	chlorine
CITES	Convention on International Trade in Endangered Species
CMV	cytomegalovirus
CN	Cetane number
CO_2	carbon dioxide
CU	Chittagong University
DAP	diaminopimelic acid
DDC	dispersion and differential centrifugation
DHF	4-dihydroxyflavone

DoE   Department of Environment
DoF   Department of Forest
DWNP   Department of Wildlife and National Parks
EC   European Community
ECAs   ecologically critical areas
EIA   environmental impact assessment
FAs   fatty acids
FAO   Food and Agriculture Organization of the United Nations
FFA   free fatty acid
FRIPGC   Fruit Tree Improvement Project Germplasm Centre
GCV   genotypic coefficient of variation
GDP   gross domestic product
GED   global ecological domains
GEZs   global ecological zones
GFW   gram of fresh weight
GSPC   global strategy for plant conservation
ha   hectare
HDL   high-density lipoprotein
HQ   headquarters
HTC   hydrothermal coefficient
IAA   indole-3-acetic acid
IARC   International Agricultural Research Centers
IBPGR   International Board for Plant Genetic Resources
ICDP   Integrated Conservation and Development Program
IgM   immunoglobulin M
IITA   International Institute of Tropical Agriculture
IOM   Institute of Medicine
IPBES   Intergovernmental Platform on Biodiversity and Ecosystem Services
ISP   International *Streptomyces* Project
ISSC-MAP   International Standard for Sustainable Wild Collection of Medicinal and Aromatic Plants
IUCN   International Union for Conservation of Nature and National Resources
JHEOA   Jabatan Hal Ehwal Orang Asli/Department of Orang Asli Development
K   potassium

KOKTAS	Koperasi Serbaguna Kakitangan Taman-Taman Sabah/ Multipurpose Cooperative of Sabah Parks Staff
LAAO	L-amino acid oxidase
LC/MS	liquid chromatography/mass spectrometry
LCFA	long-chain fatty acids
LDL	low-density lipoprotein
LLA	low linolenic acid
MAI	mean annual increment
MAR	millennium assessment report
MARDI	Malaysian Agricultural Research and Development Institute
MCA	multi-criteria analysis
MCFA	medium-chain fatty acid
MCT	medium chain triglycerides
MDGs	millennium development goals
MEA	millennium ecosystem assessment
MFF	mangrove for the future
Mg	magnesium
MGs	monoglycerides
MoEF	Ministry of Environment and Forest
MOPS	morpholinepropanesulfonic acid
MOW	mixed office wastepaper
MSTE	Ministry of Science, Technology, and the Environment
MUFA	monounsaturated fatty acid
N_2	nitrogen
Na	sodium
NACGRAB	National Centre for Genetic Resources and Biotechnology
NaOH	sodium hydroxide (caustic soda)
NBDB	National Bioresource Development Board
NBPGR	National Bureau of Plant Genetic Resources
NBSAP	National Biodiversity Strategy and Action Plan
NCSAPs	National Conservation Strategy and Action Plans
NFPTCR	National Facility for Plant Tissue Culture Repository
NH_3	ammonia
NO_3^-	nitrate
N_2O	nitrous oxide
P	phosphorus
PA	protected areas

PAH	polycyclic aromatic hydrocarbon
PGE–2	prostaglandin E-2
PGR	plant genetic resources
POPs	persistent organic pollutants
PUFA	polyunsaturated fatty acid
QTL	quantitative trait loci
RAPD	randomly amplified polymorphic DNA
RIL	recombinant inbred line
S	sulfur
SANParks	South African National Parks
SARS	severe acute respiratory syndrome
SBSTTA	subsidiary body on scientific, technological, and technical advice
SEA	strategic environmental assessment
SFA	saturated fatty acid
SFO	safflower oil
SFO-CNO	safflower oil-coconut oil
SNA	System of National Accounts
SOC	soil organic carbon
SPSS	Statistical Package for the Social Sciences
SSL	Sutera Sanctuary Lodge
SSP	seed storage protein
TEF 1-α	translational elongation factor 1-α
TFA	trans fatty acids
TNF-α	tumor necrosis factor-alpha
UNDP	United Nations Development Programme
UNEP	United Nations Environment Programme
UNESCO	United Nations Educational, Scientific, and Cultural Organization
VCO	virgin coconut oil
WCPA	World Commission on Protected Areas
WHC	World Heritage Centre
WUE	water use efficiency
WWF	World Wildlife Fund of the United Nations

PREFACE

Sustainable utilization of biodiversity resources is the key objective of the "Convention on Biological Diversity" program. It is necessary to conserve natural biodiversity by sound conservation strategies and proper utilization of its components for future generation. Towards this end, we have edited this book, Biodiversity and Conservation: Characterization and Utilization of Plants, Microbes, and Natural Resources for Sustainable Development and Ecosystem Management.

The ultimate goal of this book is to provide a pragmatic approach to preserve biological diversity. This book gathers a wide range of peer-reviewed scientific content from biodiversity researchers and conservators around the world. Collectively this book provides comprehensive knowledge and details about the present conservation status of a wide range of biological diversity that includes floral, faunal, and microbial diversity. It provides a detailed account of recent trends in conservation applications that focus mainly on agriculturally and industrially important microbes and their sustainable utilization that further enables specialist researchers to understand the potential strategies for conservation of biodiversity under changing climate conditions.

This book has been well written in simple and clear language. Tables and figures are also included in this book to make content easy to understand by the readers at any level. This book will serve as good reference material for students, scientists, researchers, and academicians working in the field of biodiversity conservation.

Chapter 1 discusses in detail biodiversity concepts, benefits, and values for economic and sustainable development. The second and third chapters extensively review the value of biodiversity in the legume family, and diversity and conservation status of spontaneous *Carum*, *Vicia*, and *Lathyrus* species in Lithuania. Plant diversity in rural home gardens of coastal taluks of Uttara Kannada in Karnataka, India has been discussed in a more informative way in Chapter 4. Chapter 5 reviews the *Primula* species as indicators of forest habitat diversity in Georgia, South Caucasus. Prospecting of coconut biodiversity for safe, healthy, and sustainable edible oil has well been explained in Chapter 6.

Chapters 7 and 8 deal extensively with diversity, metabolites, and applications of *Actinomycetes* and *Trichoderma*. Biodiversity and conservation of higher plants in Bangladesh, *in situ* and *ex situ* conservation of biodiversity in India, understanding the process of evolution, and the future of biodiversity under changing climate with special reference to infectious diseases have been discussed comprehensively in Chapter 9 to 11. Chapter 12 gives a detailed account on evolving park values among the local people in Taman Negara, and Chapter 13 discusses nature conservation in a Malaysian first-world heritage site through its recreational and tourism activities. Public awareness towards ecological services and cultural significance of mangroves in Southern Peninsular Malaysia has been well illustrated in Chapter 14.

The chapters in this book were well prepared and reviewed by many subject experts in the respective fields.

We wish to express our sincere thanks to all the authors for their valuable contributions, extraordinary revisions, and further assistance to finish this book successfully.

We also wish to thank Sandy Jones Sickels, Vice President, and Ashish Kumar, Publisher, and President, Apple Academic Press, Inc., USA for quality production and timely publication of this book.

—Jeyabalan Sangeetha, PhD
Devarajan Thangadurai, PhD
Hong Ching Goh, PhD
Saher Islam, MPhil

CHAPTER 1

BIODIVERSITY: CONCEPTS, BENEFITS, AND VALUES FOR ECONOMIC AND SUSTAINABLE DEVELOPMENT

EITIMAD HSHIM ABDEL-RAHMAN AHMED[1] and
ABDELRAHMAN ELTAHIR AHMED MUSA[2]

[1]*Biology Department, Hail University, Hail, 2400, Kingdom of Saudi Arabia*

[2]*Biology Department, Khartoum University, Khartoum, 11115, Sudan*

1.1 CONCEPTS OF BIODIVERSITY

The word "biodiversity" is a synonym for biodiversity around us, i.e., in biological systems the sum of the total of life forms at all levels of organization (Lovejoy, 1980; Norse and McManus, 1980; Wilson, 1988; Heywood, 1995). Through this, we can have an idea about the variation or variability among living organisms and of the systems of which they are part. Thus, this covers the total range of variation and variability among systems and organisms at the bioregional, landscape, ecosystem, habitat, and organismal level down to species, molecules, and genes (Heywood, 1995).

In the early 1980s, the term 'biological diversity' began to use. Lovejoy (1980) first attributed the term whoever did not provide any formal definition of biodiversity, but only focus on species richness (number of different species in a location/sample). In the same year Norse and McManus (1980), included ecological and genetic diversity. After that, Rosen used the term "biodiversity" during the first planning conference of the "National Forum on Biodiversity" at Washington, DC, on September

1986. Edward O. Wilson (1988, the entomologist) edited the proceedings of the conference titled Biodiversity, and thus popularized the concept of biodiversity as "the variety of life at every hierarchical level and spatial scale of biological organization: genes within populations, populations within spices, species within communities, communities within landscape within biomes, and biomes within the biosphere."

At the Earth Summit in Rio de Janeiro on 5 June 1992, the Convention on Biological Diversity (CBD) was opened for signature where most of the nations of the world constituted a historical commitment. The CBD provides a universal legal framework to work on biological diversity with the main goal to develop national strategies for the management and sustainable use of biological variability (CBD, 2010a). Thus, biodiversity comprehensively addressed in this global community. For the human welfare, at the same time, the genetic diversity was considered, and conservation of biodiversity was accepted as the common concern. At this conference biodiversity as defined as "the variability among living organisms from all sources, including, inter alia, terrestrial, marine, and other aquatic ecosystems and the ecological complexes of which they are part: this includes biodiversity within species, between species and ecosystems." Then, the Millennium Assessment Report (MAR) in 2005 connecting ecosystems and their services with environmental and social security for stimulating and guided action to conserve ecosystems and enhance their contribution to human well-being. Furthermore, at the 10th Conference in Nagoya, Japan (October 2010) the Parties adopted a new "Strategic Plan for Biodiversity 2011–2020 along with its 20 Aichi targets" (CBD, 2010b,c).

The Strategic Plan for Biodiversity 2011–2020 is now accepted and designated the period 2011–2020 by the United Nations (CBD, 2010b). The Aichi Biodiversity Targets (CBD, 2010c), mostly with an endpoint of 2020, and set out 20 challenging target under 5 strategic goals to stimulate the effective and urgent action towards the long-term vision of a world without biodiversity loss or degradation of ecosystems to provide essential services, thus can provide security to the planet's variety of life, and contributing to human well-being, and poverty eradication.

The goals and targets (CBD, 2010b,c) comprise both aspirations for achievement at the global level and a flexible framework for the establishment of national or regional targets of the Intergovernmental Platform on Biodiversity and Ecosystem Services (IPBES) that is going to play an important role at the science-policy interface in the future. Biodiversity, therefore, considered to be discussed at three hierarchical levels, genetic diversity, species diversity, and ecosystem diversity.

1.1.1 GENETIC DIVERSITY

The amount of genetic variation within a species that providing pheno-typic variation due to the variation in the nucleotides, genes, chromosomes within and among populations of the same species (Hamrick and Godt, 1996) is termed as genetic biodiversity. Gene is the fundamental unit of natural selection, that is why the genetic diversity considered to be the real biodiversity "fundamental currency of diversity" (Williams and Humphries, 1996). Every species need in order to maintain a good repro-ductive capacity, resistance to disease and the ability to adapt to successful survival (Kohler-Rollefson, 1997). As well as, environmental variability often increases with genetic diversity within a species (Tilman, 1999). Ultimately, this resides variations in the sequence of the four base-pairs, which are components of nucleic acids, constitute the genetic code, i.e., the key for species genetic makeup, which reflect the living organisms' ability to reproduce or to deal effectively with environmental factors such as disease, weather, predation, etc. The knowledge of the genetic struc-tures of one species, varieties regarding the amount of genetic variation is essential for a health breading population of the same species (Kohler-Rollefson, 1997). Furthermore, genetic diversity gains greater access by modern developments in biotechnology and the continuing expansion of global trade. Genetic diversity also provides biological significance to crop and livestock breeding, medicines, and so on (Bernstein and Chivian, 2008). In agriculture, genetic resources comprises of the traditional resources (wild types and the older domesticated landraces) together with modern cultivars which is vital in supporting the improvement of breeding programs for crop plants, farm animals, fisheries, and aquaculture, with a wide range of objectives for increasing yield, resistance to disease optimization of nutritional value, and adaptation to the local environment and climate change. Genetic diversity of crops increases production and decreases susceptibility to pests and climate variation (Myers, 1991), the CBD, sometimes also called the "biotrade convention", as they encourage member countries to facilitate access to genetic recourses and take measures to ensure the fair and equitable sharing of benefits derived from the use of these resources.

1.1.2 SPECIES DIVERSITY

The species biodiversity can be defined as the variation in species that may be linked as biotic communities, for examples birds, reptiles, mammals, amphibian, and so on. Species diversity increases the environmental heterogeneity (Stevens et al., 2011). The living world is considered in term of "species richness," which denotes the number of species in a site or habitat that commonly used as a synonym of species diversity (Tuomisto, 2010). Some species have disproportionate influences on the characteristics of an ecosystem (Davic, 2003). For example, keystone species, whose loss would transform or undermine the ecological processes that fundamentally change the community' species composition (Davic, 2003). Therefore, a large number of plant species reflects a greater variety of corps that will ensure natural sustainability for all life forms. The diversity of species found on Earth has increased since the origin of life of about over 600 million years ago (Gould, 1993). Although the way species are defined has different viewpoints that differ between groups and between taxonomists (Robson, 1928), but measurement of species richness and distribution are important tools in assessing the state of the environment as well as the direction and speed of change (MAR, 2005).

1.1.3 ECOSYSTEM DIVERSITY

Chapin et al. (2002) and Diaz et al. (2006) defined ecosystem diversity as it describes the variety of different natural systems found in a region and is delimited and characterized by the variations in the plants and the animals found among biological communities and the physical landscapes that support them. According to Chakhaiyar (2010), there are different ecosystems on the planet, such as deserts, rainforests, and coral reefs are all part of Earth biologically diverse. The quantitative assessment of the diversity of the ecosystem, habitat or community levels remains problematic. The scientists (Whittaker, 1972; Chapin et al., 2002; Diaz et al., 2006; Chakhaiyar, 2010) suggested that it is possible to define what is in principle meant by genetic and species diversity, and to produce various measures. Therefore, there is no unique definition and classification of ecosystems at the global level (Chapin et al., 2002; Diaz et al., 2006). The interactions among the living organisms (animals, plants, and microbial resources) of the world with the physical environment to produce a wide variety of ecosystems may be described as a range of temporal and spatial scales. Three terms for measuring biodiversity over spatial scales: alpha (within-community diversity), beta (between-community diversity), and gamma (diversity of the habitats over the total landscape or geographical area diversity) is described by Whittaker (1972).

1.2   DISTRIBUTION OF BIODIVERSITY

For nature and development, understanding the global distribution of biodiversity is one of the most significant objectives for ecologists (Benton, 2001). World ecosystem biodiversity is divided into terrestrial biodiversity (land biomes/ecosystem) and aquatic biodiversity (water biomes/ecosystem). A biome can be described as an ecological community of organisms associated with particular climatic and geographic conditions that may often contain many ecosystems (Odum, 1971; De Blij et al., 2010). The distribution of life on Earth varies greatly across the global as well as within regions, providing different patterns (i.e., the latitudinal gradient biodiversity and the hotspots biodiversity regions (Benton, 2001). Generally, there are two types of biodiversity: terrestrial biodiversity and aquatic biodiversity.

1.2.1   TERRESTRIAL BIODIVERSITY

As climate is a major factor in controlling which living organisms survive, biodiversity biomes are distributed across the Earth-based primarily on the climate (De Blij et al., 2010). Therefore, tropical areas play a central point in the understanding of the distribution of biodiversity (Heywood, 1995; De Blij et al., 2010). However, tropical areas that occur between the Tropic of Capricorn and the Tropic of Cancer are renowned for housing the most biologically diverse ecosystems on the planet. The majority of the biodiversity tends to be close to the equator, due to be the warm climate and high productivity (Mora and Robertson, 2005; De Blij et al., 2010). The diversity of all creatures depend on some factors, for example, temperature, altitude, precipitation, geography, and the presence of other species. Ecologists are not certain about the existence of latitudinal gradient (Willig et al., 2003; Cardillo et al., 2005), possibly due to the greater mean temperature at the equator compared to that of the poles. Nevertheless, species diversity for most taxa is lowest near the poles, and increases toward the tropics, reaching a peak in tropical rain forests (may contain more than half the species on Earth). This is one of the recognized patterns in ecology and called the latitudinal gradient in biological diversity (Willig et al., 2003; Cardillo et al., 2005; Mora and Robertson, 2005).

Biodiversity is constantly changing, i.e., terrestrial biodiversity is up to 25 times greater than ocean biodiversity (Benton, 2001). Although a recently discovered method puts the total number of species on Earth at 8.7 million, of which 2.1 million were estimated to live in the ocean (Mora et al., 2011). According to Food and Agriculture Organization (FAO, 2010) reported that forests, occupying about 4 billion hectares, about 31% of the earth's land surface, are believed to contain more than half the species on earth (Corlett and Primack, 2011; Mora et al., 2011).

Much of the diversity of tropical rainforests is due to the great abundance of insects; as well as, tropical forest harbor more species of birds, mammals, and plants than the most temperate areas. Furthermore, tropical areas receive intense solar energy, due to the consistent day length, resulting in warm temperatures throughout the year with minimal variation and receive large amounts of rainfall evenly distributed annually. Both of these characteristics attributes to tropical rainforest stable climate (Corlett and Primack, 2011). The tropical forest canopy structure believed to provide an array of new niches and are home to more than 50% of all the plants and animals. According to Douros and Suffness (1980), most of the plants that supply 90% of the world's food today are from the tropics, and at least 1400 plant species in tropical forests are believed to contain anticancer chemicals.

Some regions of the world that are considered to be "hotspots" for biodiversity based on the density of species. Hotspots were defined by British Ecologist Norman Myers (1988) as locations of high biodiversity that were threatened with destruction. Mittermeier et al. (1998) qualify a region as a hotspot when it contains at least 1,500 species of rare (endemic) vascular plants (i.e., > 0.5% of the world's total) and must have lost at least 70% of its original habitat. Although plants were used as the principal indicators for biological diversity, patterns of endemism and diversity were also examined for animals such as mammals, birds, reptiles, and amphibians (Reid, 1998).

Therefore, regions with a high human population tend to have the lowest number of species, the hotspot regions with many endemic species that are usually found in areas with limited human impacts (Mittermeier et al., 1998; Brooks et al., 2001; Mittermeier et al., 2011). Mostly, the hotspots exist in the tropical rain forests (Myers et al., 2000; Mittermeier et al., 2011). Earlier 25 biodiversity hotspots have been identified first, which covered most diverse terrestrial realms. After that 10 hotspots have been added, therefore resulting in a total of 35 hotspots (Myers et al., 2000; Mittermeier et al., 2011). These hotspots covering less than 2% of the world's land area and are found to have about 50% of the world's endemic plant species and 42% of the world's endemic terrestrial vertebrates (Mittermeier et al., 1998).

1.2.2 AQUATIC BIODIVERSITY

According to Elmqvist et al. (2010), aquatic ecosystem includes marine, fresh, and other aquatic habitats. Water and aquatic zones cover about the three-fourths of the earth's surface, and oceans dominate the biosphere. Temperature, dissolved oxygen content, availability of light, and nutrients necessary for photosynthesis are considered as the key factors determining biodiversity in aquatic systems. The earth's aquatic systems provide important ecological and economic services (Elmqvist et al., 2010). Ecological services include climate moderation, carbon dioxide absorption, nutrient cycling, reduced storm impact (mangrove swamps, estuaries, barrier islands), habitats, and nurseries for species (shrimp, crab, oysters, clams, fish), genetic resources, biodiversity, and so on. Beside economic services such as food, pharmaceuticals, transportation, coastal habitats and employment for human, recreation, and so on.

The major importance of marine biodiversity is in climate regulation, including biogeochemical cycling and carbon sequestration (MAR, 2005). The ocean links to the terrestrial biosphere play a huge role in the cycling of almost every material involved in biotic processes. Biodiversity influences the effectiveness of the biological pump that moves carbon from the surface ocean and sequesters it in deep waters and sediments. Some of the carbon that is absorbed by marine photosynthesis and transferred through food webs to grazers sinks to the deep ocean as fecal pellets and dead cells.

The efficiency of carbon sequestration is sensitive to the species richness and composition of the planktons community. Furthermore, key interactions in marine ecosystems can also influence ecosystem processes and the provisioning of ecological services. Certain ecosystems (coral reefs, the deep sea, and large tropical lakes) and regions contain more species than others (Mulhall, 2009).

Regarding marine ecosystem, 11 marine hotspots had been defined with hotspot characteristics in tropical coral reefs, but research is still underway (Parravicini et al., 2014). Besides, many marine hotspots extend from terrestrial hotspots; therefore an extension of present hotspots would be appropriate. Sorokin (1993) stated that marine biodiversity tends to be highest along the coasts in the Western Pacific, where sea surface temperature is highest and in a mid-latitudinal band in all oceans. Parravicini et al. (2014) defined 11 marine ecosystem hotspots that were represented by Philippines, Sundaland, Wallace, Gulf of Guinea, Southern Mascarene Island, Eastern Caribbean, Red Sea, and the Gulf of Aden. The ecosystem and market services provided by aquatic biomes had been threatened and disrupted by human activities.

1.3   BENEFITS AND VALUES OF BIODIVERSITY

The value of biodiversity can be separated into two categories: anthropocentric and ethical values (Dasgupta and Perrings, 2012). Direct and indirect economic benefits to humans that include the services of ecosystems, regulations on climate by biodiversity, generation of moisture and oxygen by plants and animals, the formation of the soil and improvement of fertility, detoxification of wastes by organisms and aesthetic and recreational benefits are considered as anthropocentric value. Aesthetic/moral and ethical/spiritual thought are composed of ethical value.

According to Elmqvist et al. (2010), human life depends on the products of living organisms worldwide, including the production of goods (i.e., food); life-support processes (i.e., water purification and storage); life-fulfilling conditions (i.e., recreation), in addition to the other values like timber, fiber, rubber, silk, wax, resin, and decorative items. Dasgupta and Perrings (2012) suggested that biodiversity has paramount importance for the social, cultural, and economic development of humankind that meets basic needs of food, shelters, and clothing by different disciplines and thought. Some parameters of the ecosystem influence human society are air quality, climate (both global and local), water purification, disease control, biological pest control, pollination, and prevention of erosion. Biodiversity is directly involved in water purification, recycling nutrients and providing fertile soils. There is a wide range from the anthropocentric benefits of biodiversity in the areas of agriculture, science, medicine, industrial materials, and ecological service (Bernstein and Chivian, 2008). There are two main types of biodiversity: direct values and indirect values on economic values, while option values are the non-use values (Elmqvist et al., 2010; Dasgupta and Perrings, 2012).

1.3.1 DIRECT VALUES

Direct values are considered to be related with the enjoyment or satisfaction received directly by biological resources, and it includes two types: consumptive use value (a non-market value that do not appear in the national and international market) and productive use value (commercial value sold in the national and international market).

The consumptive use value can be harvested directly and measure its outputs as it consumed locally, from plant sources such as crops that constitute more than two-third of the food requirements all over the world; animal sources such as fish products that have the largest source of protein in the world. Therefore, the value of nature's products that are consumed directly such as fire woods, fodder, and meat are known as consumptive value. In other words, the commodities which are consumed directly without passing through the national economy but appear in gross domestic accounts. As some wild resources can prosper through the sustainable use by the local communities. Direct use and/or consumption such as collection of berries, herbs, plants, and animals all have consumptive use values. By assigning prices, the productive use easily observed and measured (benefits/services) for its commercial value to them such as textile, leather, and paper industry. Besides the products that are commercially harvested for informal exchange markets, is known as productive use. Each species is valuable to humans (Costanza, 2000). According to MAR (2005), the global collection of biological resources provide benefits for human needs. Humans are part of the biological system, and utilize the natural resources goods and services to improve life welfare and survival. Thus this often considered the only value of biological resources that is reflected in the income accounts (Elmqvist et al., 2010).

1.3.2   INDIRECT VALUES

Costanza (2000) stated that biodiversity provides indirect benefits to human well-being which support the existence of biological life and other benefits which are difficult to quantify. It deals primarily with the natural resources and the biological diversity at the limited scale (i.e., on the local society) or at large scale (i.e., ecosystem) rather than to individuals or corporate species. Direct values often derive from indirect values that include non-consumptive values (i.e., environmental service values), social values, ethical values, besides the option values (Barbier, 1994).

1.3.3   NON-USE VALUES

Some uses or values associated with biodiversity are unmeasurable. These may include unknown genetic material, cultural or religious values. In the case of some types of biodiversity (e.g., blue whale) non-use values account for almost all of the economic value measured in a TEV calculation (Costanza, 2000).

1.4   BIODIVERSITY IN NATURE: FUNCTIONS AND SERVICES

The principal framework (Figure 1.1) is expressing the "usefulness" of biodiversity that based on the concept of ecosystem services and functions. The functions of the ecosystem are grouped into regulation, habitat, production, and information functions hat linking ecosystems to human well-being by the service-providing functions (Helliwell, 1969; de Groot et al., 2002; Howarth and Farber, 2002; Barbier et al., 2008; Elmqvist et al., 2010).

The benefits people obtain from ecosystems are known as ecosystem services (Díaz et al., 2006; Elmqvist et al., 2010). According to MAR (2005), it includes provisioning services (such as food, wood, fiber production, air, water, and fuel); regulating services (such as pest and disease control, climate regulation, water purification); cultural services (such as spiritual, aesthetic, educational, recreational benefits and a unique setting for social and cultural activities); and supporting services (such as nutrient cycling, that maintain the conditions for life on Earth, soil formation and

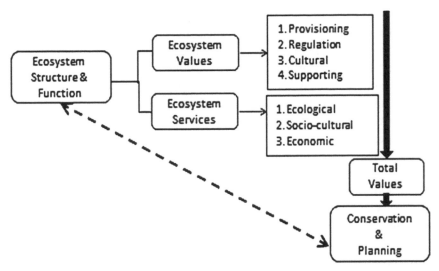

FIGURE 1.1   The principal framework of ecosystem structure and function (modified from de Groot et al., 2002).

primary production), where the so-called supporting services are regarded as the basis for the services of the other three categories, which require policy intervention with regard to the incorporation of ecosystem services into markets and payment schemes (de Groot et al., 2002).

For the survival of human being the biosphere gives the vital life support. Boumans et al. (2002) stated that biodiversity ecological services mean ecological effects of biodiversity that obtained for the flow of energy and nutrients network that affect, for example, air quality, climate, erosion, pest, and natural hazard regulations. Beside water purification and waste treatment and pollination. These ecological effects of biodiversity, in turn, affect, for example, both global warming through enhanced greenhouse gases, resource degradation, and productivity. They illustrate the link between, on the one hand, the interactions of living organisms (plants, animals, and microbes) with each other and with the nonliving components of the environment (things like air, water, and mineral soil); and on the other, the well-being of people, whether in terms of productivity, and/or stability. These living and non-living components are linked together through elements cycles and energy flows (Odum, 1971). Thus, ecosystem processes are driven by the biotic components, the density, richness, evenness of each individual species, abiotic components, and the interaction between them (Chapin et al., 2002). Therefore, biodiversity plays a vital and radical role in ecosystem resources, services, and functions (de Groot et al., 2002; Martin-Lopez et al., 2009).

All aspects of ecosystem productivity and stability dependent on biodiversity and ecosystem functions and services (Figure 1.2). According to de Groot et al. (2002), productivity is a measure of ecosystem function, i.e., the capacity of natural processes and components provide goods and services, that satisfy human needs, directly or indirectly, and stability is a measure of resilience (i.e., how an ecosystem quickly returns to an equilibrium and/or have a less chance of extinction). For example, the more diverse world is more likely to contain the most productive ecosystems that raise the total productivity of the whole community (Turner et al., 1998; Tietenberg and Lewis, 2013). Also, ecosystem processes (such as photosynthesis, microbial respiration, nitrification, denitrification, nitrogen fixation and so on) result from the life-processes of multi-species assemblages of organisms and their interactions with the biotic environment, as well as the abiotic environment itself. These processes ultimately generate services (such as primary production, decomposition, and other elements cycle) that provide utilities and benefits for humans and nature (Diaz et al., 2006). For example, forest ecosystem, not only provide the raw material for food and industries, but also maintain biological variety, regulate water flow, and absorb gases. Nevertheless, the value of biodiversity is evident in permanent grassland and pasture ecosystem functioning (Tilman et al., 1996; Diaz et al., 2006). However, analysis of the biodiversity benefits and values is essential to help economists understand and interpret the dynamics and complex interactions in ecological contexts of both the flow of energy and matter through trophic networks and the functional diversity of species within ecosystems (Turner et al., 1998; Diaz et al., 2006; Tietenberg and Lewis, 2013).

Beyond the value biodiversity has in regulating and stabilizing ecosystem processes; there are direct economic consequences of losing diversity in certain ecosystems and in the world as a whole beside deterioration lies in their partial irreversibility (Costanza and Daly, 1992). Therefore, ecosystems need to be managed to deliver multiple services in ways that avoid the passing of dangerous tipping-points (Chan, 2006; Elmqvist et al., 2010).

The loss in biodiversity, for example, depletion of natural ecosystem and species may have non-linear consequences for observable impact until a certain threshold is reached. A lake can, for example, absorb nutrients for a long time while actually increasing its productivity. However, once a certain level of algae is reached lack of oxygen causes the lake's ecosystem to break down suddenly. Therefore, losing species means losing potential genes or ecosystem, besides the other ecological consequences of changes in biodiversity all of which have a direct natural and an economic effect on people's lives (Elmqvist et al., 2010).

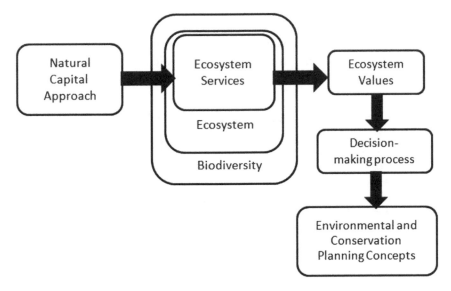

FIGURE 1.2   Natural capital function and service (modified from Martin-Lopez et al., 2009).

1.5   BIODIVERSITY FOR ECONOMIC DEVELOPMENT

Ecosystems and natural habitats management deal with the application of the principles of economics to the study of how environmental and natural resources are developed and managed (Tietenberg and Lewis, 2013).

Nevertheless, biodiversity development underpins the environment, and human well-being needs, in terms of economic development, social development and environmental protection (Elmqvist et al., 2010). Definition of the total economic value of an environmental resource (Barbier, 1994; Loomis et al., 2000) includes the use values (such as direct and indirect one) and non-use values (bequest and existence value). In order to take the non-use values properly into an account, it requires a framework for distinguishing and grouping the various values of an ecosystem (Figure 1.3).

The major contribution of environmental economists has been in the area of the valuation of environmental goods and services (de Groot et al., 2002) to commodity in a growing number of ecosystem services and to reproduce the market logic to tackle environmental problems, i.e., methods for measuring the demand curves for private and financial sectors beside goods for which there are no markets (non-market valuation) (Birol et al., 2006). Economic development has traditionally required the growth of natural capitals in the form of Gross Domestic Product (GDP).

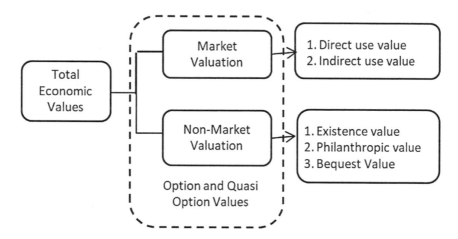

FIGURE 1.3 The total economic value of an ecosystem (modified from Turner et al., 1998 and Martin-Lopez et al., 2009).

The environmental inputs flowing into the economy form the environment can be divided into natural resources (typically mineral and biological resources) and ecosystem inputs (the water and air necessary for all life forms) indicating that biodiversity, ecosystems, ecosystem services and human well-being are linked to each other (de Groot et al., 2002). Nevertheless, genetic, species, and ecosystem biodiversity and services provide vital goods and services. Some products, such as agricultural seeds, enzymes, and microorganisms are derived from genetic resources represent the 'natural' component of the market. About 40% of world trade is based on biological products or processes (Chair's Conclusion, 2007). The economic importance of ecological services provided by associating biodiversity in terrestrial (e.g., agricultural systems) and aquatic biodiversity (e.g., fisheries biota) by implementing the CBD that directly connects economic benefits to the role of indigenous communities (CBD, 2010b). There is an increased recognition that people are part of, but not separate from, the ecosystems in which they live (Elmqvist et al., 2010), and are affected by changes in ecosystems, populations of species and genetic changes. Several years ago the CBD entered into force, with a goal of ensuring that in the process of obtaining and sharing biodiversity benefits, society would promote the conservation and sustainable use of the world's biological diversity. On the other hand, sustainable production and consumption can be stimulated, from the 'Strategic Plan for Biodiversity 2011–2020' along with its 'Aichi targets' by a combination of government incentives, including sustainable procurement policies, and harnessing market forces through a partnership with private sector. Therefore, needs for biodiversity conservation for the existence of living beings (Kohler-Rollefson, 1997; Mittermeier et al., 1998; Brooks et al., 2001) and the sustainability of resources for the promotion of natural habitats to balance the natural processes of social and economic development, will create a foundation for economic and human well-being.

Thus, biodiversity requires our attention as it provides a wide range of direct and indirect benefits to mankind, which occurs on both local and global scales for the different biodiversity levels. Therefore, assessment of progress towards the biodiversity economic benefits is based on:

1. Knowledge of ecosystem services, benefits to mankind and its current status.
2. Knowledge of market needs by reviewing the current state of the economics of ecosystems and biodiversity.
3. Quantitative review of biodiversity business risks, opportunities indicators that underlying quantitative and monetary information and help to close gaps where no such information exist.
4. National and international commitments to biodiversity need a common valuation framework and methodology.
5. The analysis of the economic impact is essential, for which economic data and gross in ecosystem services will be available.
6. Anticipate new markets and governmental policies.

7.   Strengthen existing approaches to environmental management.
8.   Improve stakeholder relationships.
9.   Demonstrate leadership in corporate sustainability.

Besides, all economic production depletes ecosystem structure and generates waste (Farley, 2012). Therefore, there is a need to allocate the finite ecosystem structure between economic production and production of life-sustaining ecosystem goods and services for solving the economic production problem with the ecological satisfaction and efficiency (MAR, 2005; CBD, 2010a,c). However, many human activities contribute to the unprecedented rate of biodiversity loss, which threatens the stability and continuity of ecosystems as well as their provision of goods and services to mankind (Diaz et al., 2006). Consequently, in recent years much attention has been focused on the analysis and valuation of the loss of biodiversity (Helliwell, 1969; Wilson and Carpenter, 1999; Costanza, 2000; Loomise et al., 2000; Birol et al., 2006). Therefore, for economic development, there is a need for natural resource (i.e., requires raw material and energy provided by nature) at different levels (species, genes, and ecosystem diversity). In addition to sustainable practices in agriculture, aquaculture, and forestry, the mapping of the ecosystem service distribution, the degree of association or interaction between ecosystem services and the spatial pattern of ecosystem services, the measure of "well-being," loss or gain; for example, during a trade that may set the stage for economic cost-benefit analysis (de Groot et al., 2002).

1.6   BIODIVERSITY FOR SUSTAINABLE DEVELOPMENT

The concept of sustainable development has emerged, as the means by which biodiversity and natural ecosystems would be saved from human over-exploitation' activates on the environment. The concept was first promoted by the World Conservation Strategy in 1980 by the International Union for Conservation of Nature and National Resources (IUCN); with the advice, cooperation, and financial assistance of the United Nations Environment Programme (UNEP) and the World Wildlife Fund (WWF) by the global conservation blueprint that grew from the United Nations Conference on the Human Environment, held in Stockholm in 1972. Sustainable development is defined as "development that meets the needs of the present without compromising the ability of future generations to meet their own needs." Therefore, societies need to handle three types of capital (natural habitats, social welfare, and market values), which may be non-substitutable and whose consumption might be irreversible (Dyllick and Hockerts, 2002).

The MAR completed in 2005 by more than 1360 scientists working in 95 countries, examined the state of 24 biodiversity services. The assessment concluded that 15 of the 24 services are in decline, including the provision of fresh water, marine fisheries production, the number and quality of places of spiritual and religious values, the ability of the atmosphere to cleanse itself of pollution, and the capacity of agro-ecosystems to provide pest control. However, it would need sound, sustainable use and management strategy governing its inputs. As compensation for loss of utility means substitutability of goods and/or biodiversity or ecosystems that can be kept within safe ecological limits (Tilman et al., 1996; Bernstein and Chivian, 2008).

Biodiversity provides the basis for ecosystems and well-being, providing products, such as water and food both wild and cultivated, besides other agricultural commodities, whose values are widely recognized (Turner et al., 1998; Bullock et al., 2007). For example, deforestation, and excessive burning of fossil fuels is altering the balance of gases in the atmospheres (Myers, 1991) as CO_2 level is building up to high levels since more CO_2 is released than the natural ecosystems can absorb it. It also reduces ground cover, disrupts water cycles as well as lead to soil erosion. The soil is washed into lakes and rivers, which silt and reduced aquatic biodiversity, among other effects (Chapin et al., 2002; MAR, 2005). Although deforestation can be controlled at a local level, the massive amount of deforestation is due to over-harvesting of trees for economic use, rather than local use. Most of the fossil fuel usage generating the CO_2 in the atmosphere stems from the industrialized nations and transition economies, but the effects are especially apparent throughout the less industrialized world.

Accordingly, the term sustainability describes how biological systems remain diverse and productive over time; in relation to humans, it refers to the potential for long-term maintenance of well-being, which in turn depends on the well-being of the environment and the responsible use of natural resources (MAR, 2005).

The interdisciplinary nature of biodiversity conservation, business, commerce, economics, biotechnology, national, and international law, social, and cultural issues was limited if not absent in most countries. This eco-efficiency is usually worked out as the economic value added by a firm in relation to its aggregated environmental impacts (Tilman et al., 1996; Bernstein and Chivian, 2008). Eco-efficiency is satisfied through the delivery of competitively priced goods and science that satisfy human demand in market sector and bring life quality that reducing natural impacts of the local communities throughout the life-span in line with the biosphere's carrying capacity (Bernstein and Chivian, 2008).

For example, the increased need for agricultural production to feed the increase in population number will likely be met largely by commercial intensification, and negative consequences for the genetic diversity of agricultural crops and livestock (Bullock et al., 2007). Concerning that as the human population grows, the demand for other stuff such as freshwater, food, and energy resources put the sustainability of the environment at risk, i.e., there is a direct and underestimated link between the current economic crisis and sustainability crisis. Noting the biodiversity is a key indicator of success in achieving sustainable development (Costanza and Daly, 1992).

Loss of biological entities affects all materials' human well-being (MAR, 2005). The progressive loss of biodiversity and threaten of cultural integrity represent obstacles towards the attainment of the Millennium Development Goals (MDGs). Hence, recognizing the objectives of the CBD to conserve biological biodiversity, sustain the use of the various components of biodiversity, and share the benefits arising from the biological biodiversity resources wisely and/or in a sustainable manner.

Acknowledging that approximately 60% of ecosystems services are being degraded or used unsustainably unless are significantly adjusted for sustainable systems (Costanza and Daly, 1992; Dyllick and Hockerts, 2002). Noting in this regards, sustainable development projects often are part of a large strategy for landscape or ecosystem-scale conservation with the change in demands and consumption pattern as well as technological advances; this value will be important in the future (Bernstein and Chivian, 2008). Biotechnology is also increasingly being used to capitalize on various biological and genetic resources. Therefore, local Biodiversity Action Plans are better coordinated with each other and with national biodiversity objectives, that is compatible with economic and conservation sustainability in local community benefits.

There is a need for a biodiversity strategy to ensure economic development and in achieving economic outcomes that should last with the feasible and sustainable natural balance (Dyllick and Hockerts, 2002; MAR, 2005; CBD, 2010a,b). Therefore, the enhancement towards a national economy may take into account all significant development programmes and biodiversity concepts, approaches, and schemes for various thematic areas, i.e., sustainable agriculture and sustainable forestry, and sustainable use of food, water, energy, residential living trends, transportation, industry, material, and waste management. Sustainable development projects typically use a number of economic tools to promote sustainability, including incentives such as certification, subsidies, and grants, or job creation in sustainable enterprises, and sometimes penalties, such fines or fees for continuing unsustainable practices. Therefore, the goals of environmental conservation and sustainable development are not conflicting and can be reinforcing each other (Barbier et al., 2008).

1.7 CONCLUSION

Biodiversity contributes to many aspects of people's economy and development, providing products from natural resources at different levels (species, genes, and ecosystem diversity), whose values are widely recognized (CBD, 2010a,b,c). Therefore, the biodiversity provides inputs that including but not limited to (1) natural resources at different categories (food, water, wood, energy, and medicines), i.e., basic material for good life, (2) natural (ecosystem) services, including water purification and storage, soil fertility, waste disposal, removal of carbon dioxide/air filtration, pollination, pest, flood, and landslide control, and (3) environmental and aesthetic pleasure, and so on.

Multiple services are produced by ecosystems, and these interact in complex ways that require a general approach to conserve the integrity and diversity of nature, in the long term and to ensure that any use of natural resources is equitable and ecologically sustainable (MAR, 2005; CBD, 2010a,b,c).

Which refers to ecological restoration and monetary value such as Cost-Benefit Analysis (CBD), combinations of CBA and Multi-Criteria Analysis (MCA), Environmental Impact Assessment (EIA) and Strategic Environmental Assessment (SEA), can be a primary component of conservation and sustainable development programmes throughout the world. Martin-Lopez et al. (2009) stated that when these functions, once performing well, deliver a healthy supply of goods and services, which in many cases accounting for national economic values by incorporating biodiversity values into planning processes and strategies to reduce poverty and integrating natural capital into the national accounts.

Whereas economics is the study of resource allocation, distribution, and consumption that maximize social welfare (CBD, 2010a,b,c; Tierenberg and Lewis, 2013), therefore, economics of biodiversity provides insight into the value of biodiversity that are already captured within the GDP such as improvement in the breed, tourism development, and economic development, and values external to the System of National Accounts (SNA) (such as firewood and ecosystem services) that generates profit (i.e., creates jobs and goods) and equitable benefits beside other unknown benefits (i.e., without biodiversity, there are no other services).

The growing commercial interest in biodiversity must be accompanied by increased investment in source conservation. However, often over-looked or poorly understood. Scientists (Tilman et al., 1996; Bernstein and Chivian, 2008) dragged a conclusion that with the continuation of sustainable relationships, adjusted to account for natural resources conservation under two measures, protect or enhance biodiversity in which living conditions and resource-use meet human needs without undermining the sustainability of natural systems and the environment for future generations to have their needs met.

In general the sustainability of conservation initiatives, particularly on those involving a combination of conservation and development (such as Integrated Conservation and Economic Development Programs), often depends on effective partnerships between government, business, and civil society (CBD, 2010a,b,c). Therefore, ecological economics is defined by its focus on nature, justice, and time, i.e., biodiversity, and sustainable development that guide ecological, economic analysis and valuation (Helliwell, 1969; Wilson and Carpenter, 1999; Costanza, 2000; Loomis et al., 2000; Birol et al., 2006). Such emphasis is about the unfolding dimensions of biodiversity that maintain ecosystem integrity, and provide a baseline for the biodiversity and key areas of development (Martin-Lopez et al., 2009; Elmqvist et al., 2010). Also, it provides a route map to the biodiversity monitoring system and sustainable approaches that moving toward conservation strategies from an economic perspective point of view. As well as, ecological restoration is a well-established practice in biodiversity conservation and ecosystem management.

For the ecosystem to sustain this level of inputs to the economy, a clear strategy is needed, regarding the following accounts on biological resources and ecosystem services. First, the environment's contribution to the economy needs to be valued based on experience gained over several decades from a sound framework of interdisciplinary collaboration between ecologists and economists. Second, national mapping and screening of biodiversity resources and environmental spatial variation need to be incorporated and clearly expressed in monetary terms/decision-making techniques such as CBD, MCA, EIA, and SEA to provide a baseline for the biodiversity and key areas of development. Third, research and structural reform may be needed in some key sectors or activities for providing ecological restorations and ecosystem self-generating processes to resume. Fourth, the strengthening of policy, legislative, and institutional frameworks and the implementation measures of intervention measures besides monitoring protocols need to be applied, follow, and evaluated continuously for adaptive management.

KEYWORDS

- **aquatic ecosystems**
- **biodiversity**
- **biodiversity and economic development**
- **biodiversity and socio-cultural development**
- **biodiversity concepts**
- **biodiversity for sustainable development**
- **biodiversity history**
- **biodiversity hotspots**
- **biodiversity non-use values**
- **biodiversity strategy and conservation**
- **biodiversity use value**
- **biodiversity values**
- **ecosystem diversity**
- **ecosystem service**
- **genetic diversity**
- **land biomes**
- **species diversity**
- **terrestrial biodiversity**
- **water ecosystems**

REFERENCES

Barbier, E. B., (1994). Valuing environmental functions: Tropical wetlands. *Land Econ.*, *70*(2), 155–173.

Barbier, E. B., Koch, E. W., Silliman, B. R., Hackery, S. D., Wolanski, E., Primavera, J., et al., (2008). Coastal ecosystem-based management with nonlinear ecological functions and values. *Science, 319*(5861), 321–323.

Benton, M. J., (2001). Biodiversity on land and in the sea. *Geolog. J.*, *36*(3/4), 211–230.

Bernstein, A., & Chivian, E., (2008). *Sustaining Life: How Human Health Depends on Biodiversity*. Oxford University Press, New York.

Birol, E., Karousakis, K., & Koundouri, P., (2006). Using economic valuation techniques to inform water resources management: A survey and critical appraisal of available techniques and an application. *Sci. Total Environ.*, *365*(1–3), 105–122.

Boumans, R., Costanza, R., Farley, J., Villa, F., & Wilson, M., (2002). Modeling the dynamics of the integrated earth system and the value of global ecosystem services using the GUMBO model. *Ecol. Econ.*, *41*, 529–560.

Brooks, T., Balmford, A., Burgess, N., Fjelda, J., Hansen, L. A., Moore, J., Rahbek, C., & Williams, P., (2001). Toward a blueprint for conservation in Africa. *BioScience, 51*, 613–624.

Bullock, J. M., Pywell, R. F., & Walker, K. J., (2007). Long-term enhancement of agricultural production by restoration of biodiversity. *J. Appl. Ecol.*, *44*(1), 6–12.

Cardillo, M., Orme, C. D., & Owens, I. P. F., (2005). Testing for latitudinal bias in diversification rates: An example using New World birds. *Ecology, 86*, 2278–2287.

CBD, (1992). *Convention on Biological Diversity*. Secretariat of the Convention on Biological Diversity, Montreal, Canada, www.cbd.int/convention/convention.shtml/.

CBD, (2010a). *Global Biodiversity Outlook 3*. Secretariat of the Convention on Biological Diversity, Montreal, Canada, https://www.cbd.int/gbo/gbo3/doc/GBO3-final-en.pdf

CBD, (2010b). *Strategic Plan for Biodiversity 2011–2020*. Secretariat of the Convention on Biological Diversity, Montreal, Canada, http://www.cbd.int/undb/media/factsheets/undb-factsheet-sp-sn.pdf

CBD, (2010c). *Aichi Biodiversity Targets*. Secretariat of the Convention on Biological Diversity, Montreal, Canada, http://www.cbd.int/sp/targets/.

Chair's Conclusion, (2010). *G–8 Ministerial Conference on the Biological Diversity*, Potsdam.

Chakhaiyar, H., (2010). *Periwinkle Environmental Education, Part XII*. Jeevandeep Prakashan Pvt Ltd., Mumbai, India.

Chan, K. M. A., Shaw, M. R., Cameron, D. R., Underwood, E. C., & Daily, G. C., (2006). Conservation planning for ecosystem services. *PLoS Biology, 4*, e379.

Chapin, F. S. III, Matson, P. A., & Mooney, H. A., (2002). *Principles of Terrestrial Ecosystem Ecology*. Springer-Verlag, New York.

Corlett, R., & Primack, R. B., (2011). *Tropical Rain Forests: An Ecological and Biogeographical Comparison*. Wiley-Blackwell Publishing, New Jersey, USA.

Costanza, R., & Daly, H. E., (1992). Natural capital and sustainable development. *Conserv. Biol.*, *6*, 37–46.

Costanza, R., (2000). Social goals and the valuation of ecosystem services. *Ecosystems, 3*, 4–10.

Dasgupta, P., & Perrings, C., (2012). Valuing biodiversity. In: Levin, S. A., (ed.), *Encyclopedia of Biodiversity* (pp. 167–179). Academic Press, New York.

Davic, R. D., (2003). Linking keystone species and functional groups: A new operational definition of the keystone species concept. *Conserv. Ecol.*, 7(1), r11, http://www. consecol.org/vol7/iss1/resp11/

De Blij, H. J., Muller, P. O., Williams, R. S., Conrad, C. T., & Long, P., (2010). *Physical Geography of the Global Environment*. Oxford University Press, New York.

De Groot, R., Wilson, M. A., & Boumans, R. M. J., (2002). A typology for the classification, description, and valuation of ecosystem functions, goods, and services. *Ecol. Econ.*, *4*, 393–408.

Díaz, S., Lavorel, S., Chapin, III, F. S., Tecco, P. A., Gurvich, D. E., & Grigulis, K., (2006). Functional diversity at the crossroads between ecosystem functioning and environmental filters. In: Canadell, J., Pitelka, L. F., & Pataki, D., (eds.), *Terrestrial Ecosystems in a Changing World* (pp. 81–91). Springer-Verlag, Berlin Heidelberg, Germany.

Douros, J., & Suffness, M., (1980). The National Cancer Institute's Natural Products Antineoplastic Development Program. In: Carter, S. K., & Sakurai, Y., (eds.), *Recent Results in Cancer Research* (Vol. 76, pp. 21–44). Springer, Berlin Heidelberg.

Dyllick, T., & Hockerts, K., (2002). Beyond the business case for corporate sustainability. *Business Strategy and the Environment*, *11*, 130–141.

Elmqvist, T., Maltby, E., Barker, T., Mortimer, M., & Perrings, C., (2010). Biodiversity, ecosystems, and ecosystem services. In: Kumar, P., (ed.), *The Economics of Ecosystems and Biodiversity: Ecological and Economic Foundations* (pp. 41–104). Earthscan, London, UK.

FAO, (2010). *Global Forest Resources Assessment 2010 – Main Report*. FAO Forestry Paper No. 163, Rome, www.fao.org/docrep/013/i1757e/i1757e00.htm.

Farley, J., (2012). Ecosystem services: The economics debate. *Ecosystem Services*, *1*, 40–49.

Gould, S. J., (1993). In: Norton, W. W., (ed.), *The Book of Life: An Illustrated History of the Evolution of Life on Earth* (pp. 6–21). New York.

Hamrick, J. L., & Godt, M. J. W., (1996). Effects of life history traits on genetic diversity in plant species. *Phil. Trans. Royal. Soc. Biol. Sci.*, *351*, 1291–1297.

Helliwell, D. R., (1969). Valuation of wildlife resources. *Regional Studies*, *3*, 41–49.

Heywood, V. H., (1995). *Global Biodiversity Assessment* (pp. 995–1027). Cambridge University Press, Cambridge, UK.

Howarth, R., & Farber, S., (2002). Accounting for the value of ecosystem services. *Ecol. Econ.*, *41*, 421–429.

Köhler-Rollefson, I., (1997). Indigenous practices of animal genetic resource management and their relevance for the conservation of domestic animal diversity in developing countries. *J. Anim. Breed. Gene.*, *114*, 231–238.

Loomis, J., Kent, P., Strange, L., Fausch, K., & Covich, A., (2000). Measuring the total economic value of restoring ecosystem services in an impaired river basin: Results from a contingent valuation survey. *Ecol. Econ.*, *33*(1), 103–117.

Lovejoy, T. E., (1980). The technical report. In: Barney, G. O., (ed.), *The Global 2000 Report to the President* (Vol. 2, pp. 327–332). Penguin.

MAR, (2005). *Ecosystems and Human Well-Being: Biodiversity Synthesis*. World Resources Institute, Washington, DC.

Martin-Lopez, B., Gomez-Baggethun, E., Gonzales, J. A., Lomas, P. L., & Montes, C., (2009). The assessment of ecosystem services provided by biodiversity: Re-thinking concepts and research needs. In: Aronoff, J. B., (ed.), *Handbook of Nature Conservation: Global, Environmental, and Economic Issues* (pp. 261–282). Nova Science Publishers, New York.

Mittermeier, R. A., Myers, N., Thomsen, J. B., Da Fonseca, G. A. B., & Olivieri, S., (1998). Biodiversity hotspots and major tropical wilderness areas: Approaches to setting conservation priorities. *Conserv. Biol., 12*(3), 516–520.

Mittermeier, C. G., Turner, W. R., Larsen, F. W., Brooks, T. M., Gascon, C., (2011). Global biodiversity conservation: the critical role of hotspots. In: Zachos, F. E., Habel, J. C., (Eds.), Biodiversity hotspots: Distribution and Protection of Priority Conservation Areas (pp. 3-22). Springer-Verlag, Berlin.

Mora, C., & Robertson, D. R., (2005). Causes of latitudinal gradients in species richness: A test with fishes of the Tropical Eastern Pacific. *Ecology, 86,* 1771–1792.

Mora, C., Tittensor, D. P., Adl, S., Simpson, A. G. B., Worm, B., & Mace, G. M., (2011). How many species are there on earth and in the ocean? *PLoS Biology, 9*(8), e1001127, https: //doi.org/10.1371/journal.pbio.1001127.

Mulhall, M., (2009). Saving rainforests of the sea: An analysis of international efforts to conserve coral reefs duke environmental law and policy forum. *Spring, 19,* 321–351.

Myers, N., (1991). Tropical forests and climate. *Climatic Change, 19,* 1, 2.

Myers, N., Mittermeier, R. A., Mittermeier, C. G., Da Fonseca, Gustavo, A. B., & Kent, J., (2000). Biodiversity hotspots for conservation priorities. *Nature, 403*(6772), 853–858.

Norse, E. A., & McManus, R. E., (1980). Ecology and living resources biological diversity. In: *Environmental Quality* (pp. 31–80). The 11th Annual Report of the Council on Environmental Quality, Washington DC, USA.

Odum, E. P., (1971). *Fundamentals of Ecology,* Saunders, New York (p. 574).

Parravicini, V., Villeger, S., McClanahan, T. R., Arias-Gonzalez, J. E., Bellwood, D. R., Belmaker, J., et al., (2014). Global mismatch between species richness and vulnerability of reef fish assemblages. *Ecol. Lett., 17,* 1101–1110.

Reid, W. V., (1998). Biodiversity hotspots. *Trends Ecol. Evolut., 13*(7), 275–280.

Robson, G. C., (1928). *The Species Problem: An Introduction to the Study of Evolutionary Divergence in Natural Populations.* Oliver and Boyd, Edinburgh (p. 283).

Sorokin, Y. I., (1993). *Coral Reef Ecology.* Springer-Verlag, Berlin Heidelberg, Germany.

Stevens, R. D., Gavilanez, M. M., Tello, J. S., & Ray, D. A., (2011). Phylogenetic structure illuminates the mechanistic role of environmental heterogeneity in community organization. *J. Anim. Ecol., 81,* 455–462.

Tietenberg, T., & Lewis, L., (2013). *Environmental and Natural Resource Economics.* Pearson, New Jersey (p. 666).

Tilman, D., (1999). The ecological consequences of changes in biodiversity: A search for general principles. *Ecology, 80,* 1455–1474.

Tilman, D., Wedin, D., & Knops, J., (1996). Productivity and sustainability influenced by biodiversity in grassland ecosystems. *Nature, 379,* 718–720.

Tuomisto, H. A., (2010). Consistent terminology for quantifying species diversity? Yes, it does exist. *Oecologia, 4,* 853–860.

Turner, R. K., Van den Bergh, J. C., Barendregt, A., & Maltby, E., (1998). *Ecological-Economic Analysis of Wetlands.* Tinbergen Institute, Amsterdam, and Rotterdam.

Whittaker, R. H., (1972). Evolution and measurement of species diversity. *Taxon, 21,* 213–251.

Williams, P. H., & Humphries, C. J., (1996). Comparing character diversity among biotas. In: Gatson, K. J., (ed.), *Biodiversity: A Biology of Number and Differences* (pp. 54–76). Blackwell Science, Oxford, UK.

Willig, M. R., Kaufmann, D. M., & Stevens, R. D., (2003). Latitudinal gradients of biodiversity: Pattern, process, scale, and synthesis. *Annu. Rev. Ecol. Syst., 34,* 273–309.

Wilson, E. O., (1988). *Biodiversity.* National Academy Press, USA.

Wilson, M. A., & Carpenter, S., (1999). Economic valuation of freshwater ecosystem services in the United States: 1971–1997. *Ecol. Appl., 9*(3), 772–783.

CHAPTER 2

THE VALUE OF BIODIVERSITY IN THE LEGUME FAMILY AND ITS BENEFITS FOR HUMAN LIVES

MIRZAEI SAEID and AZIZKHANI NEGIN

Department of Biotechnology, Institute of Science, High Technology and Environmental Sciences, Graduate University of Advanced Technology, Kerman, 7631133131, Iran

2.1 INTRODUCTION

Biodiversity is usually broad and general term that covering the great diversity of biological individuals, from gene to ecosystem. The ecosystem functioning and biological diversity are the primary key factors for the sustainability of life on the earth. In addition, biological products have commercial value and contribute considerably to the development of societies and nations. However, anthropogenic behaviors and climate change caused the loss of global biodiversity (Singh et al., 2015).

Many factors impact the ecosystems, including the availability of nitrogen, water, carbon dioxide, phosphorus, and other limiting resources (Tilman et al., 2014). Also now the fundamental issues of the community are food and energy difficulties. Recently, to feed the increasing population of the world, agriculture is based on vast populations of genetically identical and cultivars (Gresshoff et al., 2015). This human behavior causes to reduce the productivity of the ecosystems. Furthermore, there are concerns about habitat destruction and the impacts of the loss of biodiversity on ecosystem functioning and productivity. Since diversity is a vital element for productivity and stability of the ecosystem and drives to the more efficient use of limited resources, restoration, conservation, and preservation of biodiversity should be recognized as a global preference (Tilman et al., 2014). Therefore, protection of biodiversity is essential, and this will be achieved through the promotion of knowledge and identification of biodiversity whether naturally detected or produced by the experimenter.

Legume family is diverse and has 650 genera and nearly 20,000 species spread throughout the world (Doyle et al., 1994; Sprent et al., 2007). They utilized as food and animal feed products (e.g., soybean, lucerne), biofuel (e.g., *Pongamia pinnata*) supplies, ornamental (e.g., wisteria), medicine (e.g., fenugreek) or timber (e.g., acacia/wattle). This means that this family covers different aspects of human needs. Furthermore, most legumes are capable of forming nodules after microsymbiont induction, which serves as their habitat for nitrogen fixation. This symbiotic relationship provides nitrogen for plant and reducing nitrogen fertilizers demand (Humphries et al., 2013; Basu et al., 2014; Gresshoff et al., 2015). This capability has beneficial to the environment. Hence, for future global agriculture, research into the physiology, biochemistry, and genetics of legumes is inevitable.

2.2   THE POTENTIAL OF LEGUMES TO MITIGATE CLIMATE CHANGE

2.2.1   LEGUME EFFECTS ON MITIGATION OF GREENHOUSE GASES

In modern agriculture, industrial nitrogen fertilizers highly used to increase maximum crop productivity. On the other hand, production, and distribution of the nitrogen fertilizer need to a numerous deal of fossil fuel (Ferguson et al., 2010). Carbon-based fuel or fossil fuel (coal, oil, and natural gas) produces major greenhouse gases, such as carbon dioxide (CO_2), methane (CH_4) and nitrous oxide (N_2O) in the air (Stavi et al., 2013).

Also, the significant portion of applied nitrogen fertilizer–30–50%–is lost. Therefore, applying chemical fertilizers is a largely ineffective process (Ferguson et al., 2010). Loss of nitrogen from applied fertilizer (10% to 30% of the applied N) is a major cause of inefficiency in nitrogen fertilizer utilization. This loss of nitrogen can occur through many pathways: ammonia (NH_3) volatilization, nitrate (NO_3^-) leaching and gas emissions – emissions of gases such as N_2O. Losses of up to 30% of applied N as

NH_3 have been reported from pasture systems in Australia. Denitrification can also be a major pathway of nitrogen lost as N_2O, N_2 and NO_3^- during periods of irrigation or waterlog. In Australia, N_2O emissions from agriculture account for 71% of the national total (Suter et al., 2014).

Most legume species develop a relationship with soil rhizobia called symbiosis and achieve the most of the nitrogen needs through a biological nitrogen fixation (BNF) system. This mechanism reduces the demand for nitrogen fertilizers in the legume family. This is one solution and alternative to nitrogen fertilizer. BNF is performed by prokaryotes using nitrogenase enzyme complex (Ferguson et al., 2010). Nitrogen fixation is the capture of atmospheric nitrogen gas (N_2) to form NH_3, which is readily available for assimilation by plant and mycorrhiza. Most, but not all legumes able to build a mutually beneficial symbiosis with rhizobia and also there is a host specificity in this relationship (e.g., soybean, and lucerne come in to interact with *Bradyrhizobium japonicum* and *Sinorhizobium meliloti,* respectively) (Kazakoff et al., 2011; Owusu et al., 2013). In modern agriculture and silviculture, numerous legume varieties are used as the nitrogen-fixing advantage of the legume-rhizobia symbiosis. Annually, symbiotic nitrogen fixation of legumes leads to adding around 200 million tons of nitrogen to the biosphere (Graham et al., 2003).

During the symbiosis with bacteria, most legume plants form a novel organ, called the root nodule that enables symbiotic nitrogen fixation. The nodule formation systems are diverse but seem to have common mechanistic features (Gresshoff et al., 2015). This process requires an organized exchange of signals between the two symbiotic allies (Biswas et al., 2014). Early nodulation genes are key components for forming nodule in legumes. Figure 2.1 shows legume plant exudates flavonoid substances such as the daidzein (signals in soybean), genistein, and isoflavones and the flavone 7,4-dihydroxyflavone (DHF; signal in white clover), which is perceived by *Rhizobium* bacteria (mainly called *Rhizobium, Bradyrhizobium, Photorhizobium, Sinorhizobium, Azorhizobium,* and *Mesorhizobium*). This is followed by the synthesis of a return signal in bacteria, called lipochitooligosaccharide nodulation factor (Nod Factor). A receptor complex including two LysM type receptor kinases (in soybean: NFR1 and NFR5, or equivalents in other legumes) at the root epidermis is the perception at the legume root epidermis (Oldroyd et al., 2011). The downstream genes are influenced by this perception including those involved in the expression of nucleoporin, calcium spiking, cation channels, early nodule, and cytokinin signaling result in cortical and pericycle cell divisions and simultaneous bacterial infection (Biswas et al., 2014) (Figure 2.1).

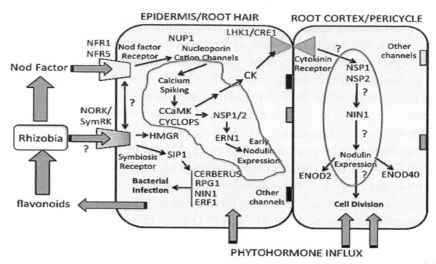

FIGURE 2.1   Model of legume nodule induction for both legumes with determinate and indeterminate nodulation pattern (Gresshoff et al., 2015).

Rhizobia through two main routes enter the plant root either via the root hair or fractures in root epidermal tissue (Oldroyd et al., 2011). The most common way is the root hair infection, and this involves the formation of infection threads. The initial visible of nod factor responses are root hair curling and formation of subsequent infection thread by which the rhizobia enter the root hair cell via redirected tip growth and new cell wall formation (in soybean: during the first 6 to 8 hours). Thinner structure and less cross-linked cell walls of the tip of emerging root hairs make them suited for redirection. This allows bacteria to be penetrated into newly induced cortical cell divisions (nodule primordia) for further infection and eventual nodule formation. Xylem-pole localized pericycle divisions, normally limited to lateral root induction, occur concurrently and allows development of nodule vascular connections (Gage et al., 2004; Ferguson et al., 2010). Fixed nitrogen (NH_3) is exported into the plant cytoplasm for assimilation by glutamine synthetase into glutamine so further moved to the plant via the xylem or converted to ureides (in soybean: as a nitrogen transport molecule) and then transported (Padilla et al., 1987).

Two major morphological types of nodules: determinate and indeterminate exist in legumes. The host plant determines the type of nodule. The site of first initial cell divisions, maintenance of a meristematic region, and the form of mature nodules are differences between the two nodule types in legume (Oldroyd et al., 2011). Legumes such as clovers and pea have indeterminate-type nodules and incomplete cell divisions lead to the formation of larger cells containing approximately 750–1000 bacteroides per infected cell, in return determinate legumes such soybean and *Pongamia pinnata* there are about 25,000 bacteroides per infected cell (Gresshoff et al., 1977; Gresshoff et al., 1978).

Therefore, legumes can participate in reducing greenhouse gases emissions and environmental concerns. Globally, the amount of CO_2 in the atmosphere is altering by the amount of CO_2 that is respired from the root systems of N_2-fixing legumes and CO_2 generated during N-fertilizer production. However, the amount of CO_2 respired from nodulated root system is higher than the other one. But, the CO_2 respired from the nodulated roots of legumes are originally captured from the atmosphere via photosynthesis, and therefore this would essentially be CO_2 neutral. By contrast, all the CO_2 released during the synthesis of N fertilizer would be derived from fossil energy and represents a net contribution to atmospheric concentrations of CO_2. Thus compared to cropping and pasture systems that are fertilized with industrial N, legumes decrease emissions of fossil-energy derived CO_2 and results in lower N_2O fluxes (Jensen et al., 2012).

Furthermore, legumes such as lucerne, clover, and various tree and shrub species (all are perennial) could also use in generating biomass for biorefineries, without the demand for nitrogen fertilization either as a monoculture or in mixtures with grasses. The potential application of legume biomass for products includes protein extraction for pharmaceuticals and feed, renewable materials, CH_4 and biofertilizer containing nutrients for recycling (Jensen et al., 2012).

2.2.2 EFFECT OF LEGUMES ON SOIL CARBON SEQUESTRATION

Carbon sequestration is a term, which explains any increase in soil organic carbon (SOC) content due to alteration in land management (Powlson et al., 2013). Legumes are important and suitable for increase sustainable and diverse cropping systems. One of the capabilities of legumes is the ability to grow on drastically marginal surplus lands and disturbed soils (Chaer et al., 2011). They also have a significant role in soil C sequestration. According to this, legumes used as cover crops in pastures, sole crops such as green (used as mulch) or brown manures, and intercrops in reduced tillage cropping systems (Baggs et al., 2003). C sequestration also mitigates climate change. It may have limitations, such that the process is reversible, the quantity of C stored in soil is finite, an increase of SOC may be changed in the fluxes of other greenhouse gases especially N_2O. Animal manure or crop residues also added to soil, as organic materials for increase SOC. However they increase SOC, but they generally do not represent an additional transfer of C from the atmosphere to land (Powlson et al., 2013). There is evidence that SOC concentrations can be increased when legumes are included in pastures in many different regions. Forage legumes (perennial legume species) through lower losses of C from their organic residues effect on SOC more than from annual legumes (Jensen et al., 2012).

2.3   SOYBEAN AS AN OIL SEED LEGUME

The soybean (*Glycine max* (L.) Merr) is a species in *Fabaceae* family native to East Asia. It is an ancestrally duplicated allotetraploid ($2n = 4\times = 40$) (Grseshoff et al., 2015). Two rounds of duplication occurred in the soybean genome (59 and 13 million years ago) resulting in a highly duplicated genome. Due to these duplication events, nearly 75% of the genes present in multiple copies (Schmutz et al., 2010). This is also correct for *GmNARK*, a gene responsible for nodulation in soybean, but the duplicated genomic region contains a homeologous copy called *GmCLV1A* (*CLAVATA1* because of the high sequence similarity to an Arabidopsis gene responsible for the stem morphology). *GmCLV1A* also has the highest similarity with *GmNARK* (over 90% identities). It is transcribed and functional, but not in the same way as *GmNARK*. It acts on shoot architecture indicating of occurring neodiversification of *CLV1* like receptor genes and generating variation (Mirzaei et al., 2014; Mirzaei et al., 2017).

This annual legume is categorized as an oilseed rather than forage. In addition, soybean has been used in China for many years as food and medicine. Today it is also the primary constituent in various processed foods and suitable substitutes for dairy products because soybean is a good source of protein and vegetable oils and also containing some of the human essential amino acids (Kanchana et al., 2015). Soybean grows in regions with hot summers having mean temperatures of 20 to 30°C. Its growth stops in temperatures of below 20°C and over 40°C. As a nitrogen-fixing legume, it can build a symbiotic association with the bacterium, *Bradyrhizobium japonicum* and fix atmospheric dinitrogen gas, which is beneficial to plant as well as the environment (Santos et al., 2013).

The world's largest soybean producers are U.S. (33%), Brazil (27%), Argentina (21%), China (7%) and India (Santos et al., 2013). Dry Soybean contains 35–40% protein; 90% of soy protein consists of two storage globulins, 11S glycinin, and 7S β-conglycinin, roughly 19% oil, of which the triglycerides are the major component. Soybean oil is characterized by relatively large amounts of the polyunsaturated fatty acids (PUFA), (i.e., Soybean oil including of five fatty acids (FA): stearic acid (4%), linolenic acid (8–10%), palmitic acid (13%), oleic acid (18%) and linoleic acid (55%) (Messina et al., 1997; Santos et al., 2013). The unsaturated fatty acid is oleic acid and saturated fatty acids (SFA) including of stearic acid, palmitic acid (Kanchana et al., 2015). Other components of dry soybean seed including 5% minerals, 35% carbohydrate, vitamins, isoflavones, and saponins (Liu et al., 1997; Bozanic et al., 2006). Whole soybeans are an excellent source of calcium, iron, zinc, phosphorus, magnesium, thiamine, riboflavin, niacin, and folic acid. Some soybean components influenced environmentally. An increase in temperature from 10°C to 40°C impacts on fatty acids and lead to a reduction in the production of linoleic (polyun-saturated oils) and linolenic (monounsaturated oils) acid and a rise in the production of oleic acid in soybeans (Schulte et al., 2013).

Oil concentration in soybean seeds is a complex quantitative trait governed by many genes mostly with small effects and under the influence of the environment. There is a negative relationship between seed oil and protein level in soybean. Therefore, it is difficult for breeders to develop high-oil soybean genotypes while retaining a high level of protein concentration due to this negative relationship. Molecular markers have been used in the past two decades to discover quantitative trait loci (QTL) and to facilitate the development of high oil genotypes. Eskandari et al. (2013) studied on the soybean oil in several environments over two years in Canada, Ontario. They found oil QTL in two recombinant inbred line (RIL) populations obtained from crosses among three soybean cultivars with moderately high oil including OAC Glencoe, RCAT Angora, and OAC Wallace. Finally, they identified oil QTL on Chromosome 9 with a significant positive impact on seed protein composition, and also observed three oil QTL on chromosomes 17, 16 and 14 and that in any of the environments had no meaningful influences on seed protein concentration. These QTL could be used in breeding programs targeted at increasing the oil concentration without changing protein concentration (Eskandari et al., 2013).

Today due to the growing world population and environmental issues, renewable energy resources are more attractive. Soybean as one of the primary raw material is suitable for biodiesel production (Santos et al., 2013). Among plant biofuel feedstocks such as corn, oil palm, sugarcane, willow, and canola, soybean is able to use atmospheric nitrogen through nitrogen fixation which brings enormous benefit to humanity and the ecosystem. Biodiesel from soybean oil is usually stated to have a Cetane number (CN) of 48–52, and very low in sulfur (Knothe et al., 2015). However, soybean seeds contain the polyunsaturated fatty acids with low oxidative stability, as a result, are not acceptable for biodiesel production. Therefore, for this purpose, it is necessary to use soybean varieties with low linolenic acid (LLA) and oxidative stability. In Brazil, over 80%, Argentina 100%, the United States about 74% and the European Union only 16% of all biodiesel produced are related to soybean. Overall, 80–85% of esters are unsaturated such as linolenic acid, linoleic acid. Hence, soybean oil with high linoleic and linolenic acid contents has high oil oxidative stability. Thus, it is efficient and convenient for biodiesel production. Soybean with low palmitic acids and high oleic could be the produced biodiesel with better qualities. It is expected that the use of soybean with LLA as raw material for biodiesel production will overcome the drawbacks of soybean biodiesel. Features of LLA soybean oil are important for use in biodiesel production. Although the total tocopherol content is decreased, the LLA content ensures good results for oxidative stability (Santos et al., 2013).

2.4   THE LEGUME BIOFUEL TREE: *PONGAMIA PINNATA*

There are multiple forms of renewable energy, such as those used for the production of heat, fuel, and electricity. Electrical energy usually is generated through various methods, including the nuclear fission, burning fossil fuels and hydroelectric power generation. But these methods are not renewable. Accordingly, many reproducible sources, varying from crop plant biofuel, plant biomass to fuel conversion, waste conversion to gas and algal biodiesel have been considered (Biswas et al., 2014).

P. pinnata is a legume tree with fast vegetative growth, oil-rich seeds (20,000 seeds per tree) and 10 to 20 m in maximum height. On average, oilseed yield of superior trees (called 'elite trees') is about 3–5 MT per hectare per year. Pongamia oil can be converted by hydrogenation to aviation A1 jet fuel (Scott et al., 2008; Klein et al., 2013). Also, it is used for ornamental purposes due to beautifully clustered flower, and biodiesel production (Kazakoff et al., 2011). It is favorable stress physiology and has a deep root system, waxy leaf, so can tolerate a broad range of rainfall (500 to 2500 mm) and minimum temperatures of –5 to 16°C and maximum temperatures of 27 to 50°C (Biswas et al., 2014). On the other hands, it is drought-tolerant: suitable for Australia and India climate; salinity-tolerant: up to an electric conductivity of 20 ds/m (grow well at 10 dS.m^{-1}). Universally it could be grown on margin lands and in soils unsuited for food production (Biswas et al., 2011; Jensen et al., 2012).

Pongamia is a true diploid ($2n = 22$), and chromosomes are nearly uniform in shape. Nuclear genome is predicted to be around 1300 Mb/ haploid genome using by flow cytometry method, making it somewhat equivalent to that of soybean (Choudhury et al., 2014). Pongamia is derived from a cross between a heterogenous mother and unknown male parents. The annual pod yield is varied from a few kilograms per tree to 20 kg per tree. About 6% of pongamia trees (derived from seed) begin to flower in year one while the major flowering development and harvest are occurring on year 4 (Jensen et al., 2012; Jiang et al., 2012). The pongamia seeds contain about 30–40% oil, which itself are rich in non-edible vegetable oil also have monounsaturated (50–55%) oleic acid (C18:1), optimal for renewable biodiesel production (Kazakoff et al., 2011; Gresshoff et al., 2015).

Studies on oil content and oil triglycerides of seeds from three selected trees in Australia after hexane extraction shown the fatty acids profile as follow: palmitic (C16:0) (~ 12%), stearic acid (C18:0) (~ 4%), linoleic acid (C18:2) (~ 20%) and oleic acid (C18:1) (~ 50%). These properties lead to the production of fatty acid methyl ester (FAME) that has important chemical features, such as a pour point of 2.1°C, a CN of 55.8, a quantity of iodine of 80.9 and viscosity at 40°C of nearly 4.3 mm^2/s. These are good according to the limits set by the North American and European Industry Standards (Scott et al., 2008). Seed oil analysis in pongamia revealed existence of variation in oil content and composition between trees and between progeny seeds of a single parent tree. For instance, individual seeds from 10 randomly selected pongamia trees showed variation in the seed mass (0.41–1.5 g), oil content (19.7–50.5%) and oleic acid/oil content (25.4–54.2%). Progeny seeds of a single parent tree showed less variation of seed oil content compared with seeds from different trees (Jiang et al., 2012).

Seed protein content is another commercially important trait because include essential amino acids. After oil extraction (due to the presence slightly toxic effects), Pongamia seed cake will be obtained and used as supplemental animal feed, second-generation bioethanol, biogas, thermo-chemical conversion and/or production of biochar (Jensen et al., 2012). The seed cake contains approximately 30% of protein with a very low rate of oil; different individual seeds differ for protein content (Scott et al., 2008). Most of the seed storage protein (SSP) of Pongamia consists of a 7S-beta-conglycinin (50,000 and 52,000 Da in size). These two proteins have a low content of essential amino acids, making the cake even less attractive as a supplementary feed source. Possibly seed cake may be used as a fertilizer for Pongamia or adjacent plantations (Gresshoff et al., 2015).

There is also a great variation in its morphological characteristics, such as leaf and pod shape, the abundance of seed formation, shape, and size and weight of seed, flowering time, growth habit and growth rate (Gresshoff et al., 2015). Moreover, as a legume, it has the ability to form nodule and fix atmospheric nitrogen, thus as a good fertilizer supplementation commonly is planted with non-legume feedstock species including switchgrass, canola, oil palm and others (Kazakoff et al., 2011).

The tree grows in various geo-climatic environments, ranging from humid tropical/sub-tropical regions (such as the coastal area of Queensland) to semiarid and cooler regions (Jiang et al., 2012). Pongamia is native to Northern Australia, India, and South-East Asia (Biswas et al., 2011). Relatives are also found in Southern China, islands north of Australia and Indochina. So far the true source of diversity is unclear. Also, Pongamia trees are seen in Dubai, Madagascar, Hawaii, and Florida (Samuel et al., 2013).

Overall, Pongamia is a valuable legume tree and compared to other biofuel feedstock species, is important due to nitrogen-fixing ability, oil composition, and high seed number. But the outcrossing nature of Pongamia prevents the establishment of uniformity that is desired by the producer. Therefore, superior trees serve as the basis for clonal propagation using different techniques such as rooted cuttings, grafts, and tissue culture techniques.

2.5 FENUGREEK IS THE LEGUME WITH MEDICINAL PROPERTIES

Plants are useful for human in various ways, and among them, some plant products are being used as a source of medicine since long ago. The efficacy and safety of herbal medicine lead to research and identify about medicinal plants (Kanchana et al., 2015). Fenugreek (*Trigonella foenum-graecum* L.) is a dicotyledonous, self-pollinated legume belonging to the family Leguminosae/Fabaceae, subfamily Papilionaceae. Stems are erect, thick, branched, green to purplish in color, length of about 30–60 cm and hairy or glabrous. Leaves are alternate, trifoliate, have three toothed ovate leaflets, with 20–25 mm long. The flowers in collections of two or solitary are white or pale yellow to light purple with a triangular shape, which are borne on leaf axils. Two types of fenugreek flowers have been described cleistogamous (closed) and aneictogamous (open) flowers. The fruits are elongated arched pods. Mature seeds are small with 5 mm long, very hard, light brown to reddish brown in color, flattened, and characterized by a groove that defines two unequal parts. The root has finger-shaped structures (Snehlata et al., 2012; Basu et al., 2014).

There are more than 260 species of fenugreek, but only 18 species of *Trigonella* are currently recognized. Some species of *Trigonella* may contain 18, 28, 30, 32 or 44 chromosomes. However, most species, such as *Trigonella foenum-graecum* L. are diploids and have $2n = 16$ chromosomes. Several Species of fenugreek are *T. foenum-graecum, T. balansae, T. corniculata, T. maritima, T. spicata, T. occulta, T. polycerata, T. calliceras, T. cretica, T. caerulea, T. lilacina, T. radiata, T. spinos*. But, the statistics of the region of cultivation is available only for *T. foenum-graecum* L. (Petropoulos et al., 2003; Snehlata et al., 2012).

As a nitrogen-fixing legume, seeds can be inoculated with suitable bacteria *Rhizobium* spp. to increase their growth potential. After inoculation, a new organ produced on root called nodule where atmospheric nitrogen is actively fixed and prepare N requirement of the plant. The fixing nitrogen mechanism subsequently reduces the application of nitrogen fertilizers (Basu et al., 2014). Genus of *Trigonella* is a reliable and effective source for improving soil characteristics and is extensively used as green manure (Naimuddin et al., 2013; Basu et al., 2014).

Fenugreek is one of the world's oldest plants. Fenugreek is grown in the continents of the west, south, Southeast Asia, Australia, Mediterranean Europe, North America, and North Africa. According to studies, India is the largest producer of Fenugreek, but it is still an unimportant crop in the continent of North America (the United States and Canada) (Acharya et al., 2008). Dry and warm conditions semi-Mediterranean climate are appropriate for proper fenugreek growth (Petropoulos et al., 2003). Water need of fenugreek is low and is counted as a rainfed crop. However, during dry phases needs watering. Drought stresses lead to the bioactive materials of fenugreek negatively affected (Basu et al., 2014).

2.5.1   GENETIC DIVERSITY OF FENUGREEK

There is a valuable genetic diversity among fenugreek genotypes for different characteristics. For instance, they are highly variable in nutritional traits. This indicates selection and crossbreeding can be used for genetic improvement of the fenugreek. Assessment of seven Iranian fenugreek genotypes showed considerable variability concerning plant nutritional minerals. In this study, several traits, including, ash, mineral content such as Mn, P, Fe, Mg, Zn, Na, Ca, K, and Cu and moisture content were measured. Studied genotypes exhibited relatively considerable differences for different parameters. The P content was ranged from 182 to 250 mg/100 gram of fresh weight (gfw) while Calcium (Ca) varied from 200.66 to 455.25 mg/100 gfw. Na and Fe content were found to be the highest in the Gaz genotype while the highest content of Mg and Mn were observed in Kashan genotype (370.1 and 0.87 mg/100 gfw, respectively). Macronutrients such as phosphorus and calcium are essential for plant growth. The Ca/P ratio of Iranian fenugreek genotypes differed from 0.80 to 2.0 that approximately is a desirable content for nutritional aspects. The highest and lowest Ca/P ratio was detected in Khansar and Ahwaz genotypes, respectively. The highest Cu and Zn content (4.1 and 2.5 mg/100 gfw, respectively) were measured in Ardestan genotype. The lowest and the highest moisture content was observed in Shahreza (80.73%) and Ardestan (86.30%) genotypes. In Mobarakeh and Khansar, the ash content was from 1.21% to 1.67% respectively (Gharneh et al., 2015).

In another study, five common fenugreek varieties of India were analyzed using molecular markers. Amplification using 9 arbitrary primers [Randomly Amplified Polymorphic DNA (RAPD) technique] revealed a total of 47 bands ranged from 200 to 5000 base pairs (bp) across studied genotypes. An average polymorphism of 62.4% was observed. Genetic diversity was also investigated by amplified fragment length polymorphism (AFLP). A total of 669 peaks were generated with 17 fluorescently labeled AFLP primer combinations. The size of the amplified fragments was from 50 to 538 bp. The mean genetic diversity across all loci was found to be 23.83 and 2.1% with RAPD and AFLP markers, respectively. Twenty-five variety-specific AFLP markers were found in all fenugreek genotypes. In this research, the RAPD technique revealed high levels of polymorphism of about 62.4% while AFLP analysis showed about 6.4% among the fenugreek varieties. AFLP marker provides reliable information and gives better coverage throughout the genome than RAPD. In general, both RAPD and AFLP techniques provide useful information on the level of polymorphism and diversity in fenugreek (Kumar et al., 2014).

A study on genetic variability and seed yield-related characteristics and direct and indirect impacts of these traits on seed yield showed that a significant amount of variability existed in morphological traits of germplasm lines such as pod length. A higher genotypic coefficient of variation (GCV) was reported for traits including test weight (14.64), primary branches per plant (15.12), total pod number per plant (18.25), secondary branches per plant (22.96), seed yield per plant (33.63) and number of pods on the main axis (51.23). This demonstrates that genetic diversity among fenugreek genotypes for these parameters is considerable and could be used for selection in breeding programs. Moreover, a high phenotypic coefficient of variation was seen in different traits. Jain et al. (2013) showed that plant height, number of pods on the main axis, total numbers of pods per plant and test weight was positively and significantly correlated with seed yield per plant. As mentioned above a high amount of diversity exists in fenugreek germplasm which is worthwhile and beneficial for breeding programs of fenugreek. Further studies are required to characterize fenugreek germplasm and identify superior genotypes.

2.5.2 APPLICATION OF FENUGREEK AS FOOD AND FEED

Fresh fenugreek leaves are used in some Indian curries and Swiss cuisines as a flavoring agent. Chopped herb greens and sprouted seeds are also used in salads (Petropoulos et al., 2003; Swaroop et al., 2014). The seeds or the extracts are utilized in gravy, sauces, meat products, relish, frozen dairy products, condiments, gelatin puddings, candy, bakery products and in alcoholic and non-alcoholic beverages. Moreover, fenugreek is utilized as a forage crop and also added to poor hay to improve its smell (fodder plant) (Nahas et al., 2009; Didarshetaban et al., 2013).

Fenugreek has high content of fibers and proteins, especially soluble dietary fiber called gum (about 20.9 g/100 g in the seed), so can alter food texture. This fiber content, in addition to the flavor ingredients, adjusts the organoleptic properties of foods. Soluble fibers also can be utilized in yogurts, nutritional, and cereal bars, nutritional beverages and dairy products. Powders of soluble fiber can be compounded with other spices and also fruit juices. Furthermore, it can be formulated as capsules or tablets with other vitamins and nutrients for direct supplements. It can also be applied in soups, sauces, sweets, milkshakes, and candies or to strengthen bakery flour for pizza, bagel, bread, noodles, cake mix, tortilla, flatbread, muffins, and baked and fried corn chips (Srinivasan et al., 2006; Im et al., 2008).

Hooda and Jood (2004) reported that wheat flour can be improved by adding 10% of fenugreek flour. This addition will promote fiber, protein content and total calcium and iron of the flour. This indicates that fenugreek can be incorporated into different dietaries as a supplement for some limiting amino acids or may also be mixed with cereals for improving their protein quality by amino acid balance. Moreover, due to antioxidant and antimicrobial effects of fenugreek seed flour, it was used in the formulation of a beef burger. Instead of soybean flour, fenugreek seed flour at 3, 6, 9 and 12% levels were used in the production of beef burgers. Substitution of soybean flour with fenugreek seed flour improved the content of essential amino acids, and also promoted retention of physiochemical quality criteria during frozen storage (Hegazy et al., 2011).

Furthermore, fenugreek seeds usage in aquaculture and fish growth will lead to improved safety and growth of fishes. Study on the Gilthead seabream (*Sparus aurata* L.) for 4 weeks revealed that fenugreek enriched diet increased the respiratory burst activity which is associated with the level of the reactive molecule containing oxygen that is toxic to for bacterial fish pathogens. Also, it served to regulate water balance by increasing the amount of fish serum proteins. The level of immunoglobulin M (IgM: safety factor molecules in the blood of fish) were increased in all fishes with a dietary regime containing fenugreek seeds. Principally fenugreek increased disease resistance and growth and survival of fishes (Awad et al., 2015).

2.5.3 CLINICAL STUDIES

The dried fenugreek seeds have been traditionally used in China, India, Egypt, and even in some parts of Europe for their beneficial health effects including antibacterial, anti-inflammatory, galactogouge, rejuvenating properties and insulinotropic (Im et al., 2008). Due to the pleasantly bitter taste of slightly sweet fenugreek seeds, they are used as a source of flavoring for foods in the forms of spice blends, teas, and curry powders. They are available in whole and ground forms (Betty et al., 2008). Medicinal values and wonderful functional of fenugreek seeds are attributed to its chemical composition. Fenugreek seeds are a rich source of protein (20–25%), dietary fiber (45–50%), mucilaginous soluble fiber (20–25%), fixed fatty acids and essential oils (6–8%), and steroidal saponins (2–5%); also some minor components such as alkaloids (trigonelline, choline, gentianine, carpaine), free unnatural amino acids (4-hydroxyisoleucine), and individual spirostanols and furastanols like diosgenin, yamogenin, and gitogenin (Trivedi et al., 2007; Didarshetaban et al., 2013). About 200 μg extracts of husk, fenugreek seed, and endosperm showed 72, 64, and 56% antioxidant activity, respectively, by free-radical scavenging method. This indicates that fenugreek has hypocholesterolemic, anti-inflammatory, antidiabetic, antioxidant, anticancer, and chemo-preventive activity due to its useful chemical constituents (Naidu et al., 2011).

2.5.3.1 CHOLESTEROL LOWERING AND ANTI-DIABETIC EFFECTS

Fenugreek seeds and leaves have long been used in Chinese medicines in the treatment of diabetes (Stark et al., 1993). Diabetes mellitus or simply diabetes is a metabolic disorder in which there are high blood sugar levels over a prolonged period. In this metabolic disorder, the body does not produce enough insulin or adequately use the hormone insulin leading to dysregulation of glucose. Besides, diabetes is a polygenic disorder, and the pathogenesis of diabetes is influenced by genetic and environmental determinants that negatively alter insulin secretion and tissue response to insulin. Now over 220 million people worldwide have diabetes also this disease is expanding at epidemic rates and is expected to double in 2030. There are many drugs to treat diabetes. However, most of these drugs have serious adverse effects (Brunetti et al., 2014). As a result, it is better to use alternative therapies such as balanced diets enriched with fresh fruits, vegetables, fiber, and antioxidants.

Studies on fenugreek seeds have demonstrated antidiabetic efficacy in different experiments. Furthermore, human studies also approved that fenugreek seeds and leaves could decrease blood cholesterol and glucose. The high insoluble fiber of seeds is effective in the treatment of diabetes by slowing digestion and intake of carbohydrate and reducing gastrointestinal absorption of glucose and helps to lower blood sugar. Fenugreek seeds are also rich sources of minerals, antioxidants, and vitamins, which assist in protecting the cells against oxidative injuries from free radical-induced (Swaroop et al., 2014).

There are several significant clinical data and scientific information on the efficiency of dietary fiber, particularly the soluble counterparts such as galactomannans or beta-glucans in the control of hypercholesterolemia. Fenugreek originated galactomannans possess the most efficacy in lowering the blood cholesterol level due to their unique structure of galactose to mannose 1:1 ratio (Brummer et al., 2003). In a study with two different dosages of fenugreek of 25 or 50 g per day of defatted fenugreek seed powder (around 50% fiber content), it has been demonstrated that fenugreek could significantly decrease blood glucose, low density lipoproteins (LDL) cholesterol and triglyceride level after 20 days (Favier et al., 1995; Prasanna et al., 2000).

Moreover, multiple human trials have uncovered that fenugreek may help lower total cholesterol in people with mild atherosclerosis or insulin or non-insulin dependent diabetes. One human double-blind trial has explained that defatted fenugreek seeds may advance the beneficial HDL cholesterol (Neelakantan et al., 2014). Furthermore, studies on animal models indicate that fenugreek may also comprise a constituent, which stimulates insulin production or sensitization. Earlier studies in animals showed that fenugreek seed extract has the potential to slow the enzymatic conversion of carbohydrates, diminished gastrointestinal absorption of glucose and consequently reduce post-prandial glucose level (Swaroop et al., 2014).

2.5.3.2 APPLICATION AS A GALACTAGOGUES

Fenugreek is one of the galactagogues that used to promote milk and stimulate breast milk secretion. It also has been proved that fenugreek has an estrogenic activity that is effective in breast milk production. Sreeja et al. (2010) suggested that fenugreek seeds include estrogen-like compounds, which raise expression of the *pS2* gene in MCF–7 cell lines (Sreeja et al., 2010; El Sakka et al., 2014). In another study, three days after childbirth, the consequences of consumption of fenugreek herbal tea on breast milk making were investigated. Results indicated that milk amount and child weight considerably differed in the fenugreek compared to control groups in the early postpartum time. Infant weights in the fenugreek group indicated a rising trend on the seventh day, and there was not any significant variation among them on the fourteenth day (El Sakka et al., 2014). The fenugreek herb helps to reduce fever and nourishes the body during an illness. Also, it can be utilized instead of the green tea to balance women's hormones. Fenugreek stimulates uterine contractions and can be helpful to childbirth. Generally, hormone precursors which are available in fenugreek seeds increase milk supply within 24 to 72 hours after first taking the herb. Fenugreek also exhibited a stimulatory impact on immune functions (Sreeja et al., 2010).

2.5.3.3 ANTIOXIDANT ACTIVITY AND ANTICARCINOGENIC EFFECTS

Antioxidant property of fenugreek is because of active phytochemicals such as phenolic and flavonoid compounds, and this has a beneficial effect on the liver and pancreas. It has been stated that fenugreek seed extract decreases lipid peroxidation and hemolysis (Kaviarasan et al., 2004). Fenugreek seed in the diet have anticarcinogenic effects; also an activity of β-glucuronidase and mucinase inhibits colon carcinogenesis. One of the causes of inducing colon cancers and irritable bowel syndromes could be low consumption of fiber in a diet. Fermentation of dietary fiber by anaerobic bacteria yields short-chain fatty acids like butyrate. Fenugreek seed with the activity of β-glucuronidase considerably reduced the free carcinogens. Mucinase helped in hydrolyzing the protective mucin, and this was correlated with the presence of fiber, flavonoids, and saponins (Devasena et al., 2003). Dixit et al. (2005) defined the aqueous fraction of fenugreek exhibits higher antioxidant activity rather than other fractions. Germinated fenugreek seeds have a significant antioxidant activity, and this may be related to the presence of flavonoids and polyphenols. Mustard and fenugreek seeds also have hypoglycemic and anti-hyperglycemic activities in diabetic mice, which could be due to the presence of antioxidant carotenoids in these spices (Dixit et al., 2005).

However, fenugreek is a natural product, but its consumption is not completely free of unfavorable effects. The most important side effects include premature labor in pregnant women and lead to hypersensitivity reactions, internal bleeding perturbations. Furthermore, diarrhea, and pyrosis, hypersensitivity reactions and hematemesis are other side effects of fenugreek. Though, several researchers expressed the antidiabetic features of fenugreek, but yet it could not be used alone to treat diabetics. It is better used as an add-on agent to the drugs for controlling diabetes (Basu et al., 2014).

2.6   LUCERNE AS A COVER CROP AND FORAGE

Cover-crops are important for providing resources for biodiversity; improving nutrient management, organic matter content and soil structures; mitigation of diffuse pollution; fixing nitrogen and suppressing weed. On the other hand, for the optimal use of land, designing multi-functional cover crops would be reasonable (Storkey et al., 2015). The lucerne (*Medicago sativa*) plant is perennial, tetraploid, and open pollinated. It has a deep root system, tolerant to grazing and its height reached to a maximum of 90 cm (Fyad-Lameche et al., 2015; Storkey et al., 2015).

Today, lucerne has been domesticated to create better patterns of recurrent harvest and regrowth. Lucerne is cultivated around the world such as New Zealand, South America, and Australia as a specialty fodder plant and used extensively for grazing (Humphries et al., 2013). Forage consumed directly by grazing animals on pasture is ~70% of the diet in Australia. The quality of Lucerne in spring and early summer increased livestock growth rates, especially for producing heavy lambs at weaning. Lucerne for dry environments is most efficient and provides the required soil moisture for forage (Watson et al., 2006; Anderson et al., 2014).

Lucerne as a legume could form nodules through an essential symbiosis with the *Sinorhizobium meliloti* in the soil. Accordingly, lucerne is a nitrogen-fixing plant with a global production of over four hundred million tons (Owusu et al., 2013). Lucerne with a deep rooting system and white clover with shallow-rooting together could fix large proportions of their N requirement symbiotically. Therefore, small amounts of soil nitrogen acquired and let spared soil N to be used by non-leguminous plants. The nitrogen fixation and deep-rooting characteristics make lucerne to be grown on low soil N levels as well as low soil water content (Pirhofer et al., 2013).

On the basis of the potential benefits to human health, there is an increased interest in producing milk containing lower-SFA and higher unsaturated FA (Vazirigohar et al., 2014). Samkova et al. (2014) found that fresh lucerne can increase the proportion of total C18 FAs in both the cow breeds (Czech Fleckvieh and Holstein) with only mild differences between the breeds. The fatty acid composition was altered after feeding with lucerne. The changes in stearic acid and oleic acid were more extensive in Holstein cows and linoleic and linolenic acids in Czech Fleckvieh cows. The different breed response on fresh lucerne was probably affected by the diverse fat yield of the breeds (Samkova et al., 2014). In addition, there have been several studies that have shown the advantage of lucerne-based pastures for enhancing meat and wool production in southern Australia (Humphries et al., 2013). CH_4 produced by cows that fed with lucerne diet is much reduced compared to grass hay. CH_4 production decrease is because of the presence of high malate content in lucerne. However, recent data showed that malate was efficient at a dose of 7.5% in the dry matter but not at lower amounts. When lucerne represents a great part of the diet or is grazed, malate in lucerne decreases enteric CH_4, not when lucerne is included in the diet as a protein source (Doreau et al., 2014).

Lucerne is an outstanding crop, mainly for organic growing systems. Due to its high nitrogen fixation and developing deep root systems, it improves soil structure, water infiltration, and soil organic matter content. Lucerne is also considered a drought-tolerant crop due to the deep root system. It can extract water from soil depths up to 2 meters and help to decrease losses of soil water by deep drainage. In crop rotations, lucerne can be used with shallow-rooted cereals and pulses, and this may improve water use efficiency (WUE) of the entire cropping system (Raza et al., 2013).

Lucern hay is a good source of slow release carbohydrates, proteins, minerals: sodium (Na), chloride (Cl), manganese (Mg), calcium (Ca), potassium (K), phosphate (P), sulfur (S); and vitamins. Lucerne is mainly used to make hay and silage, but can also be used for grazing purposes, because of its high yield, quality, and wide adaptability to different climates and soil types. The yield of crops such as potatoes, rice, cucumber, lettuce, and tomatoes often increased which are grown after lucern (Mustafa et al., 2001; Mielmann et al., 2013).

Human has also used lucerne in different forms. It has been used in the form of tea, vegetable, soup as well as raw and cooked salad. Moreover, lucerne meal was incorporated into a special "cereal mixture" and used in the feeding of small children. Moreover, lucerne was utilized to increase the protein, dietary fiber, mineral, and vitamin content of wheat flour. Lucerne is also one of the most popular sprouts available on the European markets (Mielmann et al., 2013). Lucerne biomass has considerable potential for biotechnology applications including production of soluble protein concentrates, carotenoids, vitamins, minerals, growth factors, pharmaceutical agents, transgenic enzymes and cosmetic products (Sreenath et al., 2001). Lucerne protein concentrate is a superior and recognized source of high quality for human consumptions. The limits of lucerne protein concentrate in human food are their negative sensory properties including dark color–due to polyphenols – poor solubility, granulous texture, and their grassy taste possibly because of the saponin content (D'Alvise et al., 2000).

2.7   CONCLUSION AND FUTURE PERSPECTIVES

Analysis of diversity whether existing or induced would be beneficial because it helps to identify a useful donor for some parameter for breeding programs and also select appropriate germplasm for specific environmental conditions. Here we characterized this fact through examples from legume family including soybean as an oilseed, *Pongamia pinnata* as biofuel feedstock tree, fenugreek as a medicine and vegetable plant and lucerne as a cover crop.

One component of many biological molecules is nitrogen. Thus, its availability is critical for plant growth and reproduction and, as such, is a limiting nutrient. However nitrogen gas comprises the highest amount of atmospheric composition, it is unavailable to most organisms. Many legumes overcome this limitation by initiating a symbiotic relationship with soil bacteria, broadly named as rhizobia. Therefore, nodulation, which occurred through the symbiotic process, is evolutionarily, ecologically, and agriculturally significant. In addition, compared to industrial nitrogen fertilizer, BNF is highly cost-efficient, environmentally friendly and sustainable (Reid et al., 2011; Jensen et al., 2012). Therefore, legume plants are important to agriculture and this ability confers significant advantages to leguminous species compared to non-leguminous species, in particular in nitrogen-deficient soils. Legume family is diverse and accommodates nearly 20,000 species, which are distributed throughout the world (Doyle et al., 1994; Sprent et al., 2007). Among them, soybean is a good source of protein and oilseed which has large amounts of the polyunsaturated fatty acids involved stearic acid, linolenic acid, palmitic acid, oleic acid, and linoleic acid. It is also used in many processed foods and substitutes for dairy products (Kanchana et al., 2015).

Pongamia pinnata is a legume tree by fast vegetative growth, oil-rich seeds (20,000 seeds per tree). Its oil can be converted by hydrogenation to aviation A1 jet fuel. Also, it is used in ornamental purposes and biodiesel production (Scott et al., 2008; Kazakoff et al., 2011). *Pongamia* species displays a huge range of biodiversity (under the influence of genotype and environment) as trees are outcrossing and heterozygosity is high. This is worthwhile for selection plant with outstanding traits (Gresshoff et al., 2015).

Fenugreek is a self-pollinated legume with more than 260 species (Snehlata et al., 2012). It uses as a spice, flavor enhancing, forage, and medicine. It has properties such as anti-diabetic, antioxidant, anti-carcinogenic, anti-leukemic, anti-hyperlipidemic, and anti-inflammatory (Basu et al., 2014). An enormous variation exists in fenugreek germplam, and this variation could be exploited and used in fenugreek breeding programs (Gharneh et al., 2015).

The lucerne (*Medicago sativa*) plant is perennial, with a deep root system and is appropriate and tolerant to grazing (Fyad-Lameche et al., 2015; Storkey et al., 2015). High malate content in Lucerne diet leads to a decrease in NH_4 production by cows (Mielmann et al., 2013; Samkova et al., 2014). It is an important crop, especially for organic farming systems due to its high nitrogen fixation, and improves soil structure, water infiltration and soil organic matter content (Raza et al., 2013). Human has also used Lucerne in forms of tea, vegetable, soup as well as a raw and cooked salad (Mielmann et al., 2013). Lucerne biomass also has considerable potential for biotechnology applications including production of soluble protein concentrates, carotenoids, vitamins, minerals, growth factors, pharmaceutical agents, transgenic enzymes and cosmetic products (Sreenath et al., 2001). Taken together, legume family is very diverse, and nitrogen-fixing capability confers significant advantages to leguminous species compared to non-leguminous species. Their N requirements compared to non-leguminous species are considerably low, and they could grow in nitrogen-deficient soils. This capability also has beneficial to the environment because of participating in reducing greenhouse gas emissions. They cover different aspects of human needs; used as food and animal feed crops, biofuel resources, medicine, ornamental (e.g., wisteria), or timber (e.g., acacia/wattle). Since globally, we are confronted by ever-growing food and energy challenges; legume species are a great value for the provision of these demands. Furthermore, legume species are suitable for future sustainable agriculture that requires solutions to climate change mitigation and novel systems.

KEYWORDS

- AFLP
- biodiversity
- biofuel
- climate change
- forage
- galactagogues
- genetic diversity
- legume

- **lucerne**
- **medicinal plants**
- **nitrogen fixation**
- ***Pongamia***
- **RAPD**
- **Rhizobia**
- **soybean**

REFERENCES

Acharya, S. N., Thomas, J. E., & Basu, S. K., (2008). Fenugreek an alternative crop for semiarid regions of North America. *Crop Science, 48*, 841–853.

Anderson, D., Anderson, L., Moot, D. J., & Ogle, G. I., (2014). Integrating Lucerne (*Medicago sativa, L.,*) into a high country merino system. *Proceedings of the New Zealand Grassland Association, 76*, 29–34.

Awad, E., Cerezuela, R., & Esteban, M. A., (2015). Effects of fenugreek (*Trigonella foenum graecum*) on gilthead seabream (*Sparus aurata, L.,*) immune status and growth performance. *Fish and Shellfish Immunology, 45*, 454–464.

Baggs, E., (2003). Nitrous oxide emissions following application of residues and fertilizer under zero and conventional tillage. *Plant and Soil, 254*, 361–370.

Basu, A., Basu, S. K., Kumar, A., Sharma, M., Chalghoumi, R., Hedi, A., et al., (2014). Fenugreek (*Trigonella foenum-graecum, L.,*), a potential new crop for Latin America. *American Journal of Social Issues and Humanities, 4*, 148–162.

Betty, R., (2008). The many healing virtues of fenugreek. *Spice India, 1*, 17–19.

Biswas, B., & Gresshoff, P. M., (2014). The role of symbiotic nitrogen fixation in sustainable production of biofuels. *International Journal of Molecular Sciences, 15*, 7380–7397.

Biswas, B., Scott, P. T., & Gresshoff, P. M., (2011). Tree legumes as feedstock for sustainable biofuel production: Opportunities and challenges. *Journal of Plant Physiology, 168*, 1877–1884.

Bozanic, R., (2006). Proizvodnja, svojstva i fermentacija sojinog mlijeka. *Mlj., 56*, 233–254.

Brummer, Y., Cui, W., & Wang, Q., (2003). Extraction, purification, and physicochemical characterization of fenugreek gum. *Food Hydrocolloids, 17*, 229–236.

Brunetti, A., Chiefari, E., & Foti, D., (2014). Recent advances in the molecular genetics of type 2 diabetes mellitus. *World J. Diabetes, 5*, 128–140.

Chaer, G. M., (2011). Nitrogen-fixing legume tree species for the reclamation of severely degraded lands in Brazil. *Tree Physiology, 31*, 139–149.

Choudhury, R., Basak, S., Ramesh, A. M., & Rangan, L., (2014). Nuclear DNA content of *Pongamia pinnata, L.,* and genome size stability of *in vitro*-regenerated plantlets. *Pro., 251*, 703–709.

D'Alvise, N., Lesueur–Lambert, C., Fertin, B., Dhulster, P., & Guillochon, D., (2000). Hydrolysis and large-scale ultrafiltration study of alfalfa protein concentrate enzymatic hydrolysate. *E. Mic. Tech.*, *27*, 286–294.

Devasena, T., & Menon, V. P., (2003). Fenugreek affects the activity of β-glucuronidase and mucinase in the colon. *P. Res.*, *17*, 1088–1091.

Didarshetaban, M. B., Pour, S., & Reza, H., (2013). Fenugreek (*Trigonella foenum-graecum, L.,*) as a valuable medicinal plant. *I. J. Adv. Biol. Biom. Res.*, *1*, 922–931.

Dixit, P., (2005). Antioxidant properties of germinated fenugreek seeds. *Phytotherapy Research*, *19*, 977–983.

Doreau, M., Ferlay, A., Rochette, Y., & Martin, C., (2014). Effects of dehydrated Lucerne and soya bean meal on milk production and composition, nutrient digestion, and methane and nitrogen losses in dairy cows receiving two different forages. *Animal*, *8*, 420–430.

Doyle, J. J., (1994). Phylogeny of the legume family - an approach to understanding the origins of nodulation. *Annual Review of Ecology and Systematics*, *25*, 325–349.

El Sakka, A., Salama, M., & Salama, K., (2014). The effect of fenugreek herbal tea and palm dates on breast milk production and infant weight. *J. Pedi. Sci.*, *6*, e202.

Eskandari, M., Cober, E. R., & Rajcan, I., (2013). Genetic control of soybean seed oil: I. QTL and genes associated with seed oil concentration in RIL populations derived from crossing moderately high-oil parents. *Theor. Appl. Genet.*, *126*, 483–495.

Favier, M. L., Moundras, C., Demigne, C., & Remesy, C., (1995). Fermentable carbohydrates exert a more potent cholesterol-lowering effect than cholestyramine. *B.B.A. Lip. Lip. Meta.*, *1258*, 115–121.

Ferguson, B. J., Indrasumunar, A., Hayashi, S., Lin, M. H., Lin, Y. H., Reid, D. E., & Gresshoff, P. M., (2010). Molecular analysis of legume nodule development and autoregulation. *J. Int. Plant Biol.*, *52*, 61–76.

Fyad-Lameche, F. Z., Iantcheva, A., Siljak-Yakovlev, S., & Brown, S. C., (2015). Chromosome number, genome size, seed storage protein profile and competence for direct somatic embryo formation in Algerian annual *Medicago* species. *Plant Cell, Tissue, and Organ Culture*, 124(3), 1-10.

Gage, D. J., (2004). Infection and invasion of roots by symbiotic, nitrogen-fixing rhizobia during nodulation of temperate legumes. *Microbiology and Molecular Biology Reviews*, *68*, 280–300.

Gharneh, H. A. A., & Davodalhosseini, S., (2015). Evaluation of mineral content in some native Iranian Fenugreek (*Trigonella foenum-graceum, L.,*) genotypes. *Journal of Earth, Environment, and Health Sciences*, *1*, 38.

Graham, P. H., & Vance, C. P., (2003). Legumes: Importance and constraints to greater use. *Plant Physiology*, *131*, 872–877.

Gresshoff, P. M., & Rolfe, B. G., (1978). Viability of *Rhizobium* bacteroids isolated from soybean nodule protoplasts. *Planta*, *142*, 329–333.

Gresshoff, P. M., Hayashi, S., Biswas, B., Mirzaei, S., & Indrasumunar, A., (2015). The value of biodiversity in legume symbiotic nitrogen fixation and nodulation for biofuel and food production. *Journal of Plant Physiology*, *172*, 128–136.

Gresshoff, P. M., Skotnicki, M., Eadie, J., & Rolfe, B. G., (1977). Viability of *Rhizobium* bacteroids isolated from clover nodule protoplasts. *Plant Sci. Lett.*, *10*, 299–304.

Hegazy, A., (2011). Influence of using fenugreek seed flour as antioxidant and antimicrobial agent in the manufacturing of beef burger with emphasis on frozen storage stability. *World Journal of Agricultural Sciences*, *7*, 391–399.

Hooda, S., & Jood, S., (2004). Nutritional evaluation of wheat-fenugreek blends for product making. *Plant Foods for Human Nutrition*, *59*, 149–154.

Humphries, A., (2013). Future applications of Lucerne for efficient livestock production in southern Australia. *Crop and Pasture Science*, *63*, 909–917.

Im, K. K., & Maliakel, B. P., (2008). Fenugreek dietary fiber a novel class of functional food ingredient. *Agro Food Industry Hi-Tech.*, *19*, 18–21.

Jain, A., Singh, B., Solanki, R., Saxena, S., & Kakani, R., (2013). Genetic variability and character association in fenugreek (*Trigonella foenum-graecum, L.,*). *Int. J. Seed Spices*, *3*, 22–28.

Jensen, E. S., Peoples, M. B., Boddey, R. M., Gresshoff, P. M., Hauggaard-Nielsen, H., Alves, B. J. R., & Morrison, M. J., (2012). Legumes for mitigation of climate change and the provision of feedstock for biofuels and biorefineries: A review. *Agronomy for Sustainable Development*, *32*, 329–364.

Jiang, Q., Yen, S. H., Stiller, J., Edwards, D., Scott, P. T., & Gresshoff, P. M., (2012). Biochemical and morphological diversity of the legume biofuel tree *Pongamia pinnata*. *J. Plant Genome Sci.*, *1*, 54–67.

Kanchana, P., Santha, M. L., & Raja, K. D., (2015). A review on *Glycine max* (L.) Merr. (soybean). *World Journal of Pharmacy and Pharmaceutical Sciences*, *5*, 356–371.

Kaviarasan, S., (2004). Polyphenol-rich extract of fenugreek seeds protect erythrocytes from oxidative damage. *Plant Foods for Human Nutrition*, *59*, 143–147.

Kazakoff, S. H., Gresshoff, P. M., & Scott, P. T., (2011). *Pongamia pinnata*, a sustainable feedstock for biodiesel production. *Energy Crops*, 233–254.

Klein-Marcuschamer, D., Turner, C., Allen, M., Gray, P., Dietzgen, R. G., & Gresshoff, P. M., (2013). Techno-economic analysis of renewable aviation fuel from microalgae, *Pongamia pinnata*, and sugarcane. *Biofuels, Bioproducts, and Biorefining*, *7*, 416–428.

Knothe, G., Krahl, J., & Van, G. J., (2015). *The Biodiesel Handbook*. AOCS Press, Urbana, USA.

Kumar, V., Srivastava, N., Singh, A., Vyas, M. K., Gupta, S., & Katudia, K., (2014). Genetic diversity and identification of variety-specific AFLP markers in fenugreek (*Trigonella foenum-graecum*). *African Journal of Biotechnology*, *11*, 4323–4329.

Liu, K., (1997). Soybeans: Chemistry and nutritional value of soybean components. In: Liu, K. S., (ed.), *Soybean: Chemistry, Technology, and Utilization* (pp. 25–113). Chapman, and Hall, New York.

Messina, M. J., (1997). Soyfoods: Their role in disease prevention and treatment. In: Liu, K. S., (ed.), *Soybean: Chemistry, Technology, and Utilization* (pp. 442–477). Chapman, and Hall, New York.

Mielmann, A., (2013). The utilization of Lucerne (*Medicago sativa*): A review. *British Food Journal*, *115*, 590–600.

Mirzaei, S., Batley, J., Ferguson, B. J., & Gresshoff, P. M., (2014). Transcriptome profiling of the shoot and root tips of S562L, a soybean GmCLAVATA1A mutant. *Atlas J. Biol.*, *3*, 183–205.

Mirzaei, S., Batley, J., Meksem, K., Ferguson, B. J., & Gresshoff, P. M., (2017). Neodiversification of homologous CLAVATA1-like receptor kinase genes in soybean leads to distinct developmental outcomes. *Scientific Reports, 7,* 8878. doi: 10.1038/s41598-017-08252-y.

Mustafa, A. F., Christensen, D. A., & McKinnon, J. J., (2001). Chemical composition and ruminal degradability of Lucerne (*Medicago sativa*) products. *Journal of the Science of Food and Agriculture, 81,* 1498–1503.

Nahas, R., & Moher, M., (2009). Complementary and alternative medicine for the treatment of type 2 diabetes. *Canadian Family Physician, 55,* 591–596.

Naidu, M. M., Shyamala, B. N., Naik, J. P., Sulochanamma, G., & Srinivas, P., (2011). Chemical composition and antioxidant activity of the husk and endosperm of fenugreek seeds. *LWT-Food Science and Technology, 44,* 451–456.

Naimuddin, N., Aishwath, O. P., Lal, G., Kant, K., Sharma, Y. K., & Ali, S. F., (2013). Response of *Trigonella foenum-graecum* to organic manures and *Rhizobium* inoculation in a Typic Haplustept. *Journal of Spices and Aromatic Crops, 23*(1), 110–114.

Neelakantan, N., (2014). Effect of fenugreek (*Trigonella foenum-graecum, L.,*) intake on glycemia: A meta-analysis of clinical trials. *Nutrition Journal, 13,* 1.

Oldroyd, G. E., Murray, J. D., Poole, P. S., & Downie, J. A., (2011). The rules of engagement in the legume-rhizobial symbiosis. *Annual Review of Genetics, 45,* 119–144.

Owusu-Sekyere, A., Kontturi, J., Hajiboland, R., Rahmat, S., Aliasgharzad, N., Hartikainen, H., & Seppänen, M. M., (2013). Influence of selenium (Se) on carbohydrate metabolism, nodulation, and growth in alfalfa (*Medicago sativa* L.). *Plant and Soil, 373,* 541–552.

Padilla, J. E., (1987). Nodule-specific glutamine synthetase is expressed before the onset of nitrogen fixation in *Phaseolus vulgaris, L. Plant Molecular Biology, 9,* 65–74.

Petropoulos, G. A., (2003). *Fenugreek: The Genus Trigonella.* CRC Press, Florida.

Pirhofer-Walzl, K., Eriksen, J., Rasmussen, J., Høgh-Jensen, H., & Søegaard, K., (2013). Effect of four plant species on soil 15N-access and herbage yield in temporary agricultural grasslands. *Plant and Soil, 371,* 313–325.

Powlson, D., (2013). Carbon sequestration in soils has the potential for climate change mitigation been over stated? *EGU General Assembly Conference Abstracts, 15,* 123.

Prasanna, M., (2000). Hypolipidemic effect of fenugreek: A clinical study. *Indian Journal of Pharmacology, 32,* 34–36.

Raza, A., Friedel, J. K., Moghaddam, A., Ardakani, M. R., Loiskandl, W., Himmelbauer, M., & Bodner, G., (2013). Modeling growth of different Lucerne cultivars and their effect on soil water dynamics. *Agricultural Water Management, 119,* 100–110.

Reid, D. E., Ferguson, B. J., Hayashi, S., Lin, Y. H., & Gresshoff, P. M., (2011). Molecular mechanisms controlling legume autoregulation of nodulation. *Annals of Botany, 108,* 789–795.

Samkova, E., Certikova, J., Spicka, J., Hanus, O., Pelikánová, K. M., & Kvac, M., (2014). Eighteen carbon fatty acids in milk fat of Czech Fleckvieh and Holstein cows following feeding with fresh Lucerne (*Medicago sativa, L.,*). *Animal Science Papers and Reports, 32,* 209–218.

Samuel, S., Scott, P. T., & Gresshoff, P. M., (2013). Nodulation in the legume biofuel feedstock tree *Pongamia pinnata. Agricultural Research, 2,* 207–214.

Santos, E. M., Piovesan, N. D., De Barros, E. G., & Moreira, M. A., (2013). Low linolenic soybeans for biodiesel: Characteristics, performance, and advantages. *Fuel, 104,* 861–864.

Schmutz, J., Cannon, S. B., Schlueter, J., Ma, J., Mitros, T., & Nelson, W., (2010). Genome sequence of the palaeopolyploid soybean. *Nature, 463*, 178–183.

Schulte, L. R., Ballard, T., Samarakoon, T., Yao, L., Vadlani, P., Staggenborg, S., & Rezac, M., (2013). Increased growing temperature reduces the content of polyunsaturated fatty acids in four oilseed crops. *Industrial Crops and Products, 51*, 212–219.

Scott, P. T., Pregelj, L., Chen, N., Hadler, J. S., Djordjevic, M. A., & Gresshoff, P. M., (2008). *Pongamia pinnata*: An untapped resource for the biofuels industry of the future. *Bioenergy Research, 1*, 2–11.

Singh, J. S., (2015). Biodiversity: Current perspective. *Climate Change and Environmental Sustainability, 3*, 71–72.

Snehlata, H. S., & Payal, D. R., (2012). Fenugreek (*Trigonella foenum-graecum, L.,*): An overview. *International Journal of Current Pharmaceutical Review and Research, 2*, 169–187.

Sprent, J. I., (2007). Evolving ideas of legume evolution and diversity: A taxonomic perspective on the occurrence of nodulation. *New Phytologist, 174*, 11–25.

Sreeja, S., Anju, V. S., & Sreeja, S., (2010). *In vitro* estrogenic activities of fenugreek *Trigonella foenum graecum* seeds. *Indian Journal of Medical Research, 131*, 814.

Sreenath, H. K., Koegel, R. G., Moldes, A. B., Jeffries, T. W., & Straub, R. J., (2001). Ethanol production from alfalfa fiber fractions by saccharification and fermentation. *Process Biochemistry, 36*, 1199–1204.

Srinivasan, K., (2006). Fenugreek (*Trigonella foenum-graecum*): A review of health beneficial physiological effects. *Food Reviews International, 22*, 203–224.

Stark, A., & Madar, Z., (1993). The effect of an ethanol extract derived from fenugreek (*Trigonella foenum-graecum*) on bile acid absorption and cholesterol levels in rats. *British Journal of Nutrition, 69*, 277–287.

Stavi, I., & Lal, R., (2013). Agroforestry and biochar to offset climate change: A review. *Agronomy for Sustainable Development, 33*, 81–96.

Storkey, J., Doring, T. F., Baddeley, J. A., Collins, R., Roderick, S., Stobart, R., Jones, H. E., & Watson, C. A., (2015). Framework for designing multi-functional cover crops. *Aspects of Applied Biology: Getting the Most Out of Cover Crops, 129*, 7–12.

Suter, H., (2014). Stabilized nitrogen fertilizers to reduce greenhouse gas emissions and improve nitrogen use efficiency in Australian agriculture. *International Symposium on Managing Soils for Food Security and Climate Change Adaptation and Mitigation*, Vienna, Austria.

Swaroop, A., Bagchi, M., Kumar, P., Preuss, H. G., Tiwari, K., Marone, P. A., & Bagchi, D., (2014). Safety, efficacy, and toxicological evaluation of a novel, patented anti-diabetic extract of *Trigonella foenum-graecum* seed extract (Fenfuro). *Toxicology Mechanisms and Methods, 24*, 495–503.

Tilman, D., Isbell, F., & Cowles, J. M., (2014). Biodiversity and ecosystem functioning. *Annual Review of Ecology, Evolution, and Systematics, 45*, 471.

Trivedi, P. D., (2007). A validated quantitative thin-layer chromatographic method for estimation of diosgenin in various plant samples, extract, and market formulation. *Journal of AOAC International, 90*, 358–363.

Vazirigohar, M., Dehghan-Banadaky, M., Rezayazdi, K., Krizsan, S. J., Nejati-Javaremi, A., & Shingfield, K. J., (2014). Fat source and dietary forage-to-concentrate ratio influences milk fatty-acid composition in lactating cows. *Animal, 8*, 163–174.

Watson, P., (2006). *Dairying for Tomorrow: Survey of NRM Practices on Dairy Farms.* Down to Earth Consulting, Frankston, Victoria.

CHAPTER 3

DIVERSITY AND CONSERVATION OF SPONTANEOUS *CARUM, VICIA*, AND *LATHYRUS* SPECIES IN LITHUANIA

MARŠALKIENĖ NIJOLĖ

Institute of Environment and Ecology, Aleksandras Stulginskis University, Studentų Str. 11, LT–53361 Akademija, Kaunas, Lithuania

3.1 INTRODUCTION

World plant biodiversity, its conservation, and prospects for practical use are becoming an increasingly compelling dilemma in the 21st century. The relevance of this problem is primarily determined by the rapid extinction of plant species and their populations (Raven, 2000; Crane, 2001; Sliesaravičius and Petraitytė, 2001). The prognosis delivers bad news – almost 2/3 of all organisms will have become extinct by the end of the 21st century. Over the last millennium, nearly 10% of the total number of plant species has already become extinct (Raven, 2000). Such severe extinction of species can cause ecological and biological disasters and irreversible loss of genetic resources (Sliesaravičius and Petraitytė, 2001; Petraitytė et al., 2003).

The constantly expanding number of populations calls for the booming of food and forage resources also. Besides forest felling, which results in the enlarging of the cultivated land area, the use of fertilizers and pesticides, in addition to genetically modified plant varieties, is steadily increasing. These human activities also exert a great effect on the extinction of plant species and decline their biodiversity (Brunner et al., 2001). As an alternative to the chemisation of agriculture and the spread of genetically modified varieties, organic agriculture has been rapidly developing worldwide since 1990. The purposes of organic agriculture are to produce the ecologically clean products (Lotter, 2003; Gipson et al., 2007), minimize environmental pollution, conservation of phytodiversity and use it for human needs.

Biodiversity conservation contains an extensive regime for the protection and conservation of biodiversity including the protection of species, preservation of their genetic variability, ecological communities, and landscape variety; and it needs immediate attention of the society (Balmford et al., 2005; Partel et al., 2005). The national collection, storage, reproduction, and use of national resources for rational needs in plant breeding are ones of the tasks of genetic resources. Knowledge of multiplication, ecological peculiarities, and response to changing environmental and anthropogenic factors makes the conservation of plant species possible. A promising way to conserve phytobiodiversity is an introduction, creation of collections, versatile tests of plant characteristics and selection of promising forms for breeding (Petraitytė et al., 2002; Petraitytė, 2003).

Compared to other community types, the grasslands of Europe have a rich flora, and they may develop a very high density of small-scale species (Partel et al., 1996; Partel et al., 2005). During the last hundred years, the detected decline of grassland area all across Europe has been identified (van Dijk, 1991). The main reasons of this process are the greater efficiency of the grassland farming in cultivated stands than in permanent natural grasslands, in addition to the air pollution and urbanization (Thompson and Jones, 1999; Stevens et al., 2004; Partel et al., 2005) are also playing a considerable role.

In the Republic of Lithuania, together with the decline of natural meadows, also the habitats of caraway, pea, and vetch shrink; as a consequence, the species biodiversity and genetic resources become sparse (Petraitytė, 2005; Petraitytė and Dastikaitė, 2007). In order to preserve species diversity, it is vital to accumulate the genetic fund, investigate intraspecific varying, assess, and select the most valuable samples, which would be suitable for utilizing them in breeding (Budvytytė, 2000; Petraitytė and Dastikaitė, 2007). Having accumulated the comprehensive information, the selection of the most valuable ecotypes and individuals, their introduction into selection programs become possible (Petraitytė and Dastikaitė, 2007).

National Centre for Plant Genetic Resources was founded at the Lithuanian Institute of Agriculture in 1994. The main objective was to develop the national conservation network of plant genetic resources in Lithuania (Petraityte, 2005). This project activated the coordination work and cooperation among the research institutions, which were involved in plant genetic resources conservation and investigation (Dapkevičius et al., 2008; Gelvonauskis, 2013). Since 1996, the researches of genetic resources of the economically valuable species, such as *Carum* genus: *Carum carvi* L. (caraway); *Vicia* genus: *V. angustifolia* (narrow-leaved vetch), *V. hirsuta Grey* (hairy vetch), *V. villosa* Roth (winter vetch), *V. cracca* L. (tufted vetch), *V. sepium* L. (bush vetch); *Lathyrus* genus: *L. pratensis* L. (meadow pea), *L. sylvestris* L. (flat pea), *L. palustris* L. (marsh pea) (Table 3.1) were carried out *ex situ* in the Lithuanian University of Agriculture, now – Aleksandras Stulginskis University (ASU). The aim of this work was to accumulate and analyses of the greatest possible diversity of these species. The research is the basis for the selection of valuable ecotypes that are perspective for selection, accumulation, and conservation of species genefund (Petraitytė, 2005; Petraitytė and Dastikaitė et al., 2007). The program ended in 2008 and, presently, the project on "Keeping and restoring field collections of national plant resources" is being continued.

TABLE 3.1 Seeds Samples of *Carum, Vicia,* and *Lathyrus* Spontaneous Species Collected in 1996–2000 (Sliesaravičius and Petraitytė, 2001)

Name of plant	Number of collected samples	Sown in collection
Total number of *Carum carvi* L.	127	127
Vicia genus		
V. angustifolia L.	15	15
V. hirsuta Gray	21	15
V. villosa Roth	35	35
V. cracca L.	75	75
V. sepium L.	14	13
V. pisiformis L.	1	1
Total number of *Vicia* genus:	161	149
Lathyrus genus		
L. pratensis L.	70	70
L. sylvestris L.	11	11
L. palustris L.	2	2
Total number *Lathyrus* genus:	83	83

3.2   THE NATURAL SITUATION OF LITHUANIA

Lithuania is not a big country with an area of 65,200 km², lays at the edge of the North European Plain. About 25,000–22,000 years BP (Before Present) the landscape of country was shaped by the glaciers of the last Ice Age. Lithuania's terrain is an alternation of moderate lowlands and highlands (Chomskis et al., 1958; Eidukevičienė, 2013). More than 70% of the country territory is covered by lowlands (Basalykas et al., 1965).

The landscape is punctuated by 4,433 larger or smaller lakes, and the majority of them are found in the eastern part of the country. Lithuania also has 758 rivers longer than 10 kilometers, and largest one is the Nemunas (917 km). Climate is transitional between the oceanic and continental climate (Gailiušis et al., 2001). The warm period (average temperature is > 0°C) in the biggest part of country lasts 230–250 days, in western regions (near the seaside) 260–270 days. Average day-night temperatures are –6°C in January and 16°C in July. Temperatures can reach 30 or 35°C in the summer time and –20°C in winter. Vegetation period is rather long: from 202 days in west and 169 days in the eastern part of the country. The sum of the effective temperatures over the period when the average day and night temperature tops 10°C increases going southern east: from 1950 in the north west up to 2300°C – in the southeast districts (Basalykas et al., 1965; Gailiušis et al., 2001).

The parent materials of soils in Lithuania vary in the age and genesis, the most common being Quaternary deposits. The thickness of these deposits in average 80–120 m thick and varies from less than 10 m in the northern part to 200–300 m in the Baltija and Zemaiciai Heights (Buivydaite, 2005). Glacial deposits are: morainic, glaciofluvial, and limnoglacial, and, in some places, alluvial, eolian, and organic. Preliminary estimates show that Albeluvisols occupy 30% of the country, Luvisols 27%, Cambisols (13%), Arenosols (12%), Podzols (11%), mainly in forest areas, and Gleysols and Histosols (5.3%) in the depressions. In Lithuania, about 46.3% of the soils are very close to neutral (pH_{KCl} – 6.6–6.9) or neutral (pH_{KCl} – 7.0) but more than 16% of soils are acid and very acid (Buivydaite, 2005).

From the geobotanical point of view, Lithuania's location is exceptional. Lithuania lays in both in the boreal and the broadleaved (angiosperm) forest belts (Basalykas et al., 1965). The Flora of the country is estimated to comprise about 10,600 species (1350 of these are vascular). 2/3 of Lithuania territory is covered by natural or semi-natural flora

(Balevičienė et al., 1998). Country's territory consists of 30% woodlands (Karazija, 1989).

The geographical situation, a great amount of precipitation in growth season, favorable wintering conditions, dense hydrographic net (1 km^2 – 1 km) and human activity have predetermined the prevalence of meadows on Lithuanian landscape (Balevičienė et al., 1998). Meadows and natural pastures cover 7.6% of the Lithuanian territory (Eidukeviciene, 2013). Only flood meadows near rivers and lakes and small fragments between arable fields are considered as natural ones. The Lithuanian meadow communities are ascribed to 35 associations included into 5 classes: *Cl. Asteretea tripolii*, *Cl. Molinio-Arrhenatheretea*, *Cl. Festuco-Brometea*, *Cl. Trifolio-Geranietea*, *Cl. Nardetea* (Rašomavičius, 1998).

3.3 DIVERSITY OF CARAWAY (*CARUM CARVI* L.) POPULATION IN LITHUANIA

3.3.1 CHARACTERISTICS OF CARAWAY (*CARUM CARVI* L.)

There are known about 30 species of *Carum* genus, which differ by the form of their fruits, size, the color of stem, form of leaves and leafstalks and other features. Caraway (*Carum carvi* L.) is mostly predominant, and nowadays it has the greatest practical importance among the plants that concentrate the essential oil in its fruits (Petraitytė, 2005). Caraway fruits are one of the main and most widely used wild spice plants in Lithuania and north Europe (Dastikaitė, 1997; Petraitytė and Dastikaitė, 2007). Caraway, European-Asian species having disjunctive-European-Asian habitat, which spreads from the West to the East for more than 5000 kilometers. Caraway found itself in Europe in the Neolithic Period, with many other ancestors of cultivated plants, native from the Mediterranean, brought by old inhabitants of the Central and Northern Europe. The oldest caraway fruits, which were found in Europe (nowadays, the territory of Switzerland), are exactly from the Neolithic Age (Sinskaya, 1969; Petraitytė, 2005; Petraitytė and Dastikaitė, 2007).

Caraway is a biennial or annual (in rare cases – perennial) plant (Wulf and Maleeva, 1969; Sklyarevsky and Gubanov, 1986). According to the vegetation period, they are divided into annual (*Carum carvi* L. f. *annuum*) and biennial (*Carum carvi* L. f. *biennial*) forms. In recent years, there has been a growing interest in annual caraways, which wild forms were spread in the territory of the Asia Minor.

The most valuable materials, accumulated in the caraway fruits, are their essential oils. Their composition consists of hydrocarbons, terpenes, aldehydes, cetones, phenols, and organic acids. The essential oils actively participate in the metabolism processes of the plant (Vokk and Loomaegi, 1998). Some scientists consider that essential oils protect the plants from diseases and pests; they act as antiseptics and stimulate the convalescence of the damaged plant. The essential oils distinguish themselves by bactericidal and fungicidal functions, which partly protect the plant (Toxopeus and Bouwmeester, 1993). The main constituents of the caraway essential oils are compounds belonging to monocyclic terpenes – carvone and limonene (Bouwmeester and Smid, 1993). The carvone in caraway fruits make up to 60%, and limonene – 40% (Bouwmeester, 1998).

The caraway fruits have been used already for a long time to improve the flavor and value of food because of their winning distinction by antimicrobial and antioxidant properties. Caraway is characterized by its antiseptic, pain sedative, antispasmodic, depletive, antimicrobial, and antioxidant features (Nikolčiuk and Žigar, 1996; Vokk and Loomaegi, 1998; Petraitytė 2005). The essential oil of caraway fruits and its main component carvone holds a high economic potential. Recently, the greater attention has been paid to the investigations of terpenes, including carvone, in order to protect and cure human organism against oxidant stresses, as well as cardio diseases and cancerous (Wagner and Elmadfa, 2003). Amongst monoterpenes, carvone is the most important one among potato sprouts inhibitors, and it holds a forth hope to use it in practice (Kleinkopf et al., 2003; Carvalko and Foncea, 2006, Petraitytė, 2005). In Lithuania, caraways are one of most widely grown medicinal and aromatic plants and in 2005 were included into the European research of ten priority plants (Radušienė and Janulis, 2004).

3.3.2 GEOBOTANIC ASSESSMENT OF HABITATS IN SITU

During the years from 1996 to 2000, while performing the expeditions, 123 *Carum carvi* habitats were found and described *in situ*. The investigations were carried out in meadows of different economic activities: meadows (used for mowing) and pastures (used for grazing). Descriptions of the geobotanic community were analyzed according to the Braun-Blanquet cover-abundance scale (Braun-Blanquet, 1964) used to analyze vegetation in ecological studies. Community taxonomic dependence was established according to the international and Lithuanian plant classifications (Rothmaler et al., 1988; Balevičienė, 1998; Petraitrytė, 2005).

The most frequently caraway habitats were found in the south and southeeast meadows of Lithuania. In this region because of laky and hilly relief, sandy soils not very attractive for intensive agriculture, still rich in natural meadows. The number of habitats in the central and north of Lithuania is limited because of intensive agriculture, and natural or semi-natural meadows are not plentiful (Petraitrytė, 2005; Maršalkienė, 2011).

The majority of studied caraway habitat communities to widely spread Lithuanian alliances of the class of *Molinion–Arrhenatheretea elatioris* R. Tx., namely: to *Cynosurion cristati* R. Tx. – 48%, to *Arrhenatherion elatioris* (Br.-Bl. 1925) W. Koch – 36.6%, to *Molinion caeruleae* W. Koch – 2.4%. 6.5% of the described communities belonged to the alliance *Trifolion medii* Th. Müller of the *Trifolio-Geranietea sanguinei* Th. Müller class. Those communities were found on edges of woods and in the woods meadows (Petraitrytė, 2005; Maršalkienė, 2011).

4.9% of communities were ascribed to the class of the *Festuco-Brometea erecti* Br.-Bl. et R. Tx., all found in the south and southeast of Lithuania, on slopes of rivers and hills, in locations, which were non-suitable for land cultivation. Two (2.6%) communities belonged to *Agropyretalia repentis* Obert., Th. Müller et Görs Oberd et al. (Petraitrytė, 2005; Maršalkienė, 2011).

61% of investigated habitats were ascribed to pastures, and 39% – to hayfields (especially, the communities from the *Arrhenatherion elatioris* class) and total, 168 plant species were identified (Petraitrytė, 2005; Maršalkienė, 2011).

Caraway was most abundant in the communities of the *Cynosurion cristati* confederation, especially in the locations, where no other activities were carried out, with the exception of pasturages. Systematic grazing ensured the more favorable conditions for caraway surviving. In this type habitats, the caraway grew in the first layer of sward, by tags and small isles and this must be explained by seed spreading around a mother plant. In pasture type communities, the caraway grew smaller, but it was more productive, had more inflorescences and higher fruit yield, if compared with hayfields (Figure 3.1) (Petraitrytė, 2005; Maršalkienė, 2011).

The least richness of caraway plants, in comparison with other types of habitats, was found in the communities of dwarfish *Festuco-Brometea erecti* clas (Petraitrytė, 2005; Maršalkienė, 2011). Almost all such class community meadows were used as pasture, but caraway spreading was limited there, restricted by unsuitable soil and moisture conditions. In the unexploited meadows, the caraway plants became rare and after 5–7 years they just vanished. Especially in the unexploited meadows, the invasion of nitrofilic plants, especially *Cirsium arvence,* from the besides of the cultivated fields, negatively influenced on caraway survival.

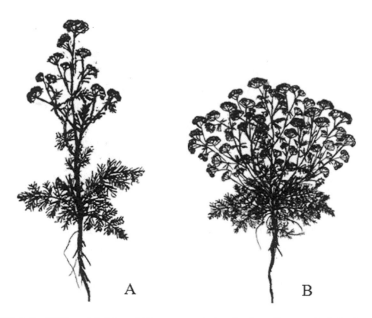

FIGURE 3.1 Different habits of *Carum carvi*: A: single-stalker, typical for hayfields; B: procumbent, multi-stalk – typical for a pasture.

3.3.3 SOIL ANALYSES OF HABITATS

The soil of each habitat has been analyzed for pH, mobile phosphorus, potassium, and total nitrogen (N). Phosphorus in the form of P_2O_5 has been determined by the photoelectric colorimeter method with molybdate; potassium in the form of K_2O – using a frame photometer by the A-L method (ISO 10694, 1995), total N – by the Kjeldahl method (ISO 11261, 1995), and pH in KCl – potentiometrically (ISO 10390, 1994). The soil was classified by its texture: clay, loam, and sandy loam.

The performed soil analyses of caraway habitats indicated that, in the most frequent cases, the caraways grew in neutral – alkaline and close to neutral, loamy, fertile, humus, potassium, and phosphorus-rich soils. Also, there were found some habitats in sand and peat soils (pH 5.3–6.5) with low levels of mobile potassium and phosphorus, and hummus, not typical for caraway. According to the data of analyses, soils of pastures were more fertile (Table 3.2) and soil texture varied more widely in comparison to meadows. Most prevailing soils in pastures were clay soils (27%), sandy loams (24%) and loamy sands (21%), in meadows – sandy loams (28%) and clay loams (65%) (Petraitytė, 2005).

TABLE 3.2 Agrochemical Soil Characterization of Pastures and Meadows (Petraitytė, 2005)

Habitat type	Agrochemical indicators	x	min	max	V%
Pasture	P_2O_5, mg/kg	184.0	59.0	580.0	65.8
	K_2O, mg/kg	199.7	81.0	332.0	34.6
	Humus,%	2.5	1.1	5.3	47.3
	pH_{KCL}	7.0	5.3	7.8	6.7
Meadow	P_2O_5, mg/kg	162.3	35.0	640.0	82.5
	K_2O, mg/kg	171.0	15.0	364.0	40.9
	Humus,%	2.1	0.66	4.34	42.9
	pH_{KCL}	6.9	5.8	7.6	6.2

3.3.4 ASSESSMENT OF MORPHOMETRICAL CARAWAY TRAITS EX SITU

During the period of 2000–2004, the collections of 107 wild caraway samples (ecotypes) were grown, and field trials were carried out at the experimental station of the ASU (Petraitytė, 2005; Petraitytė and Dastikaitė, 2007). The morphological and productivity traits: stalk height, branching height, the quantity of inflorescences, the weight of 1000 seeds and seed yield per plant, were used for plant assessment. At different caraway development stages, the thermal and humidity conditions were characterized by Selianinov's hydrothermal coefficient (HTC). HTC – the ratio of precipitation and the sum of above zero temperatures: HTC = P/0.1 T, where P is the total amount of precipitation over a period, and T is the sum of positive temperatures over the same period (Kudakas and Urbonas, 1983). For the show of meteorological conditions of caraway vegetation period, HTC was calculated for period April 1st – July 20th (Table 3.3). When HTC is up to 0.3 – the investigated period are very dry, 0.4–0.5 – dry, 0.6–0,7 – arid, 0.8–1.0 – insufficiently wet, 1–1.5 – sufficiently wet, 1.5 and > – wet.

The study results demonstrated a high variation of *Carum carvi* L. ecotypes in biometrical traits. The height of caraway stalks was 76.5 cm in average and variation of stem heights – 14.6%. The maximal height of some caraway ecotypes plant stalk was 1.9 times higher than the minimum. Stalk height was also affected by meteorological conditions. A week positive dependence of caraway stalk height on the ratio of precipitation and air temperature (HTC) was estimated (r = 0.3638). The quantity of inflorescences was 68.6 pieces in average. A high variation of the quantity of inflorescences was detected – 74.6%. The maximal quantity of inflorescences was 21.1 times higher than the minimum. The weight of 1000 caraway fruits was 2.1 g in average. The lowest variation indicated of the weight of 1000 caraway fruits – 5%. In *ex situ* conditions, based on average data, the caraways delivered 3.7 g of fruit yield. In prosperous years some caraway populations gave a 5-fold higher yield (19.1 g) than the average yield received during the study years. The variation of this trait was among the highest ones – 86.6% (Petraitytė, 2005; Petraitytė and Dastikaitė, 2007).

TABLE 3.3   Hydrothermal Coefficient (HTC) of Caraway Vegetation Period (2000–2004) (Petraitytė, 2005)

Investigation year	2000	2001	2002	2003	2004	Multiannual value
Carum carvi vegetation						
4–7 months	1.0	1.7	1.4	1.7	2.0	1.8
Vicia and *Lathyrus* vegetation						
4–6 months	-	1.3	1.2	1.3	1.2	1.5
7–8 months	-	1.7	0.5	1.5	1.7	1.5
9–10 months	-	2.4	3.9	2.2	1.8	1.9

3.3.5 ESSENTIAL OIL CONTENT IN FRUITS

The characteristic chemical composition of plants (species, subspecies, kind) is constant, and little changeable, but the influence of various environmental factors have an impact on the composition of secondary metabolites (and terpenoids, similarly). The composition of the essential oil is influenced by these environment factors: air temperature, precipitation, light/illumination, chemical content of the soil, mechanical damage and diseases. These factors can have an impact both on quality and composition of essential oils. Different chemotypes (races) that are not related to morphological chemotypes can also exist (Sur, 1993).

The analysis of the amount of the essential oils in caraway fruits was done by water steam distillation method. Carvone amount in essential oil was carried out by FISONC GC gas chromatographer. The essential oil amount in caraway fruits varied from 3.2 to7.0% and was 4.8% in average. A negative correlation was established between caraway essential oil amount and HTC (r = –0.6553). The most favorable year for essential oil synthesis in caraway fruit was the hot and dry weather during summer of 2000 (Table 3.3), and most unfavorable – the cool and wet year of 2004 (Petraitytė, 2005). Dependence of the essential oil synthesis on environmental conditions is mentioned by other authors as well (Bouwmeester and Smid, 1995). Amount of carvone in essential oil varied from 49.6 to 60.7% and was 52.9% in average. There was no correlation estimated between the amount of carvone and HTC (Petraitytė, 2005).

3.3.6 VARIATION IN COLORS OF CARAWAY PETALS AND FOLIAGE

Plant species and ecotypes are differently adapted to low positive, temperatures, and frost stresses (Kratsch and Wise, 2000). When the mentioned stresses could arise the various changes of gene expressions (Guy, 1990), such as increased hairiness of leaf (Roy, 1999), amount of anthocyanins (Larcher, 1995) or carotenoids (Haldimann, 1998). Temperature is one of the main factors that allow showing the diversity of foliage coloration (Zhang et al., 2004). It is considered, that the different coloration of petals of the same species ecotypes or individuals is related to resistance to low temperatures (Warren and Mackenzie, 2001). During the study years, estimations of visual coloration of different caraway coenopopulations blossom petals and foliage have been performed. Qualitative evaluation was performed (in 2001) using 5-point system: color of petals (5 – fuchsia; 4 – dark rose, 3 – rose, 2 – pink, 1 – white); color of foliage (5 – blue-green, 4 – grey green, 3 – dark green, 2 – green; 1 – light green); earliness of ecotypes (5 – ultra early, 4 – early, 3 – average early, 2 – late, 1 – ultra late) (Petraitytė, 2005; Petraitytė and Dastikaitė, 2007).

Caraway with pink-tinged blossoms was more generous in the years with changeable spring weather, when warm spells interchanged with cold. Clear caraway petal coloration was detected in 2001, when in spring, after a rather warm period, in the first decade of April, then the temperature was 28.3°C, a cold spell followed, and for couple night frosts occurred. In 2001, more like 90% of caraway ecotypes petals of blossom varied from pink to dark rose and some had fuchsia color (Figure 3.2). The diversity of caraway foliage coloration was also more recurrent and continual. On the basis of petal color intensity, there were recognized the white, pink, rose, and fuchsia caraway forms (Petraitytė, 2005; Petraitytė and Dastikaitė, 2007).

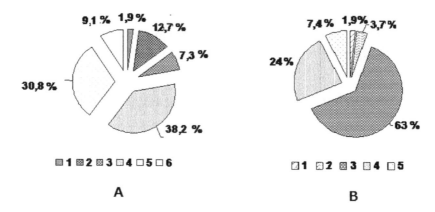

FIGURE 3.2   Petals (A) and leaves (B) color of different *Carum carvi* coenopopulations (2001): A – 1 – Fuchsia; 2 – Rose - pink, 3 – Pink>70%-white; 4 – Pink>50%-white; 5 – Pink>30%-white; 6 –White, and, B - 1- Blue-green, 2 – Grey-green, 3 – Dark green, 4 – Green; 5 – Light green.

3.3.7 BIOCHEMICAL FOLIAGE ASSESSMENT IN DIFFERENT ECOTYPES

In meadow phytocenoses, the individual ecotypes of the same species adapt different light intensity and differ in foliage pigment composition and amount (Protczuk et al., 2002). Foliage pigment content and composition indicate the biochemical diversity within the plant species, as well as their tolerance to intensive light or shade (Thayer and Bjorkman, 1990; Murchie and Horton, 1997; Rosevear, 2001).

The content of such pigments like chlorophylls and carotenoids in the caraway foliage of different ecotypes was investigated in *ex situ* collection, applying the colorimetric method and in a 100% extract of acetone, using a spectrophotometer Beckman DU–40. In May 2001, the rosette stage caraway foliage pigments, like chlorophylls *a*, *b*, and carotenoids, of the investigated ecotypes were analyzed and hierarchic – cluster analysis was performed, and three clusters were traced (Petraitytė, 2005; Petraitytė and Dastikaitė, 2007).

In analyzed caraway foliage of cluster I, the highest volume of total chlorophylls and carotenoids were detected, and the lowest ratio of chlorophylls *a* and *b* were found (Figure 3.8). This cluster consisted of 20.1% investigated ecotypes. The caraway foliage of ecotypes of cluster II had the lowest volume of chlorophylls and carotenoids and by the ratio of chlorophylls *a* and *b*, took the intermediary position between other clusters. The highest ratios of chlorophylls *a* and *b* were found in the caraway ecotypes of cluster III. In cluster III, the total volume of chlorophylls was higher by 20%, and the volume of carotenoids – by 22%, if compared with the results of cluster II. Meanwhile, the total volume of chlorophylls was lower by 35%, and the volume of carotenoids – by 25%, if compared with the results of cluster I. Almost 36% of the investigated ecotypes belonged to cluster III (Petraitytė, 2005; Petraitytė and Dastikaitė, 2007).

According to the results of chlorophylls *a* and *b* ratio, the caraway of cluster I could be ascribed to the shade tolerating ecotypes. Majority of these caraways *in situ* grew in the more or lower shaded habitats (parks, orchards, wood meadows) and of fertile communities of high growing *Arrhenatherion elatioris* meadows (Petraitytė, 2005). The higher volume of carotenoids in the foliage of the ecotypes of this cluster could be a protective function, while acclimatizing in more intensive lighting in *ex situ* conditions, in comparison with more shady *in situ* habitats (Rasevear et al., 2001; Li et al., 2002).

The ecotypes of cluster II, which amounted nearly to a half (45.5%) of the investigated collection, occupied the intermediate position according to the obtained research data. Because of texture and pH value soils of habitats of cluster II *in situ* were rather rich and favorable for caraway growth (Petraitytė and Dastikaitė, 2007).

According to the ratios of chlorophylls *a* and *b*, the caraway of cluster III could be ascribed to the ecotypes that tolerate intensive lighting. The soils of cluster habitats were among the poorest and most unfavorable, because of pH value like 6.3 or 7.8 and texture – coherent sands, heavy clays. Habitats of cluster III included a big part of *Cynosurion cristate* communities and the majority of *Festuco-Brometea erecti* (Petraitytė, 2005). Meanwhile, in such habitats, the microclimate, caused by heat, draught or frost, are more sharp (Stoutjiesdijk and Barkman, 1987).

Ultra-early ecotypes with colored petals and ultra-late ecotypes with white petals had higher levels of leaf pigments (Petraitytė and Dastikaitė, 2007). According to literature data, such ecotypes have a higher cold resistance (Haldimann, 1998). According to the data of other authors, a high chlorophyll content directly correlates with high plant productivity (Kabanova and Chaika, 2001). Leaf color and chlorophyll ratios had no correlations in between (Petraitytė and Dastikaitė, 2007).

3.3.8   CLUSTER ANALYSIS BASED ON CARAWAY AGROBIOLOGICAL TRAITS

In order to summarize the research data of the investigated diversity of caraway ecotypes *ex situ*, a hierarchic cluster analysis was performed according to morphometric, productivity, and phenologic traits (during period 2000–2004), as well as to research data of petal and foliage colors (Figure 3.3) (Petraitytė, 2005; Petraitytė and Dastikaitė, 2007). The phenologic differences, in this case, earliness of blooming, had an impact on their morphometric and productivity traits. The ecotypes with the most intensive colored petals and foliage were attributed to the cluster I (Figure 3.4). The ecotypes of this cluster lag other clusters by all morphometric and productivity traits (Figures 3.5 and 3.6), but exceeded the others (Figure 3.7) by the amount of essential oils in fruits and amount of carvone in the essential oil. In comparison to clusters I and IV, the caraways of "medium early" ecotypes of cluster II and "medium late" ecotypes of cluster III had average data of the morphologic and productivity traits. The late ecotypes of cluster IV had the biggest heights of stem and stem branching point, the highest number of blossoms, the mass of 1000 fruits and the highest yield from one plant (Figure 3.6), also these caraway ecotypes had the highest amount of essential oil (Figure 3.7) (Petraitytė, 2005; Petraitytė and Dastikaitė, 2007). The main substantial productivity traits of caraways are the amount of the essential oils in fruits and amount of carvone in these essential oils (Warren and Mackenzie, 2001). According to these two main traits, most productive caraway was of medium late and late ecotypes groups of cluster III and IV. "Early-season" ecotypes showed the lowest height, fruit weight, and total yield, but they had the highest carvone content in essential oils of fruits. Pink petals were characteristic of early-season caraway coeno-populations. The medium-late and late forms were characterized by their white petal color, the higher pigment content in leaves, the higher fruit productivity and the higher output of their essential oils (Petraitytė, 2005; Petraitytė and Dastikaitė, 2007). A number of inflorescences and total fruit yield had the highest liability. A weight of 1000 fruits and content of carvone in essential oil were the most stable parameters (Petraitytė, 2005; Petraitytė and Dastikaitė, 2007) (Figure 3.8).

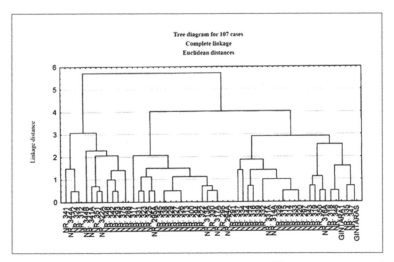

FIGURE 3.3 Dendrogram indicating similarities among wild caraway coenopopulations by petals, leaf color, and phenological dates.

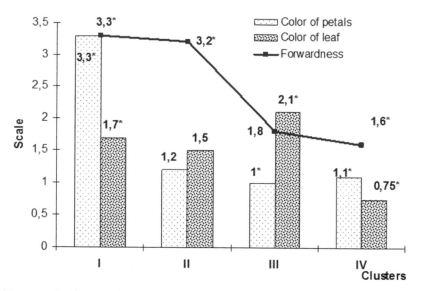

*Means are significantly different at P≤ 0.05

FIGURE 3.4   Earliness, petal, and leaf color traits characteristics of different caraway cenopopulation clusters.

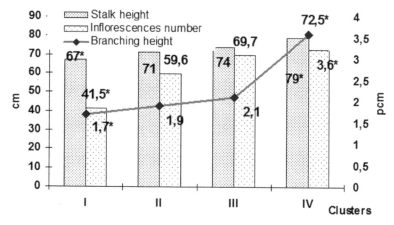

*Means are significantly different at P≤ 0.05

FIGURE 3.5   Height of stalks and branching, inflorescence traits characteristics of different caraway cenopopulations clusters.

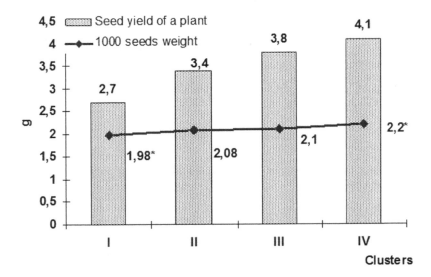

*Means are significantly different at P≤ 0.05

FIGURE 3.6 Weight of thousand seeds and yield of one plant characteristic of different caraway cenopopulation clusters.

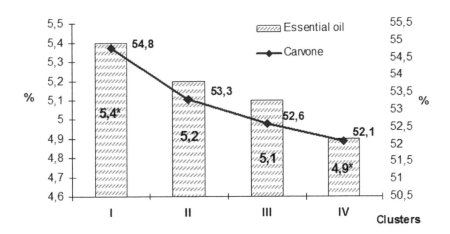

*Means are significantly different at P≤ 0.05

FIGURE 3.7 Content of essential oil and carvone characteristic of different caraway cenopopulation clusters.

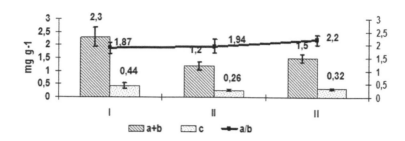

FIGURE 3.8 Different clusters of caraway cenopopulations by leaf pigments (a+b - chlorophylls a and b content; c - carotenoids content; a/b - the ratio of chlorophylls a and b).

3.4 DIVERSITY OF *VICA* AND *LATHYRUS* SPECIES POPULATION IN LITHUANIA

3.4.1 *VICIA AND LATHYRUS GENUS SPECIES*

Some 50,000 well-grazed spontaneous plant species grow around the world. Only negligible portion has been tamed (Loftas, 1995). Legumes (pulses), as a source of proteins and energy, are one of the major plants used for food and forage since olden times (Sliesaravičius and Petraitytė, 2001). Nutritionally, they are on the second place after the representatives of the *Poaceae* family. Legumes are valued for their ability to fix biological nitrogen and protect the soil from erosion. Legumes are also used for alternative cropping systems as living mulches (Sliesaravičius and Petraitytė, 2001; Baresel et al., 2002; Sheaffer and Seguin, 2003).

Worldwide, Fabaceae (*Leguminosae*) with 800 genera and 20,000 species, is the third largest family of flowering plants, after the *Orchidaceae* and *Asteraceae* (Lewis et al., 2005). *Vicia* contains several economically important food and forage species, which have a center of distribution in the Eastern Mediterranean (Maxted, 1995; Smykal et al., 2015). *Lathyrus* genus is considered to be younger and progressive one (Attokurov, 1993). Due to their current economic value and potential for future utilization, *Vicia*, and *Lathyrus* genus had a high priority for collection, conservation, and forage development (Sliesaravičius and Petraitytė, 2001).

One hundred sixty species of only *Vicia* genus are used for food and fodder. Only a small part of these species has been domesticated. Extinction of species and populations promotes search and selection activities of other valuable spontaneous species (Sliesaravičius and Petraitytė, 2001). The species under investigation are assessed not only as forage (Sheaffer and Sequin, 2003; Fraser et al., 2004), but also according to their ability to adapt to changing climatic conditions, genetic stability of major quality parameters, disease, and pest resistance (Skvorcov et al., 2002). The main object of domestication is not the species as a whole, but its individual ecotypes. The priority work is an investigation of species biodiversity and selection of promising ecotypes.

About 39 tribes of the *Fabaceae* family, some 115 species, among which the spontaneous species make 13 species of *Vicia* (vetches) and 9 species of *Lathyrus* genus (sweet peas), can be found in Lithuania (Natkevičaitė-Ivanauskienė, 1983). Perennial, long-living (20 years and more) species of *Fabaceae* family *Vicia* (*V. cracca*, *V. sepium*) and *Lathyrus* (*L. pratensis*, *L. palustris*, *L. sylvestris*) genuses, which are of good fodder value, grow in natural meadows of Lithuania (Stancevičius, 1971). Due to anthropogenic and technogenic factors, the areas under natural meadows are vanishing. For this reason, the cenopopulations of these species are lost, and the biodiversity of the species is getting poor (Balevičienė et al., 1998). Since 1998, in the framework of the research program *Genofund*, the collection of *in situ* growing *V. cracca*, *V. sepium*, *L. pratensis*, *L. palustris*, and *L. sylvestris* coenopopulations has been accumulated at the ASU. Simultaneously, the cenopopulations of short–living *V. villosa*, *V. angustifolia* and *V. hirsuta* genuses, which are of a high fodder value and characteristic to agrophytocoenoses, have also been collected (Table 3.1) (Sliesaravičius and Petraitytė, 2001).

3.4.2 INVESTIGATION OF ANNUAL VETCH

The annual vetch *V. sativa* L., *V. angustifolia* L., *V. villosa* Roth., and *V. hirsuta* Grey are typical plants of agrophytocenoses. *V. hirsuta* and *V. angustifolia* are attributed to the weed group. *V. sativa* and *V. villosa* are domesticated. *V. villosa* is semi-natural, grows in phytocenoses as a weed. The objective of the study was to estimate and investigate the capacity of adaptability of annual vetches, under the equal agrotechnical conditions, seeking to conserve the genefund and select the promising forms suitable for practical use (Maršalkienė, 2015). The seed specimens of *V. angustifolia*, *V. hirsuta*, and *V. villosa* were collected from different geographical origins in Lithuania during the period from 1998 to 2001. *V. sativa* was grown for the sake of comparison. The annual spring vetches (*V. angustifolia* and *V. hirsuta*) were sown *ex situ* (2001–2004) in spring, on the 20th–30th of April and wintering *V. villosa* – on the 1st–10th of September. The vetches were grown in a particular agrophytocenosis of the crops, characteristic to these species (*V. angustifolia* and *V. hirsuta* – with barley; and *V. villosa* – with rye) (Maršalkienė, 2015).

Plant assessment was based on measurements of such traits, like stem height, a branching point, number of branches, number of pods, quantity of seeds per pod, the weight of 1000 seeds, fresh above-ground mass. The volume of crude protein content was determined in air-dried mass (Kjeldahl method); the content of carotenoides – using a colorimetric method, and disease resistance were also estimated. In order to estimate the weather data, the hydrothermal coefficient (HTC) was calculated (Kudakas and Urbonas, 1983) for three calendar stages: April–June, July–August, and September–October (Maršalkienė, 2015).

3.4.3 ECOGEOGRAPHICAL DISTRIBUTION OF INVESTIGATED ANNUAL VETCHES

The accessions of *V. angustifolia* – narrow-leaved and *V. hirsuta* – hairy vetch, both collected during expeditions, often grew together in spring cereal phytocenoses, less often – in winter cereal phytocenoses, waste or abandoned the land, less frequently – in grasslands, sandy loam and light loam soils (pH 5.8–7.0). These species are rather frequent on the whole territory of Lithuania, especially in the south and southeast regions of the country, which are characterized by sandy soils, and less frequent – in northern Lithuania (Sliesaravičius et al., 2005; Maršalkienė, 2015). The noticeable decline in the species and population induces the expanded search and selection of valuable spontaneous species (Ckvorcov and Kuklina, 2002).

V. villosa (winter vetch) is thought to have originated on the Mediterranean littoral. It belongs to the European flora element (Stancevičius, 1971). *V. villosa* was specific to rye agrophytocenoses, it occurred less frequently on waste or abandoned the land and seldom in spring cereals and potatoes. Winter vetch grows on different acidity (pH 5.1–6.9) sandy loams, in some sites even on sandy soils. *V. villosa* is the most widespread plant in the south-east and east of Lithuania, far less frequent – in south-west and center of Lithuania, it was not found at all in northern Lithuania. Only one accession of this plant was found in the northwestern part of the country (Maršalkienė, 2015a; Maršalkienė, 2015b).

3.4.4 *VICIA ANGUSTIFOLIA L.*

V. angustifolia L. (narrow-leaved vetch) is a pioneer of the genus (Sinskaya, 1969), another Latin name – *V. sativa* ssp. *Nigra* L. (L.) (Maxted, 1995). A range of intermediate forms between *V. angustifolia* and *V. sativa* (Sinskaya, 1969) was identified. The phylogenetic relationships test between individual *Fabaceae* species revealed *V. angustifolia* to possess one of the highest genetic polymorphism levels (Nickrent and Patrick, 1998), which partly confirm the theory of the origin of species of *Vicia* genus. *V. angustifolia* belongs to the elements of the west-Asian flora and grows almost in all Europe, America, Australia, and South Africa. Narrow-leaved vetch is an annual plant, terophyte, and rarely overwinters (Stancevičius, 1971). In Lithuania, *V. angustifolia* is rather common and grows on various soils, in cereal crops as a weed, dry grasslands, forests, and bush (Sliesaravičius et al., 2005; Maršalkienė, 2015a).

Having estimated phonological stages of the *V. angustifolia ecotypes*, four groups of earliness were distinguished according to the beginning of blooming and length of the growing season in *ex situ* collection (Table 3.4). The largest number of the ecotypes (42%) was attributed to an early group, 26.3% – to a medium-early group, and 16.3% – to a late group ecotypes. To an ultra-early group was attributed to one ecotype (5.4% of all ecotypes). The groups, which were distinguished according to the length of the growing season, differed not only by the beginning of blooming but also by other traits (Table 3.4). The late ecotypes were characterized by the taller plants, greater mass, and weight of 1000 seeds, the volume of crude protein and clearly, expressed leaf polymorphism (Figure 3.9). Leaves accounted for 55.6% of the total above-ground plant mass of *V. angustifolia* (Figure 3.12) (Sliesaravičius et al., 2005; Maršalkienė, 2015).

Vetch of the ultra-early ecotype significantly diverged in habit. It is a non-branchy, single-stemmed vetch form (Figure 3.9). The plants of this ecotype distinguished by the highest seed number per pod (Table 3.4). The early medium ecotypes could be most valuable in terms of practical relevance: average stem branchiness, foliage mass percentage, protein content were found to be the highest values. The ecotypes, which were attributed to a late group, branched from enlobe more than the vetches of the other groups; the traits of their pods parameters and weight of 1000 seeds were the highest. The ecotypes of narrow-leaved were also characterized also by the seed coats color (Figure 3.10). The seed coats color varied from greenish grey with black pigment spots to black. Some ecotypes had yellowish brown seed with indistinct pigment spots. A lighter seed color and a smaller number of pigment spots were more specific to the medium early and late branchy ecotypes with wider and glosser leaves (Sliesaravičius et al., 2005; Maršalkienė, 2015).

3.4.5   VICIA VILLOSA ROTH.

Hairy (winter) vetch (*V. villosa*) provides a good soil cover and is used as to control weeds in alternative cropping systems (Anugroho et al., 2009; Mischler et al., 2010; Halde et al. 2015) and used for forage, hay, silage, and green manure (Viesermana and Leon, 2013; Renzi et al., 2017). Hairy vetch is well adapted to organic cultivation and grows well on a wide range of soil types – on sandy, nitrogen-depleted, and light acidity soils (Sheafer and Seguin, 2003; Dastikaitė et al., 2009). In this sense *V. villosa* is of interest to agricultural science and practice. The wild populations of hairy vetch have shown a high level of variability of agronomic characteristics and may be considered as having a potential for its utilization in breeding and the development of new cultivars (Mihailovic et al., 2008).

V. villosa in Lithuania is an archaeophyte and arrived together with winter cereals (Gudžinskas, 1999). In Lithuania, hairy vetch (as is the wild type) is a rather rare plant and mostly grows in agrocoenoses as a weed (Maršalkienė, 2015). The selection job was carried out from 1934 to 1952. Over 30 accessions of local winter vetch were collected and several winter hardly numbers and variety *"Pūkiai"* were released (Lazauskas and Dapkus, 1992; Petraitytė et al., 2007).

TABLE 3.4 Earliness groups of *Vicia Angustifolia, V. Hirsuta* and *V. Villosa* Cenopopulations (Maršalkienė, 2015)

Earliness type	Length of growing period (days)	Number of samples, %	Stem height, cm	Green mass of one plant, g	Branch number From enlobe	Branch number From the main stem	Number of pods per plant	Seeds set in a pod	Weight of 1000 seeds, g
V. angustifolia									
Ultra-early	75–80	9	35.1 ± 5.4	2.3 ± 0.7	1.9 ± 0.2	2.0 ± 0.2	7.1 ± 0.9	7.6 ± 1.1	7.5 ± 0.9
Early	81–85	22	42.6 ± 7.2	3.2 ± 1.1	2.3 ± 0.4	2.5 ± 0.3	9.4 ± 1.2	8.0 ± 1.1	14.5 ± 1.8
Average early	86–95	60	48.7 ± 9.6	3.6 ± 1.6	1.8 ± 1.0	4.2 ± 0.6	13.2 ± 1.9	7.5 ± 1.6	19.2 ± 2.1
Late	<96	9	54.3 ± 6.4	4.0 ± 0.8	0.9 ± 0.4	4.8 ± 0.3	11.7 ± 1.2	5.8 ± 0.9	22.3 ± 1.4
Average			4.5 ± 12.1	3.3 ± 1.8	1.9 ± 1.3	2.8 ± 1.5	11.6 ± 2.2	7.4 ± 2.1	17.1 ± 3.2
V. hirsuta									
Early	65–75	20	70.2 ± 5.8	1.1 ± 0.2	–	0.9 ± 0.2	54.1 ± 8.2	2.0 ± 0.00	6.3 ± 1.1
Average early	76–90	57	86.4 ± 7.9	1.9 ± 0.3	–	1.2 ± 0.4	56.5 ± 11.9	2.0 ± 0.01	7.1 ± 1.4
Late	<91	24	91.2 ± 5.3	2.1 ± 0.2	–	1.5 ± 0.3	57.0 ± 7.4	1.9 ± 0.01	7.4 ± 1.2
Average			84.4 ± 14.4	1.7 ± 0.5	–	1.3 ± 0.8	56.8 ± 138	2.0 ± 0.01	7.3 ± 1.6
V. villosa									
Ultra-early	100–105	7	78.4 ± 13.4	21.5 ± 5.6	2.6 ± 0.4	1.2 ± 0.3	14.6 ± 4,1	4.4 ± 0.3	23.3 ± 2.3
Early	106–110	19	101.5 ± 12.1	28.8 ± 4.2	1.8 ± 0.3	1.9 ± 0.3	23.5 ± 8.6	4.7 ± 0.5	28.9 ± 3.1
Average early	111–120	44	121.8 ± 18.2	32.2 ± 7.6	1.0 ± 0.1	3.1 ± 0.9	29.0 ± 11.8	4.5 ± 0.8	30.2 ± 3.4
Late	121–130	16	135.2 ± 16.5	34.4 ± 5.4	1.0 ± 0.1	3.5 ± 0.5	33.8 ± 4.9	4.5 ± 0.5	31.4 ± 2.8
Ultra late	<131	14	154.2 ± 17.2	33.2 ± 6.2	1.0 ± 0.0	3.8 ± 0.4	36.2 ± 6.3	4.3 ± 0.3	29.2 ± 2.0
Average			124.4 ± 28.2	32.6 ± 9.9	1.3 ± 0.7	3.2 ± 1.6	30.5 ± 16.3	4.5 ± 0.9	30.4 ± 4.1
V. sativa 'Tverai'									
Average early	105		84.0 ± 5.9	17.5 ± 2.4	1.0 ± 0.6	3.9 ± 1.2	16.5 ± 3.2	5.2 ± 0.5	54.0 ± 1.2

All the ecotypes tested by us were varied in terms of phenologic and morphometric traits. According to all traits, the plants of very early ecotypes lagged behind the other groups most considerably (Table 3.4). The number of seeds per pod of all the traits tested was found to be the least variable. The foliage and stem ratio (Figure 3.12) was similar in the aboveground mass. The total value of above-ground plant mass was increased by the mass of inflorescences that contains the largest amount of crude protein (Figure 3.13) (Maršalkienė, 2015a). In different years, the ecotypes of hairy vetch varied according to all biometric traits. This suggests that this species is polymorphic and can adapt well to changing ecological conditions, without losing the parameters, specific to the ecotypes (Petraitytė et al., 2007).

Hairy vetch, among others grown in collection annual vetch species, distinguishes by the longest growing season and was most heavily affected by diseases and pests. The end of the growing season was influenced by the spread and affect of diseases. In the year 2001 and 2003, when the weather was unusually hot and wet, powdery mildew caused by *Erysiphe communis* Grev. f. *vicea* Jacz. was identified. About 50% of ecotypes were found to be affected, because of wet and cool weather favorable to the spread of *Ascochyta* sp. Disease resistance of different cenopopulations was diverse. Nine (or 16%) cenopopulations, resistant to powdery mildew, and seven populations (12.5%) resistant to *Ascochyta* were discriminated. During 2001–2004, there followed the assessments on *Bruchus atomarius* L. damage on leaves, flowers, and pods. Late *V. villosa* forms were most heavily affected by this pest. Especially heavy pest damage was identified in 2002, when the weather during the July–August period was exceptionally dry and hot (Figure 3.3) (Petraitytė et al., 2007).

Hairy vetch is well adapted to a wide range of soil types, especially to sandy and light acidity soils (Dastikaitė et al., 2009). The analysis of adaptability of the *V. villosa* to soil pH was performed in the greenhouses and phyto-cameras during. The experiments were conducted with hairy vetch variety *Pūkiai* and wild population sample No 34, which covered investigation of vetch sensibility to the substrates with pH from 6.5 down to 3.3. The greatest hairy vetch viability and productivity were observed in the substrates with pH values from 5.8 to 5.5. Hairy vetch viability and productivity were mostly inhibited in the substrates with pH values from 3.3 to 3.5. Vetch number 34 tolerated substrates with pH 5.8–5.2 better then vetch *Pūkiai*. Whereas, vetch *Pūkiai* tolerated substrates with pH 5.8 and 3.3 relatively better than vetch number 34 (Dastikaitė et al., 2009).

3.4.6 VICIA HIRSUTA GREY

During the experimental years, the first early forms of *V. hirsuta* started blooming at the end of June (Table 3.4), and one week later – the plants of late forms. The traits of ecotypes of this species were characterized by a relatively low diversity. The populations dominated with the height variations from 70.2 to 91.2 centimeters. The mass of plant leaves accounted for 45.6% of the air-dried mass of the plant (Figure 3.12). Leaves contained 26.6% of crude proteins, i.e., comparing the quantity of proteins in leaves; it was similar to sand vetches and surpassed other species (Figure 3.13). The deviation of protein content from the average value was relatively small. *V. hirsuta* was noted for a relatively weak response to diverse ecological conditions of different years (Maršalkienė, 2015a).

3.4.7 VICIA SATIVA L.

This species was found on the whole territory of Lithuania (Gudžinskas, 1999). It was formed through a range of transitional forms from *V. angustifolia* (Sinskaya, 1969). It was the most widespread species of the cultivated (domesticated) vetch. Originally, it was a food crop known in Europe since the Neolithic times; later it became a forage and green manure crop (Sinskaya, 1969). In our experiments, the *V. sativa* variety *Tverai* was grown as a control plant. The seed-ripening period of this variety coincided with that of *V. villosa* early forms (Table 3.4). The stem height was equal to that of medium early *V. hirsuta*, and plant mass was lower than that of *V. villosa*. However, it was several times higher than that of *V. angustifolia* and *V. hirsuta*. According to pod productivity, this vetch lagged behind *V. villosa* and *V. hirsuta* and was distinguished by the seed number per pod. According to this parameter, *V. sativa* was identical to *V. angustifolia* and surpassed all species, which were tested by weight of 1000 seeds. Both *V. sativa* and *V. angustifolia* had a similar number of sets of seeds per pod. Since the number of seeds is genetically determined and is only weakly

affected by the environmental conditions, it is a relatively stable parameter, indicating the relatedness of these two species (Maršalkienė, 2015a).

3.4.8 ASSESSMENT OF ABOVE-GROUND MASS AND ITS PROTEIN CONTENT OF SELF-SEEDING VICIA AND LATHYRUS SPECIES

The mass productivity traits studies of *Vicia* and *Lathyrus* species were
carried out in 2004–2008, using the collection of the Experimental Station.
Five spontaneous *Vicia* genus species and three species of the *Lathyrus*
genus were studied: *V. angustifolia, V. hirsuta, V. villosa* L., *V. cracca* L.,
V. sepium, L. pratensis L., *L. sylvestris* L., and *L. palustris* L. The seeds
of the perennial spontaneous vetches and sweet peas of the studied species
were collected from different geographical origins during 1998–2002.
Investigated species of perennial vetches (*V. cracca, V. sepium*) and sweet
peas (*L. pratensis, L. sylvestris* and *L. palustris*) were sown on the 5th of
May in 2003, using the field boxes of two square meters in size, together
with the wild perennial grasses, particular for phytocenosis plants (*Phleum
pratense* L., *Lolium perenne* L., *Festuca pratensis* Huds.).

The evaluation of above-ground mass and protein content was based
on testing of the following parameters: the average above-ground mass
of one plant, ratio of foliage, the percentage of blossoms and inflorescences in the total mass, the crude protein content of the individual plant
ground mass ant parts like: foliage, stems, flowers. Ten plants at their early
button-flowering stage were taken from each field box for this research.
The morphometric analysis of specimens was performed in the laboratory.
The crude protein analysis was carried out using the Kjeldahl method.

Among the tested species, the maximum stem height and the greatest
aboveground mass was observed in the species of *L. sylvestris* L. and *V.
villosa* (Figure 3.11). The vetches of *V. angustifolia* and *V. sepium* had
the shortest plant stems. The aboveground mass weight of the plants of
the tested species correlated with the stem height. It was determined the
positive linear relationship between stem height and aboveground mass
weight (r = 0.880). However, the aboveground mass was also dependent
on the morphometricl traits of the tested species: size of the leaves and
blossoms, their branchiness. The lowest aboveground mass was observed

in the vetches of the *V. hirsuta*, whose stem height was one of the highest ones, if compared with other species.

The largest portion in the aboveground mass of the tested species (excluding *V. villosa*) consisted of foliage –average 52.4% (Figure 3.12). The leaf mass of different species ranged from 42.1% to 56.0%. The largest foliage mass was observed in the plants of the species of *L. pratensis, V. cracca* and *V. angustifolia*. The stem mass of all species tested was smaller than the foliage mass, and ranged from 41.5 to 45.7%; an average value was 43.7%. The mass of inflorescences in the above-ground mass ranged from 2.3 to 2.5%, because during the research years the plants of the species *V. villosa* used to have averagely 14.5% blossoms of the general above-ground mass, it means 5.8–6.3 times more than other tested species of vetches and sweet peas (Figure 3.12).

The highest protein content in the aboveground mass consisted of blossoms (Figure 3.13). The protein content of the blossoms of the tested species ranged from 33.1% (*V. angustifolia*) up to 40.1% (*V. sepium*), and it reached 33.4% on average (Figure 3.13). The protein content of blossoms of *V. sepium* was especially high – 6.7% higher than the general average of the species tested. The crude protein content in foliage ranged from 21.4% (*L. pratensis*) to 26.9% (*V. villosa*). The total foliage protein content of the tested vetch species was on average 25.62%, and a little less in the sweet pea species – on average 22.13% (Figure 3.13). The average volume of crude protein of the stems of the studied species was 9.8%. The highest volume of crude protein of the stems (on average 13.3%) was observed in the annual vetches (*V. hirsuta, V. angustifolia,* and *V. villosa*). *V. villosa* (21.9%), *L. sepium* (21.9%) and *V. angustifolia* (20.3%) were character-ized by the highest volume of crude protein in the above-ground grass mass. *L. pratensis* and *V. cracca* were described by the lowest volume of crude protein.

The experimental findings indicated that variable weather conditions affected phenologic, morphometric, productivity, and biochemical traits of the ecotypes of all investigated species. The relationship between the tested parameters and the location of geographical origin was not found during the experimental period.

FIGURE 3.9 Different forms of *V. angustifolia* in 2002: (a) – late, (b) – average early, (c) –ultra early.

FIGURE 3.10 Seed diversity of different *V. angustifolia* coenopopulations.

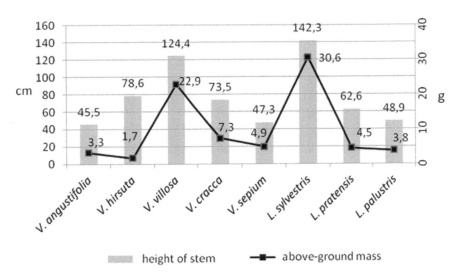

FIGURE 3.11   Stem height and aboveground mass weight of vetches and sweet peas.

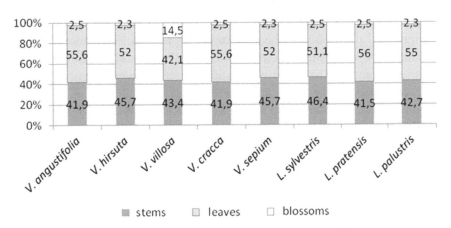

FIGURE 3.12   Ratio of the stems, leaves, and blossoms in the total plant mass.

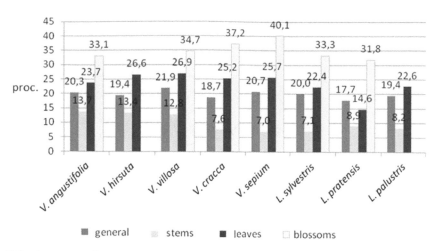

FIGURE 3.13 The protein content of vetches and sweet peas in the aboveground mass.

3.5 CONSERVATION ACTIVITIES

The developments in the European Agricultural Policy have caused some large changes in the European farming and agricultural research systems (Selge and Lillak, 2005). The new strategies and socio-economic factors needed to be increasingly taken into account by farmers, scientists, advisors, environmentalists, and other stakeholders. Nevertheless, grassland will continue to play an indispensable role in obtaining high-quality products from ruminant farm animals (Selge and Lillak, 2005; Luscher et al., 2014).

Ex situ and *in situ* research supplements each other, reveals species polymorphism and adaptive properties (Marum, 1999). Plant growth in the collection and storage in cold chambers provide the conditions to stabilize original genotypic traits (Lemežienė and Kanapeckas, 2000). Meanwhile, growth *ex situ* cannot replace growth *in situ* because the impact of various environmental conditions is eliminated and species evolution is inhibited.

Wild plant populations are still being formed by such evolution forces like natural selection, mutation, migration, and hybridization. More stable survival potential is detected in wild plant populations with more individuals, because this provides a better chance to form genotype combinations, which are adapted to the changing environment. Therefore, species genetic diversity is a guarantee of its survival. Population heterogeneity but not area determines its survival, therefore, at the declining area of natural meadows it is important to preserve genetic *Carum*, *Vicia*, and *Lathyrus* biodiversity both *in situ* and *ex situ* (Sliesaravičius and Petraitytė, 2001; Petraitytė, 2005).

Long-term seed storage for plant genetic resources was established in 1997 in the National Plant Genetic Resources Coordinating Centre, located at the Lithuanian Institute of Agriculture. The agricultural crops were presented by the largest number (63%) of accessions at the long-term seed storage of the Plant Gene Bank. Another large group of seed accessions represented the forest trees. Seed samples were dried for 2–3 months under specific conditions: of temperature from 15°C to 20°C, and relative air humidity at the drying chamber – 10–15%. Such conditions allowed achieving a seed moisture content at 3–5% (6–6.5% for pea, bean). After seeds were packed in the airtight aluminum foil bags and stored at –18°C. The first seed germination test was carried out after 5 years of storage at a temperature of –18°C. Long-term storage conditions guaranteed seed survival for decades as only very limited metabolism could occur there (Gelvonauskis, 2013).

Phenologic observations, morphometric, and biochemical research indicated that the spontaneous *Vicia* and *Lathyrus* were distinguished by their rich morphological diversity, different beginning of blossoming, different number of inflorescences and seeds, amounts of produced mass and raw proteins (Sliesaravičius and Petraitytė, 2001; Sliesaravičius et al., 2005; Petraitytė et al., 2007; Maršalkienė, 2015a). According to qualitative and quantitative parameters of *Carum carvi*, the ecotypes were selected that were outstanding by the quantity of essential oils in their fruits, high carvone content in essential oils, fruit weight and one plant yield. They also could be described as early, resistant to temperature stresses, having pink flowers and late highly productive cenopopulations (Petraitytė, 2005; Petraitytė and Dastikaitė, 2007). In order to practice a long-term seed storage, there were submitted 57 *Carum carvi* samples, 20 of *Vicia* and 22 *Lathyrus* genus species.

3.6 CONCLUSION

In view of quality fodder production, all researched species of perennial *Lathyrus* and *Vicia* are critical, because they are vanishing due to the consequences from the biotic and abiotic factors. This negative development is much accelerated in the habitats of rare (limited) species. However, *L. palustris* and *L. pisiformis* (both are included in *Red Data Book of Lithuania*) should be preserved as rare and highly valuable species. The succession of communities, caused by water level regulation and transfer from intensive exploitation of meadow to extensive one has extruded those species. A very close cooperation between conservation managers and livestock farmers is necessary in order to achieve successful grassland biodiversity conservation. The effective grassland farming and biodiversity conservation could be combined while applying the scientific insight on evolution and ecology of grassland biodiversity.

KEYWORDS

- agrobiological assessment
- biodiversity
- *Carum carvi*
- carvone
- conservation activities
- ecotypes
- essential oil
- gene fund
- habitat
- *Lathyrus palustris*
- *Lathyrus pisiformis*
- Lithuania
- phytodiversity
- protein content
- Pūkiai
- spontaneous species
- *Vicia angustifolia*
- *Vicia sativa*
- *Vicia villosa*

REFERENCES

Anugroho, F., Kitou, M., Nagumo, F., Kinjo, K., & Tokashiki, Y., (2009). Growth, nitrogen fixation, and nutrient uptake of hairy vetch as a cover crop in a subtropical region. *Weed Biology and Management, 9*(1), 63–71.

Attokurov, K., (1993). Biology of juvenile age of some species of the genus *Lathyrus* L. *Introdukcia i Aklimatizatcija, 26,* 95–98 (in Russian).

Balevičienė, J., Kizienė, B., Lazdauskaitė, Z., Patalauskaitė, D., Rašomavičius, V., Sinkevičienė, Z., Tučienė, A., & Venckus, Z., (1998). Vegetation of Lithuania: *1. Meadows*: Cl. *Asteretea tripolii*, Cl. *Molinio-Arrhenatheretea*, Cl. *Festuco-Brometea*, Cl. *Trifolio-Geranietea*, Cl. *Nardetea*. Šviesa Publishers, Kaunas – Vilnius, pp. 1–269.

Balmford, A., Bennun, L., Brink, B. T., Cooper, D., Cote, I. M., Crane, P., et al., (2005). The convention on biological diversity's 2010 target. *Science, 307,* 212–213.

Baresel, J., Schenkel, W., & Reents, H., (2003). Screening of leguminous plant species to assess their usability for green manuring and for mixed cropping in organic farming. *Bundesanstalt für Landwirtschaft und Ernährung, Bonn, Geschäftsstelle Bundesprogramm Ökologischer Landbau*, 42 (in German).

Basalykas, A., (1965). *Physical Geography of Lithuanian SSR.* Mintis, Vilnius, (in Lithuanian).

Bouwmeester, H. J., & Kuijpers, A. M., (1993). Relationship between assimilate supply and essential oil accumulation in annual and biennial caraway (*Carum carvi,* L.,). *Journal of Essential Oil Research, 9*(2), 143–152.

Bouwmeester, H. J., Gershenzon, J., Konings, M. S., & El-Feraly, F. S., (1998). Biosynthesis of the monoterpenes limonene and carvone in the fruit of caraway. *Plant Physiology, 117,* 901–912.

Braun-Blangue, J., (1964). *Planzensoziologie, Grundzuge der Vegetationskunde.* Wien, New York.

Brunner, A. M., Li, J., Di Fazio, S. P., Shevchenko, O., Montgomery, B. E., Mohamed, R., et al., (2007). Genetic containment of forest plantations. *Tree Genetics and Genomes, 3,* 75–100.

Budvytytė, A., (2000). Lithuanian plant genefund and its preservation. *Žemdirbystė, 72,* 229–238 (in Lithuanian).

Buivydaite, V. V., (2005). *Soil survey and available soil data in Lithuania.* In: Jones, R. J. A., Houskova, B., Bullock, P., & Montanarella, L., (eds.), *Soil Resources of Europe* (Vol. 9, pp. 211–223). European Soil Bureau Research Report, Italy.

Chomskis, V., (1958). *Geographical Chronicle.* Vilnius, (in Lithuanian).

Crane, P., (2001). Botanic gardens for the 21st century. *Garden Wise, 15,* 4–7.

Dabkevičius, Z., Gelvonauskis, B., & Leistrumaitė, A., (2008). Investigation of genetic resources of cultivated plants in Lithuania. *Biologija, 54*(2), 51–55.

Dastikaitė, A., (1997). *Caraway (Biology, Agricultural Engineering, Industrial-Use).* Kaunas (in Lithuanian).

Dastikaitė, A., Sliesaravičius, A., & Maršalkienė, N., (2009). Sensibility of two hairy vetch (*Vicia villosa* Roth.) genotypes to soil acidity. *Agronomy Research, 7*(1), 233–238.

De Carvalho, C. C. R., & De Foncea, M. M., (2006). Carvone: Why and how should one bother to produce this terpene. *Food Chemistry, 95,* 413–422.

Eidukevičiene, M., (2013). *Natural Geography of Lithuania*, Klaipėda (in Lithuanian).

Gailiušis, B., Jablonskis, J., & Kovalenkovienė, M., (2001). *Rivers of Lithuania, Hydrography, and Outflow*. Kaunas (in Lithuanian).

Gelvonauskis, B., (2013). Preservation of plant genetic resources in Lithuania. *Mokslas Ir Technika.*, *3*, 38–39 (in Lithuanian).

Gibson, R. H., Pearce, S., Morris, R. J., Symondson, W. O. C., & Memmott, J., (2007). Plant diversity and land use under organic and conventional agriculture: A whole-farm approach. *Journal of Applied Ecology*, *44*, 792–803.

Gudžinskas, Z., (1999). Conspectus of alien species of Lithuania. 10. Fabaceae. *Botanica Lithuanica*, *5*(2), 103–114.

Guy, C. L., (1990). Cold acclimation and freezing stress tolerance: Role of protein metabolism. *Annual Review of Plant Physiology and Plant Molecular Biology*, *41*, 187–223.

Halde, C., Bamford, K. C., & Entz, M. H., (2015). Crop agronomic performance under a six-year continuous organic no-till system and other tilled and conventionally-managed systems in the northern Great Plains of Canada. *Agriculture, Ecosystems, and Environment*, *213*, 121–130.

Haldimann, P., (1998). Low growth temperature-induced changes to pigment composition and photosynthesis in *Zea mays* genotypes differing in chilling sensitivity. *Plant, Cell, and Environment*, *21*(2), 200–214.

Kabanova, S. N., & Chaika, M. T., (2001). Correlation analysis of triticale morphology, chlorophyll content, and productivity. *Journal Agronomy and Crop Science*, *186*, 281–285.

Karazija, S., (1989). *Lithuanian Forest Types*. Vilnius.

Kleinkopf, G. E., Oberg, N. A., & Olsen, N. L., (2003). Sprout inhibition in storage: Current status, new chemistries, and natural compounds. *American Journal of Potato Research*, *80*(5), 317–327.

Kratsch, H. A., & Wise, R. R., (2000). The ultrastructure of chilling stress. *Plant, Cell, and Environment*, *23*, 337–350.

Kudakas, V., & Urbonas, R., (1983). Selianinov's hydrothermal coefficient. *Žemės Ūkis*, *12*, 25–26 (in Lithuanian).

Larcher, W., (1995). *Physiological Plant Ecology*. Springer, Germany.

Lazauskas, J., & Dapkus, R., (1992). *Field Crop Breeding in Lithuania*. Vilnius (in Lithuanian).

Lemežienė, N., & Kanapeckas, J., (2000). Lithuania perennial herb wild ecotypes - the collection of genetic creation and breeding. *Žemdirbystė*, *72*, 182–195.

Lewis, G., Schrire, B., Mackinder, B., & Lock, M., (2005). *Legumes of the World*. Royal Botanic Gardens, Kew, UK.

Li, H. A., Radunz, P. H., & Schmid, G. H., (2002). Influence of different light intensities on the content of diosgenin, lipids, carotenoids, and fatty acids in leaves of *Dioscorea zingiberensis*. *Zeitschrift fur Naturforschung C*, *57*, 134–143.

Lotter, D. W., (2003). Organic agriculture. *Journal Sustain Agriculture*, *21*(4), 59–128.

Luscher, A., Mueller-Harvey, I., Soussana, J. F., Rees, R. M., & Peyraud, J. L., (2014). Potential of legume-based grassland–livestock systems in Europe: A review. *Grass and Forage Science*, *69*(2), 206–228.

Maršalkienė, N., (2011). Phytocenological variation and effect of human activity on caraway (*Carum carvi* L.) survival in natural habitats. *NJF Report*, 7(1), 39–44.

Maršalkienė, N., (2015a). Investigation of some wild annual vetch (*Vicia*, L.,). *Biologija*, 61(1), 15–24.

Maršalkienė, N., (2015b). Locations of hairy vetch (*Vicia villosa* Roth) in Lithuania. *Scripta Horti Botanici*, 14, 60–65 (in Lithuanian).

Marum, P., (1999). Should *in situ* conservation replace *ex situ* conservation of forage crops? *Botanica Lithuanica*, 2, 99–104.

Maxted, N., (1995). *An Ecogeographical Study of Vicia Subgenus Vicia*. IPGRI, Rome, Italy.

Mihailović, V., Mikić, A., Vasiljević, S., Katić, S., Karagić, D., & Ćupina, B., (2008). Forage yields in urban populations of hairy vetch (*Vicia villosa* Roth) from Serbia. *Grassland Science in Europe*, 13, 281–283.

Mischler, R. D., Duiker, S. W., Curran, W. S., & Wilson, D., (2010). Hairy vetch management for no-till organic corn production. *Agronomy Journal*, 102, 355–362.

Murchie, E. H., & Horton, P., (1997). Acclimation of photosynthesis to irradiance and spectral quality in British plant species: Chloroplast content, photosynthetic capacity, and habitat preference. *Plant, Cell, and Environment*, 19, 1083–1090.

Natkevičaitė-Ivanauskienė, M., (1983). *Basics of Botanical Geography and Phytocoenology*. Mokslas, Vilnius (in Lithuanian).

Nickrent, D. L., & Patrick, J. A., (1998). The nuclear ribosomal DNA intergenic spacers of wild and cultivated soybean have low variation and cryptic subrepeats. *Genome*, 41, 183–192.

Nikolaiciuk, L., & Zigar, M., (1996). *Medical Plants*. Varpas, Kaunas (in Lithuanian).

Partel, M., Bruun, H. H., & Sammul, M., (2005). Biodiversity in temperate European grasslands: Origin and conservation. *Grassland Science in Europe*, 10, 1–14.

Partel, M., Zobel, M., Zobel, K., & Van der Maarel, E., (1996). The species pool and its relation to species richness: Evidence from Estonian plant communities. *Oikos*, 75, 111–117.

Petraitytė, N., & Dastikaitė, A., (2007). Agrobiological assessment of wild *Carum carvi* cenopopulations biodiversity *ex situ*. *Biologija*, 53(4), 74–79.

Petraitytė, N., (2003). Investigation of common caraway (*Carum carvi* L.) morphobiochemical properties stability *ex situ*. *Ekologija*, 1, 43–46.

Petraitytė, N., (2005). Phenotypic and genetic diversity of caraway (*Carum carvi* L.) population in Lithuania. *Summary of Doctoral Dissertation, Biomedical Sciences*, Kaunas-Akademija.

Petraitytė, N., Dastikaitė, A., & Sliesaravicius, A., (2002). Investigation of common caraway (*Carum carvi* L.) morphobiochemical properties stability *in situ*. *Žemdirbystė*, 78(2), 274–282.

Petraitytė, N., Karklelienė, R., & Maročkienė, N., (2003). Caraway, carrot, and another Umbellifer genetic resources in Lithuania. *ECP/GR Vegetables Network Meeting and Joint Ad Hoc Meeting on Leafy Vegetables*, 75–77.

Petraitytė, N., Sliesaravičius, A., & Dastikaitė, A., (2007). Potential reproduction and real seed productivity of *Vicia villosa*, L., *Biologija*, 53(2), 48–51.

Radusiene, J., & Janulis, V., (2004). Trends of investigation, use, and conservation of medicinal, aromatic plants of diversity. *Medicina*, 40(8), 705–709 (In Lithuanian).

Rašomavičius, V., (2001). *Habitats of European Importance in Lithuania.* Vilnius.

Raven, P. H., (2000). Plant conservation globally and locally. *Rhodora, 102,* 243–245.

Renzi, J. P., Chantre, G. R., & Cantamutto, M. A., (2017). *Vicia villosa* ssp. *villosa* Roth field emergence model in a semiarid agroecosystem. Grass and Forage Science, *66* (in press).

Rosevear, M. J., Young, A. J., & Jonson, G. N., (2001). Growth conditions are more important than species origin in determining leaf pigment content of British plant species. *Functional Ecology, 15,* 474–480.

Rothmaler, W., Schubert, R., Werner, K., & Meusel, H., (1998). *Exkursionsflora.* Berlin.

Roy, B. A., Stanton, M. L., & Eppley, S. M., (1999). Effects of environmental stress on leaf hair density and consequences for selection. *Journal Evolution Biology, 12,* 1089–1103.

Selge, A., & Lillak, R., (2005). Integration efficient grassland farming and biodiversity. *Grassland Science in Europe, 10,* 1–14.

Sheaffer, C. C., & Seguin, P., (2003). Forage legumes for sustainable *cropping* systems. *Journal of Crop Production, 8*(1/2), 187–216.

Sinskaya, E. N., (1969). *Historical Geography of Cultivated Flora.* Kolos, Leningrad (in Russian).

Sklyarevsky, L. J., & Gubanov, I. A., (1986). *Medicinal Plants in the Home.* Roselhozizdat, Maskva (in Russian).

Sliesaravičius, A., & Petraitytė, N., (2001). Accumulation and research of the Lithuanian fodder legume genera *Vicia,* L., and *Lathyrus,* L., genetic resources. *Biologija, 4,* 61–65.

Sliesaravičius, A., Petraitytė, N., & Dastikaitė, A., (2005). The study of phenotypical diversity in wild narrow-leafed vetch (*V. angustifolia, L.,*). *Biologija, 3,* 31–35.

Smýkal, P., Coyne, C. J., Ambrose, M. J., Maxted, N., Schaefer, H., Blair, M. W., et al., (2015). Legume crops phylogeny and genetic diversity for science and breeding. *Critical Reviews in Plant Sciences, 34,* 1–3.

Sokolov, M. S., & Marchenko, A. I., (2002). Potential risk of transgenic plants cultivation and their crop consumption. *Selskokhozyaistvennaya Biologiya, 5,* 3–23 (in Russian).

Stancevičius, A., (1971). *Flora of Lithuanian SSR.* Vilnius (in Lithuanian).

Stevens, C. J., Dise, N. B., Mountford, J. O., & Gowing, D. J., (2004). Impact of nitrogen deposition on the species richness of grasslands. *Science, 303,* 1876–1879.

Stoutjesdisk, P. H., & Barkman, J. J., (1992). *Microclimate Vegetation and Fauna.* Opulus Press, Uppsala.

Sur, C. B., (1993). Composition of essential oils of medicinal plants. *Plant Resources* (in Russian).

Thayer, S. S., & Bjorkman, O., (1990). Leaf xanthophylls content and composition in sun shade determined by HPLC. *Photosynthesis Research, 23,* 331–343.

Thompson, K., & Jones, A., (1999). Human population density and prediction of local plant extinction in Britain. *Conservation Biology, 13,* 185–189.

Toxopeus, H., & Bouwmeester, H. J., (1993). Improvement of essential oil and carvone production the Netherlands. *Industrial Crops and Products, 1,* 295–301.

Trusov, Y., Bogdanova, V. S., & Berdnikov, V. A., (2004). Evolution of the regular zone of histone H1 in Fabaceae plants. *Journal of Molecular Evolution, 59*(4), 546–555.

Van Dijk, G., (1991). The status of semi-natural grasslands in Europe. In: Goriup, P. D., Batten, L. A., & Norton, J. A., (eds.), *The Conservation of Lowland Dry Grassland Birds in Europe* (pp. 15–36). Joint Nature Conservation Committee, Peterborough.

Vokk, R., & Loomaegi, T., (1998). Antimicrobial and antioxidative properties of kitchen herbs. *B-Tecniskas Zinatnes., 289*(12), 94–96.

Wagner, K. H., & Elmadfa, I., (2003). Biological relevance of terpenoids - overview focusing on mono-, di-, and tetraterpenes. *Annals of Nutrition and Metabolism, 47*(3/4), 95–106.

Warren, J., & Mackenzie, S., (2001). Why are all color combinations not equally represented as flower-color polymorphisms? *New Phytologist, 151*(1), 237–244.

Wettstein, D., (1957). Chlorophyll letale und der submikroskopische Formwechsel der Plastiden. *Experimental Cell Research, 12,* 427–434.

Wiersemana, J. H., & Blanca, L., (2003). *World Economic Plants: A Standard Reference* (2nd edn.). CRC Press.

Wulf, E., & Maleeva, O., (1969). *Global Resources of Useful Plants.* Nauka, Leningrad (in Russian).

Zhang, P., Jin, G. Q., Zhou, Z. C., Yu, L., & Fan, H. H., (2004). Provenance difference and geographic variation pattern for seedling trait of *Shima superb. Forest Research, 17*(2), 192–198.

CHAPTER 4

PLANT DIVERSITY IN THE RURAL HOME GARDENS OF COASTAL TALUKS OF UTTARA KANNADA IN KARNATAKA, INDIA

SHIVANAND S. BHAT[1], L. RAJANNA[2], ABHISHEK MUNDARAGI[3], and DEVARAJAN THANGADURAI[3]

[1]*Government Arts and Science College, Karwar, Karnataka, 581301, India*

[2]*Department of Botany, Bangalore University, Bengaluru, Karnataka, 560056, India*

[3]*Department of Botany, Karnatak University, Dharwad, Karnataka, 580003, India*

4.1 INTRODUCTION

Homegardens are best representatives of agroforestry systems and generally perform a wide array of functions. They are often described as land use system involving multipurpose plants such as shrubs and trees along with agricultural crops. Typically, grown across an individual house and maintained and managed by family labor. Homegardens have several features which make them a viable and important option for *in situ* conservation. They can serve as refuges for different plant varieties that were once more widespread in the larger agro-ecosystems. Homegardens are sites for experimentation and introduction of new cultivars arising from the exchange and interaction between cultures and communities. They are also important systems for the study of the evolution of plant genetic resources because of the complex species diversity and interactions which take place in homegardens and are being replaced by modern commercial cultivars elsewhere. It is, therefore, crucial to understanding their dynamics so that they can take their place as a component of *in situ* conservation of global agro-biodiversity. Socio-economic aspects of homegardens such as economic benefits received from the produce of the garden, gender participation in homegarden management, the relationship between the socio-cultural status of families and biological composition of homegardens would reveal the contribution of homegardens to rural socio-economic sustainability.

During ethnobotanical field trips to remote villages of coastal districts of Karnataka State, a rich divergent plant varieties were observed in the backyard or homegardens of almost every single rural homes. Further, a literature survey on homegarden biodiversity indicated that they play an instrumental role *in situ* conservation of agrobiodiversity and have been found to contribute significantly towards food security and socio-economic sustainability among these regions. Across globally, home gardening is considered as oldest land use activity and has been found to be evolved through agricultural intensification in response to an increased shortage of available land. Nevertheless, homegardens have received a great deal of attention these days as they significantly contribute to sustainable agroecosystems approach and exhibit several important key features such as nutrient recycling, soil conservation and improved management practices with minimized external inputs and most importantly they are eco-friendly (Torquebiau, 1992; Jose and Shanmugaratnam, 1993).

Among the available data extensive research has been carried out across tropical and sub-tropical regions of Asia, Africa, and Meso-America, North America and Europe (Nair and Kumar, 2006), In India, scientific studies on homegardens have been mainly carried out across states such as Kerala, Assam, and Andaman islands (Kumar et al., 1994; Puskaran, 2002; Das and Das, 2005; Pandey et al., 2006, 2007). However, availability of scientific data on homegardens of Karnataka State is scarce. Among few reports available, Shastri et al. (2002) has described village ecosystems. Thus, an attempt was made to enrich the current knowledge and compile the floristic composition of selected homegardens across Coastal taluks of Uttara Kannada district of Karnataka, India. This book chapter brings an overview and strongly focus on following attributes (1) to inventorise species diversity of plants maintained in the homegardens, (2) to analyse the various local benefits and usage values of homegardens such as source of food, medicine, ornamental plants, fodder plants and others, (3) to analyse the various economic roles of homegardens, mainly as a source of family income and food, and, (4) to study the gender participation and social aspects connected with homegardens.

4.2   METHODOLOGY AND STUDY AREA

Initially different village ecosystems have been chosen so as to represent the different farming systems, ethnic diversity and biological diversity of Uttara Kannada. Regular field visits were undertaken to gather information about plant diversity across homegardens during 2013–2015. Around 150 homegardens were chosen and were carefully surveyed. Homegardens of these selected localities were investigated for floral diversity, economic, and management aspects. The intensive survey was made on the plant species collected from the homegardens of each household (each homegarden as a single sampling unit). The senior individual of household members of every homegarden was interviewed in the local language (Kannada, Konkani) to collect ethnobotanical relevance of plant species and their economic benefits. Further, plant species were characterized depending on their mode of usage and categorized into fruits, vegetables, ornamental, medicinal, and multipurpose plant species (comprising multiple benefits). Identification and authentication of the plant species were made citing available flora (Cooke, 1967; Bhat, 2003). Further information related to gender participation, management practices and economic outputs of the homegardens was also surveyed. Appropriate ecological and statistical tools were used for data analysis and drawing conclusions.

Uttara Kannada is also known as North Canara, and it is a district in the state of Karnataka. It is bordered by Belgaum District to the north, Dharwad, Shimoga, and Udupi Districts to the south, Haveri District to the east and to the western part with the Arabian Sea. Karwar is an administrative headquarters of the district. It comprises 11 taluks, with 5 in the coastal region, *viz.* Bhatkal, Honnavar, Kumata, Ankola, and Karwar (Figure 4.1). Totally 150 potential homegardens of 5 coastal taluks of Uttara Kannada district have been selected for the present study.

FIGURE 4.1   (See color insert.) Overview of some of the traditional homegardens in the study area.

4.3   SPECIES DIVERSITY

Homegardens proved to be a basic agroforestry system mainly for food, medicine, and for the protection of individual compounds of the houses. All the plant species documented during the study period is utilized for several purposes except the weeds in the homegardens. The size of the gardens studied were ranged from 0.01 to 0.05 ha and the average size being 0.02 ha. During the study period at different seasons, as many as 362 plant species were recorded from 150 homegardens. Interestingly, 9 homegardens exhibited a great diversity of plants comprising nearly 100 species (Table 4.1) following 66 homegardens which showed approximately 71–80 plant species for every household. Among the different habits of plants studied, a total of 108 tree species were present contributing to a total of 35.29% to the plant diversity. Further, tree species such as mango, jack, and tamarind forms a canopy and offered shade to smaller plants such as ginger, turmeric, and colocasia plant species. Second abundant plant habit recorded was shrubs with a total of 101 plant species (25.49%), followed by 92 herb plants (23.21%) and 61 climbers species (15.7%) (Figure 4.2). Most of the plants belonged to families such as *Fabaceae* (30 spp.), *Apocynaceae* (27 spp.), *Lamiaceae* (21 spp.), *Acanthaceae* (15 spp.), *Malvaceae* (15 spp.), *Cucurbitaceae* (13 spp.), *Euphorbiaceae* (12 spp.), *Araceae* (11 spp.), *Asteraceae* (11 spp.), and *Phyllanthaceae* (11 spp.) in order of decreasing number of species (Table 4.2). The least number of plant species recorded in a single homegarden was 54 while 141 plant species were found highest. Sixty-six homegardens exhibited 44% of species diversity and ranged from 71 to 80 plant species for every individual homehgarden (Table 4.3). Based on the homegarden plants' utility they can be grouped as medicinal (16.03%), ornamental (34.81%), vegetables (12.16%), fruit yielding (11.33%), miscellaneous uses (8.84%) and other minor categories (Table 4.4). Other previous studies on homegardens indicated that trees are a most dominant group (Das and Das, 2005). Further, plant species surveyed in the present study was comparatively higher than that of previously reported by several authors across other parts of India (Nair and Shreedharan, 1986; Kumar et al., 1994; John and Nair, 1999). However, higher plant diversity have been reported across homegardens of Northern Thailand (230 species, Black et al., 1996), Nicaragua (324 species, Mendez et al., 2001) and West Java (602 species, Karyona, 1990). Among the individual plant species studied most common plants found were areca palm (*Areca catechu*), followed by coconut palm (*Cocos nucifera*), mango tree (*Mangifera indica*), banana (*Musa paradisiaca*), hibiscus (*Hibiscus rosa-sinensis*) and holy basil (*Ocimum tenuiflorum*). Three plants were found in common among all the homegardens, *viz.*, Areca palm, coconut palm, and banana. Jose and Shanmugaratanam (1993) reported similar observation from the homegardens of coastal Kerala, followed by Das and Das (2005) and Pandey et al. (2006) across

Assam and Andaman, respectively. Holy basil is a sacred plant of the Hindu community, and it is found in front of most of the houses visited in the study area (93%).

Socioeconomic aspects of the homegardens have been studied. Species richness in the gardens of Havyaka, Halakki, Karivokkaliga, and Namadhary community has been analyzed. Specially role of gender in preharvesting and post-harvesting is also recorded. Species diversity is rich in the homegardens of Havyaka and Halakki community. Ornamental, medicinal, and fruit yielding plants are dominant in the homegardens of Havyaka community whereas Halakki's prefer vegetables and fruits in their garden because the home garden is a source of income for them. Thus, homegardens carry vast potential in providing food, natural medicine, and other basic household necessities further thereby aiding in the conservation of the plant genetic diversity (Bhat and Rajanna, 2016).

TABLE 4.1   List of Plants in the Homegardens of Uttara Kannada in Karnataka, India*

Botanical name	Family	Local name	Habit	Uses
Abelmoschus esculentus (L.) Moench	Malvaceae	Bende	Shrub	Vegetable
Abrus precatorius L.	Fabaceae	Gulgunji	Climber	Miscellaneous
Abutilon indicum (L.) Sweet.	Malvaceae	Shree mudre	Herb	Ornamental
Acacia auriculiformis Benth.	Fabaceae	Acacia	Tree	Multipurpose
Acacia farnesiana (L.) Willd.	Fabaceae	Kastoori jaali	Tree	Ornamental
Acacia mangium Willd.	Fabaceae	Mangium	Tree	Multipurpose
Acalypha hispida Burm. f.	Euphorbiaceae	Bekkina baala	Shrub	Ornamental
Acalypha wilkesiana Muell. Arg.	Euphorbiaceae	Katri gida	Shrub	Ornamental
Acorus calamus L.	Acoraceae	Bajae	Herb	Medicinal
Aechmea gamosepala Wittm.	Bromaliaceae	Benki kaddi	Herb	Ornamental
Aegle marmelos (L.) Correa	Rutaceae	Bilva	Tree	Multipurpose
Aerva lanata (L.) Juss	Amaranthaceae	Bili hindi	Herb	Medicinal

TABLE 4.1 *(Continued)*

Botanical name	Family	Local name	Habit	Uses
Agave sisalana Perrine ex Engelm.	Asparagaceae	Kattaale	Shrub	Ornamental
Ailanthus triphysa (Dennst.) Alston	Simaroubaceae	Guggula dhupa	Tree	Miscellaneous
Albizia lebbeck (L.) Benth.	Fabaceae	Kalbhaage	Tree	Miscellaneous
Albizia saman (Jacq.) Merr.	Fabaceae	Devdaaru	Tree	Miscellaneous
Allamanda cathartica L.	Apocynaceae	Methai huvu	Climber	Ornamental
Allamanda schottii Pohl	Apocynaceae	Haladi kote	Shrub	Ornamental
Alocasia macrorrhizos (L.) G. Don	Araceae	Marasanige	Herb	Vegetable
Aloe vera (L.) Burm. f.	Xanthorrhoeaceae	Lole sara	Herb	Medicinal
Alpinia calcarata (Haw) Roscoe.	Zingiberaceae	Rasana gida	Herb	Medicinal
Alpinia galangal (L.) Willd.	Zingiberaceae	Kallu shunti	Herb	Medicinal
Alstonia scholaris (L.) R. Br.	Apocynaceae	Saptaparni	Tree	Medicinal
Alternanthera bettzickiana (Regel) G. Nicholson	Amaranthaceae	Bannada harive	Herb	Ornamental
Alternanthera brasiliana (L.) Kuntze	Amaranthaceae	Bili gonde Bannada harive	Herb	Ornamental
Amaranthus hybridus L.	Amaranthaceae	Dantu harige	Herb	Vegetable
Amaranthus tricolor L.	Amaranthaceae	Harive	Herb	Vegetable
Amorphophallus paeoniifolius (Dennst.) Nicolson var. *campanulatus*	Araceae	Suvarnagadde	Herb	Vegetable
Anacardium occidentale L.	Anacardiaceae	Geru	Tree	Fruit
Ananas comosus (L.) Merr.	Bromaliaceae	Ananaas	Herb	Fruit

TABLE 4.1   *(Continued)*

Botanical name	Family	Local name	Habit	Uses
Andrographis paniculata (Burm. f.) Nees	Acanthaceae	Kiraayti kaddi	Herb	Medicinal
Anethum graveolens L.	Apiaceae	Sabbasige	Herb	Vegetable
Angelonia salicariifolia Bonpl.	Plantaginaceae	Aame huvu	Herb	Ornamental
Annona muricata L.	Annonaceae	Hanumana phala	Tree	Fruit
Annona reticulata L.	Annonaceae	Ramphala	Tree	Fruit
Annona squamosa L.	Annonaceae	Sitaphala	Tree	Fruit
Antigonon leptopus Hook. & Arn.	Polygonaceae	Papermint	Climber	Ornamental
Areca catechu L.	Arecaceae	Adike	Tree	Masticatory
Artabotrys hexapetalous (L. f.) Bhandari	Annonaceae	Manoranjini	Shrub	Ornamental
Artocarpus altilis L.	Moraceae	Beru halasu	Tree	Vegetable
Artocarpus camansi Blanco	Moraceae	Neeru halasu	Tree	Vegetable
Artocarpus gomezianus Wall. ex Trec.	Moraceae	Vaate huli	Tree	Spice
Artocarpus heterophyllus Lam.	Moraceae	Halasu	Tree	Fruit
Asclepias curassavica L.	Apocynaceae	Chaduranga	Herb	Ornamental
Asparagus racemosus Willd.	Asparagaceae	Shataavari	Climber	Medicinal
Asystasia gangetica (L.) T. Anderson	Acanthaceae	Maithaal kaddi	Herb	Ornamental
Averrhoa bilimbi L.	Oxalidaceae	Bimbalu	Tree	Multipurpose
Averrhoa carambola L.	Oxalidaceae	Karamalu	Tree	Fruit
Azadirachta indica A. Juss.	Meliaceae	Kahi bevu	Tree	Medicinal

TABLE 4.1 *(Continued)*

Botanical name	Family	Local name	Habit	Uses
Bambusa bambos (L.) Voss	Poaceae	Bidiru	Tree	Miscellaneous
Barleria cristata L.	Acanthaceae	Gorante	Shrub	Ornamental
Barleria prionites L.	Acanthaceae	Shaastri bale	Shrub	Ornamental
Barleria strigosa Willd.	Acanthaceae	Gentige huvu	Shrub	Ornamental
Basella alba L.	Basellaceae	Basale	Herb	Vegetable
Bauhinia acuminata L.	Fabaceae	Bili mandara	Shrub	Ornamental
Bauhinia tomentosa L.	Fabaceae	Mani Mandara	Shrub	Ornamental
Bauhinia variegata L.	Fabaceae	Mara mandara	Tree	Ornamental
Benincasa hispida (Thunb.) Cogn.	Cucurbitaceae	Budugumbala	Climber	Vegetable
Beta vulgaris L.	Chenopodiaceae	Beetroot	Herb	Vegetable
Bixa orellana L.	Bixaceae	Sindhoori kaayi	Tree	Multipurpose
Boerhavia diffusa L.	Nyctaginaaceae	Punarnava	Herb	Medicinal
Bougainvillea buttiana Holttum & Standl.	Nyctaginaaceae	Kagadada huvu	Shrub	Ornamental
Brassica caulorapa L. var. *gongylodes*	Brassicaceae	Knol khol	Herb	Vegetable
Breynia retusa (Dennst.) Alston	Phyllanthaceae	Beli gida	Shrub	Fence
Breynia vitis-idaea (Burm. f.) C.E.C. Fisch	Phyllanthaceae	Beli gida	Shrub	Fence
Bridellia retusa (L.) A. Juss	Phyllanthaceae	Kove mullu	Tree	Miscellaneous
Bridellia stipularis (L.) A. Juss	Phyllanthaceae	Beli gida	Shrub	Fence
Brophyllum pinnatum (Lam.) Oken	Crassulaceae	Kaadu basale	Herb	Medicinal
Butea monosperma (Lam.) Taub.	Fabaceae	Muttuga	Tree	Multipurpose

TABLE 4.1 *(Continued)*

Botanical name	Family	Local name	Habit	Uses
Cactus repandus (L.) Mill.	Cactaceae	Papasu kalli	Shrub	Fence
Caesalpinia bonduc (L.) Roxb.	Fabaceae	Gajjuga	Climber	Miscellaneous
Caesalpinia pulcherrima (L.) Sw.	Fabaceae	Meese huvu	Tree	Ornamental
Caladium bicolor (Aiton) Vent.	Araceae	Bannada kesu	Herb	Ornamental
Caladium humboldtii (Raf.) Schott	Araceae	Bili bannada kesu	Herb	Ornamental
Calathea zebrine (Sims) Lindl.	Maranthaceae	Show arrowroot	Herb	Ornamental
Calliandra haematocephala Hassk.	Fabaceae	Sanna bottle brush	Shrub	Ornamental
Calophyllum inophyllum L.	Calophyllaceae	Sura hone	Tree	Multipurpose
Calotropis gigantean (L.) Dryand.	Apocynaceae	Ekke	Shrub	Medicinal
Cananga odorata (Lam) Hook. f. & Thomson	Annonaceae	Apurva chmpaka	Tree	Ornamental
Canavalia rosea (Sw.) DC.	Fabaceae	Samudra katti avare	Climber	Vegetable
Canavalia ensiformis (L.) DC.	Fabaceae	Katti avare	Climber	Vegetable
Canavalia gladiata (Jacq.) DC.	Fabaceae	Katti avare (pink seed)	Climber	Vegetable
Canna indica L.	Cannaceae	Kaabaale	Herb	Ornamental
Capsicum annuum L.	Solanaceae	Kempu menasu	Shrub	Condiment
Capsicum frutescens L.	Solanaceae	Nuchu menasu	Shrub	Condiment
Cardiospermum helicacabum L.	Sapindaceae	Agniballi	Climber	Medicinal
Carica papaya L.	Caricaceae	Pappayi	Tree	Fruit
Carissa spinarum L.	Apocynaceae	Kavali	Shrub	Fruit

TABLE 4.1 *(Continued)*

Botanical name	Family	Local name	Habit	Uses
Caryota urens L.	Arecaceae	Baini mara	Tree	Miscellaneous
Cascabela thevetia (L.) Lippold.	Apocynaceae	Karaveera	Tree	Ornamental
Casurina equisetifolia L.	Casuarinaceae	Gaali mara	Tree	Multipurpose
Catharanthus roseus (L.) G. Don	Apocynaceae	Nitya pushpa	Herb	Ornamental
Ceiba pentandra (L.) Gaertn.	Malvaceae	Bili burga	Tree	Miscellaneous
Celosia argentia L. var. *argentea*	Amaranthaceae	Kolijuttu	Herb	Ornamental
Centella asiatica (L.) Urb.	Apiaceae	Ondelaga	Herb	Medicinal
Cestrum nocturnum L.	Solanaceae	Raatri raani	Shrub	Ornamental
Cheilocostus speciosus (J. Koenig) C. Specht	Costaceae	Narikabbu	Herb	Medicinal
Chrysanthemum morifolium Ramat.	Asteraceae	Sevantige	Herb	Ornamental
Cinnamomum verum J. Presl	Lauraceae	Daalchini	Tree	Spice
Cissus quadrangularis L.	Vitaceae	Sandu vaata balli	Climber	Medicinal
Citrullus lanatus (Thunb.) Matsum & Nakai	Cucurbitaceae	Kallangadi	Climber	Fruit
Citrus aurantiifolia (Christm.) Swingle	Rutaceae	Nimbu	Shrub	Fruit
Citrus aurantium L.	Rutaceae	Huli kanchi	Tree	Fruit
Citrus limon (L.) Burm.f.	Rutaceae	Gaja nimbe	Shrub	Fruit
Citrus maxima (Burm.) Osbeck	Rutaceae	Sakre kanchi	Tree	Fruit
Citrus medica L.	Rutaceae	Maadalu	Shrub	Fruit
Citrus sinensis (L.) Osbeck	Rutaceae	Kittale	Tree	Fruit

TABLE 4.1   *(Continued)*

Botanical name	Family	Local name	Habit	Uses
Cleome gynandra L.	Cleomaceae	Meese huvu	Herb	Ornamental
Clerodendrum aculeatum (L.) Griseb.	Lamiaceae	Beli mallige	Shrub	Fence
Clerodendrum calamitosum L.	Lamiaceae	Bili gonchalu	Shrub	Ornamental
Clerodendrum chinense (Osbeck) Mabb.	Lamiaceae	Parimala gundu gonchalu	Shrub	Ornamental
Clerodendrum incisum Klotzsch.	Lamiaceae	Sugandhi mallige	Shrub	Ornamental
Clerodendrum indicum (L.) Kuntze.	Lamiaceae	Kari kaalu mallige	Shrub	Ornamental
Clerodendrum inermae (L.) Gaertn.	Lamiaceae	College mallige	Shrub	Fence
Clerodendrum infortunatum L.	Lamiaceae	Beli gida	Shrub	Fence
Clerodendrum paniculatum L.	Lamiaceae	Teru huvu	Shrub	Fence
Clerodendrum splendens G. Don	Lamiaceae	Kempi huvina balli	Climber	Ornamental
Clerodendrum thomsoniae Balf.f.	Lamiaceae	Rakta balli	Climber	Ornamental
Clitoria ternatea L.	Fabaceae	Shanka pushpa	Climber	Ornamental
Coccinia grandis (L.) Voigt	Cucurbitaceae	Tonde kaayi	Climber	Vegetable
Cocos nucifera L.	Arecaceae	Tengu	Tree	Fruit
Codiaeum variegatum (L.) Rumph. ex A. Juss.	Euphorbiaceae	Croton	Shrub	Ornamental
Coffea arabica L.	Rubiaceae	Coffee	Shrub	Miscellaneous
Colocasia esculenta var. *aquatica* (L.) Schott	Araceae	Bilu kesu	Herb	Vegetable
Colocasia esculenta var. *esculenta* (L.) Schott	Araceae	Kesu	Herb	Vegetable

TABLE 4.1 *(Continued)*

Botanical name	Family	Local name	Habit	Uses
Cordia dichotoma Forst. f.	Boraginaceae	Challe hannu	Tree	Multipurpose
Cordyline fruiticosa (L.) A. Chev.	Asparagaceae	Dracaena	Shrub	Ornamental
Coriandrum sativum L.	Apiaceae	Kottumbari	Herb	Condiment
Cosmos sulphureus Cav.	Asteraceae	Ketaki	Herb	Ornamental
Costus pictus D. Don	Costaceae	Diabetes gida	Herb	Medicinal
Couroupita guianensis Aubl.	Lecythidaceae	Naga linga pushpa	Tree	Ornamental
Crossandra infundibuliformis (L.) Nees	Acanthaceae	Abballi	Shrub	Ornamental
Cryptolepis dubia (Burm. f.)	Apocynaceae	Kare balli	Climber	Medicinal
Cucumis melo L.	Cucurbitaceae	Ibbudalu	Climber	Fruit
Cucumis melo L.	Cucurbitaceae	Moge kaayi	Climber	Vegetable
Cucumis sativus L.	Cucurbitaceae	Southe kaayi	Climber	Vegetable
Cucurbita moschata (Duchesne ex Lam) Duchesne	Cucurbitaceae	Putla kaayi	Climber	Vegetable
Cuphea hyssopifolia Kunth.	Lythraceae	Putti gulaabi	Herb	Ornamental
Curcuma amada Roxb.	Zingiberaceae	Ambe kombu	Herb	Condiment
Curcuma longa L.	Zingiberaceae	Arishina	Herb	Medicinal
Cymbopogon citratus (DC.) Stapf.	Poaceae	Majjige hullu	Herb	Medicinal
Cymbopogon flexuosus (Nees ex Steud.) W. Watson	Poaceae	Enne majjige hullu	Herb	Miscellaneous
Cymopsis tetragonoloba (L.) Taub.	Fabaceae	Chaulikaai	Herb	Vegetable

TABLE 4.1   *(Continued)*

Botanical name	Family	Local name	Habit	Uses
Cynodon dactylon (L.) Pers.	Poaceae	Garike hullu	Herb	Miscellaneous
Dalhea tuberosa L.	Asteraceae	Dere	Herb	Ornamental
Delonix regia (Hook.) Raf.	Fabaceae	May flower	Tree	Miscellaneous
Dendranthema indicum (L.) De Moulins	Asteraceae	Bottu sevantige	Herb	Ornamental
Dichrostachys cinerea (L.) Wight & Arn.	Fabaceae	Shame patre	Tree	Multipurpose
Dieffenbachia seguine (Jacq.) Schoot	Araceae	Bannada kesu	Herb	Ornamental
Dillenia indica L.	Dilleniaceae	Betta kanaglu	Tree	Miscellaneous
Dioscorea alata L.	Dioscoriaceae	Mandigenasu	Climber	Vegetable
Dioscorea bulbifera L.	Dioscoriaceae	Heggenasu	Climber	Vegetable
Dombeya burgessiae Gerrard ex Harv. & Sond.	Malvaceae	December huvu	Shrub	Ornamental
Dracaena braunii Engl.	Asparagaceae	Dracaena	Shrub	Ornamental
Dracaena fragrans (L.) Ker Gawl.	Asparagaceae	Dracaena	Shrub	Ornamental
Dracaena reflexa Lam.	Asparagaceae	Patte dracena	Shrub	Ornamental
Dregea volubilis (L.f.) Benth. ex Hook. f.	Apocynaceae	Hegele balli	Climber	Miscellaneous
Duranta erecta L.	Verbenaceae	Durantha	Shrub	Ornamental
Eclipta prostrata (L.) L.	Asteraceae	Bhrangaraja	Herb	Medicinal
Epiphyllum oxypetalum (DC.) Haworth	Asteraceae	Brahma kamala	Herb	Ornamental
Epipremnum aureum (L.) Engl. cv. 'Aureum'	Araceae	Sanna money plant	Climber	Ornamental
Epipremnum pinnatum (L.)	Araceae	Dodda money plant	Climber	Ornamental

TABLE 4.1   *(Continued)*

Botanical name	Family	Local name	Habit	Uses
Eryngium foetidum L.	Apiaceae	Raakshsa kottumbari	Herb	Condiment
Erythrina variegata L.	Fabaceae	Channe	Tree	Miscellaneous
Euphorbia milii Des Moul	Euphorbiaceae	Show kalli	Shrub	Ornamental
Euphorbia neriifolia L.	Euphorbiaceae	Kodukalli	Shrub	Fence
Euphorbia pulcherrima Willd. ex Klotzsch	Euphorbiaceae	Show gida	Shrub	Ornamental
Euphorbia tirucalli L.	Euphorbiaceae	Milky bush	Shrub	Ornamental
Ficus benghalensis L.	Moraceae	Aalada mara	Tree	Miscellaneous
Ficus elastica Roxb. ex Hornem	Moraceae	Rubber	Tree	Ornamental
Ficus hispida L. f.	Moraceae	Geretle	Tree	Miscellaneous
Ficus racemosa L.	Moraceae	Atti mara	Tree	Multipurpose
Ficus religiosa L.	Moraceae	Ashwatha, arali mara	Tree	Miscellaneous
Foeniculum vulgare Mill.	Apiaceae	Bade soppu	Herb	Spice
Garcinia indica (Thouars) Choisy	Clusiaceae	Murugalu	Tree	Fruit
Gardenia jasminoides J. Ellis.	Rubiaceae	Nanjatle	Shrub	Ornamental
Getonia floribunda Roxb.	Combretaceae	Girgitti	Climber	Fence
Gliricidia sepium (Jacq.) Walp	Fabaceae	Gobbara soppu	Tree	Multipurpose
Gloriosa superba L.	Colchicaceae	Gouri huvu	Climber	Ornamental
Gmelina arborea Roxb.	Lamiaceae	Shivani	Tree	Miscellaneous
Gomphrena globosa L.	Amaranthaceae	Umi gonde	Herb	Ornamental
Gossypium barbadense L.	Malvaceae	Hatti	Shrub	Miscellaneous

TABLE 4.1 *(Continued)*

Botanical name	Family	Local name	Habit	Uses
Hedychium coronarium J. Koenig.	Zingiberaceae	Sugandha pushpa	Herb	Ornamental
Helianthus annuus L.	Asteraceae	Suryakaanti	Shrub	Ornamental
Heliconia psittacorum L. f.	Strelitziaceae	Hakki kaabaale	Herb	Ornamental
Heliconia rostrata Ruiz & Pav.	Strelitziaceae	Meenu kaabale	Herb	Ornamental
Helicteres isora L.	Malvaceae	Edamuri	Shrub	Miscellaneous
Hemelia patens Jacq.	Rubiaceae	Hemelia	Shrub	Ornamental
Hemidesmus indicus (L.) R. Br. ex Schult.	Apocynaceae	Sogade beru	Climber	Medicinal
Hibiscus mutabilis L.	Malvaceae	Chandra kaanti	Shrub	Ornamental
Hibiscus radiates Cav.	Malvaceae	Mullu daasavala	Shrub	Ornamental
Hibiscus rosa-sinensis L.	Malvaceae	Daasavala	Tree	Ornamental
Hibiscus schizopetalus (Dyer.) Hook. f.	Malvaceae	Gante daasavala	Tree	Ornamental
Hibiscus syriacus L.	Malvaceae	Neeli daasavala	Shrub	Ornamental
Hippeastrum puniceum (Lam.) Voss	Amaryllidaceae	Phonehuvu	Herb	Ornamental
Holarrhena pubescens (Buch.-Ham.) Wall. ex G. Don	Apocynaceae	Kodasiga	Tree	Medicinal
Holigarna arnottiana Hook.f.	Anacardiaceae	Holageru	Tree	Miscellaneous
Hymenocallis littoralis (Jacq.) Salisb.	Amaryllidaceae	Jedara lily	Herb	Ornamental
Ichnocarpus frutescens (L.) W.T. Aiton	Apocynaceae	Gauri, vanamaali balli	Climber	Medicinal
Impatiens balsamina L.	Balsaminaceae	Gouri huvu	Herb	Ornamental
Impatiens walleriana Hook.f.	Balsaminaceae	Gouri huvu	Herb	Ornamental

TABLE 4.1 *(Continued)*

Botanical name	Family	Local name	Habit	Uses
Ipomoea batatas (L.) Poir.	Convolvulaceae	Genasu/Gonne	Climber	Vegetable
Ipomoea cairica (L.) Sweet	Convolvulaceae	Railway creeper	Climber	Ornamental
Ipomoea fistulosa Mart. ex Choisy	Convolvulaceae	Beli gutta	Shrub	Fence
Ipomoea quamoclit L.	Convolvulaceae	Baaglu torana	Climber	Ornamental
Ipomoea muricata (L.) Jacq.	Convolvulaceae	Lavanga badane	Climber	Vegetable
Ixora chinensis Lam.	Rubiaceae	Ixora	Shrub	Ornamental
Ixora coccinea L.	Rubiaceae	Kusumale	Shrub	Ornamental
Ixora finlaysoniana Wall. ex G. Don	Rubiaceae	Bili ashoka chandu	Shrub	Ornamental
Jasminum grandiflorum L.	Oleaceae	Jaaji mallige	Climber	Ornamental
Jasminum multiflorum (Burm. f.) Andrews	Oleaceae	Moggu mallige	Climber	Ornamental
Jasminum sambac (L.) Sol.	Oleaceae	Gundu mallige	Climber	Ornamental
Jatropha curcas L.	Euphorbiaceae	Avadakalu	Shrub	Fence
Jatropha gossypifolia L.	Euphorbiaceae	Kaadu avdalu	Shrub	Fence
Justicia adhatoda L.	Acanthaceae	Aadusoge	Shrub	Medicinal
Justicia betonica L.	Acanthaceae	Bannada gida	Shrub	Ornamental
Justicia carnea Lindl.	Acanthaceae	Famingo huvu	Shrub	Ornamental
Justicia gendarussa Burm. f.	Acanthaceae	College aadusoge	Shrub	Fence
Kaempferia galanga L.	Zingiberaceae	Kachchoora	Herb	Medicinal
Kopsia fruticosa (Roxb.) A. DC.	Apocynaceae	Bannada huvu	Shrub	Ornamental
Lablab purpureus (L.) Sweet	Fabaceae	Avare	Climber	Vegetable

TABLE 4.1   *(Continued)*

Botanical name	Family	Local name	Habit	Uses
Lagenaria siceraria (Molina) Standl.	Cucurbitaceae	Sore kaayi	Climber	Vegetable
Lantana camara L.	Verbenaceae	Chadurangi	Shrub	Ornamental
Lantana montevidensis (Spreng.) Briq.	Verbenaceae	Sanna chadurangi	Shrub	Ornamental
Lawsonia inermis L.	Lythraceae	Madrangi	Shrub	Multipurpose
Leptadenia reticulata (Retz.) Wight & Arn.	Apocynaceae	Jeevanthi	Climber	Medicinal
Leucas lavandulifolia Sm.	Lamiaceae	Tumbe	Herb	Medicinal
Luffa acutangula (L.) Roxb.	Cucurbitaceae	Heere kaayi	Climber	Vegetable
Luffa cylindrica (L.) M. Roem.	Cucurbitaceae	Bolu here	Climber	Vegetable
Lycopersicon esculentum Mill.	Solanaceae	Tomato	Herb	Vegetable
Macaranga peltata (Roxb.) Muell. Arg.	Euphorbiaceae	Chandkalu	Tree	Miscellaneous
Madhuca longifolia (L.) J.F. Macbr.	Sapotaceae	Hippe	Tree	Miscellaneous
Magnolia champaca (L.) Baill.	Magnoliaceae	Sampige	Tree	Ornamental
Malvaviscus penduliflorus DC.	Malvaceae	Cheepu daasavala	Shrub	Ornamental
Mammelia suriga (Buch-Ham. ex Roxb.) Kosterm.	Calophyllaceae	Suragi	Tree	Multipurpose
Mangifera indica L.	Anacardiaceae	Maavu	Tree	Fruit
Manihot esculenta Cranz.	Euphorbiaceae	Mara	Tree	Multipurpose
Manilkara zapota (L.) P. Royen	Sapotaceae	Chikku	Tree	Fruit
Marantha arundinaceae L.	Maranthaceae	Aaraaroot	Herb	Medicinal
Marsdenia sylvestris (Retz.) P.I. Forst.	Apocynaceae	Madhu nashini	Climber	Medicinal

TABLE 4.1 *(Continued)*

Botanical name	Family	Local name	Habit	Uses
Melia azedarach L.	Meliaceae	Hucchu bevu	Tree	Medicinal
Memecylon edule Roxb.	Melastomataceae	Adchare	Tree	Medicinal
Mentha piperita L.	Lamiaceae	Pudina	Herb	Vegetable
Millingtonia hortensis L.f.	Bignoniaceae	Mugilu mallige	Tree	Ornamental
Mimusops elengi L.	Sapotaceae	Ranjalu	Tree	Multipurpose
Mirabilis jalapa L.	Nyctaginaaceae	Madyanna mallige	Herb	Ornamental
Momordica charantia L.	Cucurbitaceae	Haagalu	Climber	Vegetable
Momordica dioica Roxb. ex Willd.	Cucurbitaceae	Beli southe	Climber	Vegetable
Morinda citrifolia L.	Rubiaceae	Noni	Shrub	Medicinal
Moringa oleifera Lam.	Moringaceae	Nuggi kaayi	Tree	Vegetable
Morus alba L.	Moraceae	Hippu nerale	Shrub	Fruit
Muntingia calabura L.	Muntingiaceae	Garden cherry	Tree	Fruit
Murraya koenigii (L.) Spreng.	Rutaceae	Karibevu	Tree	Multipurpose
Musa acuminate Colla.	Musaceae	Cavendish	Tree	Fruit
Musa paradisiaca L.	Musaceae	Baale mara	Tree	Fruit
Myristica fragrance Houtt.	Myristicaceae	Jaayi kaayi	Tree	Spice
Narium oleander L.	Apocynaceae	Kanagile	Shrub	Ornamental
Neolamarckia cadamba (Roxb.)	Rubiaceae	Apathya mara, kadamba mara	Tree	Medicinal
Nephelium lappaceum L.	Sapindaceae	Rambuton	Tree	Fruit
Nyctanthus arbor-tristis L.	Oleaceae	Paarijata	Tree	Ornamental

TABLE 4.1   *(Continued)*

Botanical name	Family	Local name	Habit	Uses
Ocimum gratissimum L.	Lamiaceae	Rama tulsi	Shrub	Medicinal
Ocimum tenuiflorum L.	Lamiaceae	Tulasi	Shrub	Medicinal
Opuntia dillenii (Ker Gawl.) Haw.	Cactaceae	Kalli	Shrub	Fence
Pachystachys lutea Nees.	Acanthaceae	Candy huvu	Shrub	Ornamental
Pandanus amaryllifolius Roxb.	Pandanaceae	Biryaani ele	Herb	Condiment
Passiflora edulis Sims	Passifloraceae	Sharbat balli	Climber	Fruit
Peltophorum pterocarpum (DC.) K. Heyene	Fabaceae	Haladai gulmoher	Tree	Miscellaneous
Pentas lanceolata (Forssk.) Deflers	Rubiaceae	Show huvu	Herb	Ornamental
Persea americana Mill.	Lauraceae	Benne hannu	Tree	Fruit
Persicaria chinensis (L.) H. Gross	Polygonaceae	Kanne kudi	Herb	Vegetable
Phyllanthus acidus (L.) Skeels	Phyllanthaceae	Raja nelli	Tree	Fruit
Phyllanthus amarus L.	Phyllanthaceae	Nelanelli	Herb	Medicinal
Phyllanthus debilis Klein ex Willd.	Phyllanthaceae	Nelanelli	Herb	Medicinal
Phyllanthus emblica L.	Phyllanthaceae	Nelli	Tree	Fruit
Phyllanthus myrtifolius (Wight) Muell. Arg.	Phyllanthaceae	Show gida	Shrub	Fence
Phyllanthus reticulatus Poir.	Phyllanthaceae	Shayi kaayi	Shrub	Fence
Pimenta dioica (L.) Merr.	Myrtaceae	Sakala sambaara	Tree	Spice
Piper betle L.	Piperaceae	Veelyadele	Climber	Masticatory
Piper longum L.	Piperaceae	Hippli	Climber	Medicinal

TABLE 4.1   *(Continued)*

Botanical name	Family	Local name	Habit	Uses
Piper nigrum L.	Piperaceae	Kari menasu	Climber	Spice
Pistia stratoides L.	Araceae	Neeru gulaabi	Herb	Ornamental
Pithecellobium dulce (Roxb.) Benth.	Fabaceae	Hulse	Tree	Fruit
Plectranthus amboinicus (Lour.) Spreng.	Lamiaceae	Dodda patre/ Sambar soppu	Herb	Medicinal
Plectranthus scutellarioides (L.) R.Br.	Lamiaceae	Bannada tulasi	Herb	Ornamental
Plumbago indica L.	Plumbaginaceae	Kempu, chitra mula	Shrub	Medicinal
Plumbago zeylanica L.	Plumbaginaceae	Bili chitra mula	Shrub	Medicinal
Plumeria obtusa L.	Apocynaceae	Bili gosampige	Tree	Ornamental
Plumeria rubra L.	Apocynaceae	Gosampige	Tree	Ornamental
Polyalthia longifolia (Sonn.) Thwaites	Annonaceae	Madras ashosa	Tree	Ornamental
Portulaca grandiflora Hook.	Portulacaceae	Chigare huvu	Herb	Ornamental
Pseuderanthemum bicolor (Schrank) Radlk. ex Lindau	Acanthaceae	Motimallige	Shrub	Ornamental
Psidium guajava L.	Myrtaceae	Perale	Tree	Fruit
Psophocarpus tetragonolobus (L.) DC.	Fabaceae	Matti avare	Climber	Vegetable
Punica granatum L.	Lythraceae	Daalimbe	Shrub	Fruit
Quisqualis indica L.	Combretaceae	Bombay mallige	Climber	Ornamental
Raphanus sativus L.	Brassicaceae	Mulangi	Herb	Vegetable
Rauvolfia serpentina (L.) Benth. ex Kurz	Apocynaceae	Sarpagandha	Shrub	Medicinal

TABLE 4.1   *(Continued)*

Botanical name	Family	Local name	Habit	Uses
Rauvolfia tetraphylla L.	Apocynaceae	Dodda sarpagandha	Shrub	Medicinal
Ravanala madagascariensis Sonn.	Strelitziaceae	Fan bale	Tree	Ornamental
Ricinus communis L.	Euphorbiaceae	Haralenne gida	Shrub	Miscellaneous
Rosa indica L.	Rosaceae	Gulaabi	Shrub	Ornamental
Ruellia tuberosa L.	Acanthaceae	Beli gadde gida	Herb	Medicinal
Ruta graveolens L.	Rutaceae	Naga daali	Herb	Medicinal
Saccharum officinarum L.	Poaceae	Kabbu	Shrub	Multipurpose
Salacia chinensis L.	Celastraceae	Eknayaka	Shrub	Medicinal
Salvia coccinea Buchoz ex Etl.	Lamiaceae	Salvia huvu	Shrub	Ornamental
Sansevieria trifasciata Prain.	Asparagaceae	Haav naaru	Herb	Ornamental
Santalum album L.	Santalaceae	Shreegandha	Tree	Multipurpose
Sapindus trifoliatus L.	Sapindaceae	Atle kaayi	Tree	Medicinal
Saraca asoca (Roxb.) Willd.	Fabaceae	Ashoka	Tree	Medicinal
Sauropus androgynous (L.) Merr.	Phyllanthaceae	Vitamin soppu	Shrub	Medicinal
Scadoxus multiflorus (Martyn) Raf.	Amaryllidaceae	Bhu-chakra	Herb	Ornamental
Sesamum indicum L.	Pedaliaceae	Yallu	Herb	Miscellaneous
Sida rhombifolia L.	Malvaceae	Kadlangadle	Herb	Medicinal
Solanum melongena L.	Solanaceae	Badane	Shrub	Vegetable
Solanum torvum Sw.	Solanaceae	Gulla badane	Shrub	Vegetable
Spathodea campanulata P. Beauv.	Bignoniaceae	Neeru kaayi mara	Tree	Ornamental

TABLE 4.1   *(Continued)*

Botanical name	Family	Local name	Habit	Uses
Spilanthes mauritiana L.	Asteraceae	Gantalu gonde	Herb	Medicinal
Spondias dulcis Parkinson.	Anacardiaceae	Sihi amate	Tree	Fruit
Spondias indica (Wight & Arn.)	Anacardiaceae	Kadu amate	Tree	Miscellaneous
Spondias pinnata (L. f.) Kurz	Anacardiaceae	Huli amate	Tree	Fruit
Stachytarpheta jamaicensis (L.) Vahl.	Verbenaceae	Nili uttarani	Herb	Ornamental
Streblus asper Lour	Moraceae	Ganchi mara	Tree	Miscellaneous
Strelitzia reginae L.	Strelitziaceae	Hakki kaabaale	Herb	Ornamental
Strychnos nux-vomica L.	Loganiaceae	Kaasarka	Tree	Miscellaneous
Syzygium aromaticum (L.) Merr. & L.M. Perry	Myrtaceae	Lavanga	Tree	Spice
Syzygium caryophyllatum (L.) Alston	Myrtaceae	Kuntu	Tree	Fruit
Syzygium cumini (L.) Skeels	Myrtaceae	Nerale	Tree	Multipurpose
Syzygium jambos (L.) Alston	Myrtaceae	Punnerale	Tree	Fruit
Syzygium malaccensis (L.) Merr. & L.M. Perry	Myrtaceae	Jambu	Tree	Fruit
Syzygium samarangense (Blume) Merr. & L. M. Perry	Myrtaceae	Jambe	Tree	Fruit
Tabernaemontana divaricata (L.) R.Br. ex Roem. & Schult.	Apocynaceae	Nandibattalu	Shrub	Ornamental
Tagetes erecta L.	Asteraceae	Gonde huvu	Herb	Ornamental
Talinum triangulare (Jacq.) Willd.	Talinaceae	Bombay basale	Herb	Vegetable

TABLE 4.1   *(Continued)*

Botanical name	Family	Local name	Habit	Uses
Tamarindus indica L.	Fabaceae	Hunase	Tree	Condiment
Tectona grandis L. f.	Lamiaceae	Saagavaani	Tree	Timber
Terminalia arjuna L.	Combretaceae	Arjuna mara	Tree	Timber
Terminalia catappa L.	Combretaceae	Kaadu badami	Tree	Miscellaneous
Terminalia chebula Retz.	Combretaceae	Anale	Tree	Medicinal
Ternera ulmifolia L.	Passifloraceae	Haldi huvu	Shrub	Ornamental
Theobroma cacao L.	Malvaceae	Koko mara	Tree	Fruit
Thespesia populnea (L.) Sol. ex Correa	Malvaceae	Hoovarasi	Tree	Ornamental
Thunbergia erecta (Benth.) T. Anderson	Acanthaceae	Krisna huvu	Shrub	Ornamental
Thunbergia fragrans Roxb.	Acanthaceae	Bili Krishna huvu	Climber	Ornamental
Thunbergia grandiflora Roxb.	Acanthaceae	Aakasha krishna	Climber	Ornamental
Tinospora cordifolia (Willd.) Miers	Menispermaceae	Amruta balli	Climber	Medicinal
Tithonia rotundifolia (Mill.) S.F. Blake	Asteraceae	Jenia huvu	Shrub	Ornamental
Trichosanthes cucumerina L.	Cucurbitaceae	Padvaala kaayi	Climber	Vegetable
Tylophora indica (Burm. f.) Merr.	Apocynaceae	Kaphada balli	Climber	Medicinal
Uvaria narum (Dunal)	Annonaceae	Kakke balli	Climber	Medicinal
Vanilla planifolia Jacks.	Orchidaceae	Venilla	Climber	Condiment

TABLE 4.1 *(Continued)*

Botanical name	Family	Local name	Habit	Uses
Vigna unguiculata (L.) Walp. subsp. *cylindrica* (L.) Verdc.	Fabaceae	Bavade kaalu	Climber	Vegetable
Vigna unguiculata (L.) Walp. subsp. *unguiculata*	Fabaceae	Halasande	Climber	Vegetable
Vitex negundo L.	Lamiaceae	Nukki soppu	Shrub	Medicinal
Vitex trifolia L.	Lamiaceae	Lakki soppu	Shrub	Fence
Wedelia trilobata (L.) Hotche	Asteraceae	Bottu sevantige	Climber	Ornamental
Woodfordia fruiticosa (L.) Kurz	Lythraceae	Dhataki	Shrub	Fence
Xanthosoma sagittifolium (L.) Schott	Araceae	Budukesu	Herb	Vegetable
Zanthoxylum rhetsa (Roxb.) DC.	Rutaceae	Jummana mara	Tree	Condiment
Zingiber officinale Roscoe	Zingiberaceae	Shunti	Herb	Medicinal
Ziziphus mauritiana Lam.	Rhamnaceae	Bugari mara	Tree	Fruit
Ziziphus oenoplia (L.) Mill.	Rhamnaceae	Choori mullu	Shrub	Fence
Ziziphus rugosa Lam.	Rhamnaceae	Bili mulluhannu	Shrub	Fence

*Prepared on the basis of APG III system of classification; Fence: planted to form a barrier or to mark the boundary of a home garden, Miscellaneous: not deliberately cultivated but used by the family members; Multipurpose: more than two uses

FIGURE 4.2   (See color insert.) Vegetables, fruits, and beans maintained in the homegardens of coastal taluks of Uttara Kannada.

TABLE 4.2   Family-Wise Distribution of Plants in the Homegardens of Uttara Kannada in Karnataka, India

Family	Number of species
Acanthaceae	15
Acoraceae	1
Amaranthaceae	7
Amaryllidaceae	3
Anacardiaceae	6
Annonaceae	7
Apiaceae	5
Apocynaceae	27
Araceae	11
Arecaceae	3
Asparagaceae	7
Asteraceae	11
Balsaminaceae	2
Basellaceae	1
Bignoniaceae	2
Bixaceae	1
Boraginaceae	1
Brassicaceae	2
Bromaliaceae	2
Cactaceae	2
Calophyllaceae	2
Cannaceae	1
Caricaceae	1
Casuarinaceae	1
Celastraceae	1
Chenopodiaceae	1
Cleomaceae	3
Clusiaceae	1

TABLE 4.2 *(Continued)*

Family	Number of species
Colchicaceae	1
Combretaceae	5
Convolvulaceae	5
Costaceae	2
Crassulaceae	1
Cucurbitaceae	13
Diascoriaceae	2
Dilleniaceae	1
Euphorbiaceae	12
Fabaceae	30
Lamiaceae	21
Lauraceae	2
Lecythidaceae	1
Loganiaceae	1
Lythraceae	3
Magnoliaceae	1
Malvaceae	15
Maranthaceae	2
Melastomaceae	1
Meliaceae	2
Menispermaceae	1
Moraceae	10
Moringaceae	1
Phyllnthaceae	11
Piperaceae	3
Plantaginaceae	1
Plumbaginaceae	2
Poaceae	5

TABLE 4.2 *(Continued)*

Family	Number of species
Polygonaceae	2
Portulachaceae	1
Rhamnaceae	3
Rosaceae	1
Rubiaceae	9
Rutaceae	10
Santalaceae	1

TABLE 4.3　Species Range in the Homegardens of Uttara Kannada in Karnataka, India

Species range	Number of homegardens
50–60	20
61–70	22
71–80	66
81–90	26
91–100	10
100>	6

TABLE 4.4　Usage Category of Plants in the Homegardens of Uttara Kannada in Karnataka, India

Usage category	Number of species	% of species
Ornamental	126	34.81
Medicinal	58	16.03
Vegetable	44	12.16
Fruit yielding	41	11.33
Miscellaneous	32	8.84
Fence plants	21	5.81
Multipurpose	20	5.53
Condiments	9	2.49
Spice	7	1.94
Masticatory	2	0.55
Timber	2	0.55

4.3.1   FRUIT YIELDING PLANTS

A total number of fruit yielding species reported from the home gardens of the study area is 41. Among this, 1 (2.43%) species is herb, 3 (7.31%) species are shrubs, 6 (14.64%) species are climbers, and 31 (75.61%) species are trees. Households of this region cultivate 14 (34.15%) species of fruit yielding plants commercially and 27 (65.85%) species of fruit yielding species cultivated in the home gardens for their own use.

4.3.2   MEDICINAL PLANTS

A total number of medicinal plant species reported from the home gardens of the study area is 58. Among this, 25 species are herbs, 12 species are climbers, 12 species are shrubs and 9 species are trees. Medicines are prepared from leaf (15 spp.), root (13 spp.), rhizome/stem (8 spp.), fruit/flower (6 spp.), and bark (4 spp.). In the remote villages, people without hospital facility are totally depending on the plant-based medicines.

4.3.3   ORNAMENTAL PLANTS

Most of the homegardens comprised ornamental plants which are generally grown for decoration. However, some of the ornamental plants cultivated carried medicinal value. Ornamental plants exhibited great variation with respect to size, shape, and color. A total of 126 species of ornamental plants were recorded, in which 39 species of herbs, 52 species of shrubs, 18 species of climbers and 17 species of trees. Previous studies indicate similar observation, increased cultivation of ornamental and commercial plants across homegardens is an indicative of urbanization of families practicing home gardening (Karyona, 1990; Drescher, 1996; Bhat and Rajanna, 2016).

4.3.4   VEGETABLES

One of the main reasons to grow plants in the home garden is to get vegetables for their own use or also as a source of family income. In the present study, 44 species of vegetables have been recorded. In which 27 vegetables are cultivated for their family income (commercial) in a small scale and 17 species of vegetables are noncommercially used for their own purpose. Growth habit analysis of vegetables reveals that herbs – 17 species, shrubs – 3 species, climbers – 22 species and only 2 tree species. Most of the climbers like cucumber, bitter guard, ridge guard, snake guard, pumpkin, and ash guard are cultivated during the rainy season, but Halakki and Kari-vokkalu people grow all such vegetables based on the demand in the local market.

4.3.5   MISCELLANEOUS PLANTS

Some plant species grow naturally and rapidly. In this population of plants, some are used for particular purposes, and some are of no use, but still, they will be kept in the garden. These plants are called miscellaneous plants. A total of 31 species of miscellaneous plants has been recorded. Lifeform indicates the presence of herbs (3 spp.), shrubs (4 spp.), trees (22 spp.) and climbers (3 spp.).

4.3.6   MULTIPURPOSE PLANTS

These are the plants which are protected but not deliberately cultivated sometimes. These plants give shade, protection from wind, soil erosion, used as medicinal, fruit yielding, firewood, manure, and as ornamental. Single species shows many uses and such plants are categorized as multipurpose plants. A total of 20 species of multipurpose species were recorded, in which 18 species are trees and 2 are shrubs.

4.3.7   FENCE PLANTS

Nowadays cement compounds are common in cities and even villages also. This is the impact of modernization. But in most of the villages home, gardens are protected by natural fences. Some of the plants in the fences are thorny, and some are bitter in taste to the animals. These plants will grow in the form of a bush. Hence this type of natural fences will protect useful plants in the home garden from animals. A total number of fence species reported from the home gardens of the study area are 21. Among this, 1 (4.76%) species is climber and 20 (95.24%) species are shrubs.

4.3.8   SPICES

A total of 7 species of spices were recorded from the study area. This region is very famous for black pepper and cardamom since before 1947. Before independence Gerusoppa was very famous for its quality black pepper. Black pepper is grown on aecanut, coconut, jackfruit and also on mango trees in the home garden. Cardamom plant is more common in the upper ghat region, and black pepper is seen in both coastal and upper ghats of Uttara Kannada. Black pepper is the major spice yielding plant of this region. Lifeform of spices is classified into herbs (1 spp.), climbers (1 spp.) and trees (5 spp.).

Socioeconomic aspects of the homegardens have also been studied. Species richness in the gardens of Havyaka, Halakki, Karivokkaliga, and Namadhary community has been analyzed. Especially, the role of gender in preharvesting and post-harvesting is also recorded. Species diversity is rich in the home gardens of Havyaka and Halakki community. Ornamental, medicinal, and fruit yielding plants are dominant in the home gardens of Havyaka community whereas Halakki's prefer vegetables and fruits in their garden because the home garden is a source of income for them.

4.4   TRADITIONAL MANAGEMENT PRACTICES

Traditional management practices have played cornerstone in enhanced diversity of plants across homegardens of Uttara Kannada especially annual and perennial herbs, shrubs, and woody perennials. Animal waste such as cow dung and litter are common fertilizers for homegardening, and in some cases, chemical fertilizers are used across urban areas to increase soil fertility. People involving both genders actively participate in home gardening activities and generally perform activities such as land preparation, sowing, fertilizing, and weeding followed by material procurement and harvesting of tuberous crops, fruits of large trees like mango, jackfruit and breadfruit (Table 4.5). The information related to increased interest in conservation of homegardens, Family size of the respondents and occupation of the respondents, have been furnished in Tables 4.6, 4.7, and 4.8, respectively. Homegardening also functions as an economic source of income for the women belonging to rural communities. Most often, people sell fresh vegetables and fruits on the roadside as places like Gunavante,

Kasaragod, Karki, Handigon, Baggon, Banki kodlu, Bavikodlu, Chendiya villages are nearby to National Highway 66 and taluk center to maintain semi-commercial homegardens. However, the plant crops are generally dominated by the market demands, for instance, seasonal crops such as breadfruit, tuber, and leaf petioles of colocasia, drumstick, radish, jackfruit, papaya, cucumber, ridge guard, lady's finger, pumpkin, bottle guard, amaranthus, and basella are few among the other that are most often cultivated at homegardens because of increased demand. In most of the rural household, women sell the homegarden products while men take care of homegarden. In the majority of the cases, except weeding and watering rest of the pre-harvest work will be done by men. Even children involved in the pre and post harvesting work during their vacation in semi-commercial homegardens of Halakki vokkalu and Karivokkalu communities (Bhat and Rajanna, 2016).

TABLE 4.5   Percentage-Wise Gender Role in Homegarden Activities

Activity	Men	Women	Children
PRE HARVEST			
Plot Selection	94	56	-
Clearance	94	46	7
Compost preparation	91	50	3
Fencing	85	39	6
Land preparation	82	18	10
Planting	62	30	1
Crop selection	59	91	-
Fertilizing	71	30	9
Nursery preparation	80	20	4
Watering the plants	51	89	6
Supporting the climbers to spread	81	42	7
Weed removal	06	104	10
Pest control	95	15	-
Harvesting	62	70	16
POST HARVEST			
Sorting for seeds	8	92	42
Marketing	41	105	4
Seed storage	11	139	-

TABLE 4.6   Reasons for Having Interest in Conservation of Homegardens

Reasons	Number of respondents
Source of food	46
Save money	39
Source of income	65
Soil stabilization	-
Shade	-

TABLE 4.7   Family Size of the Respondents

Category	Number of homegardens	Percentage
2–4 members	58	38.67
5–10 members	91	60.67
>10 members	1	0.66

TABLE 4.8   Occupation of the Respondents

Category	Partial involvement	Complete involvement
Agriculture	39	61
Business	19	-
Daily labor	20	-
Government employee	11	-

4.5   CONCLUSION AND FUTURE TRENDS

Indeed, homegardens prove to be a major contributor of food, medicine, and in the conservation of the plant genetic diversity. In the present study, it is notable to mention that 34.84% of the recorded species were found to be ornamental out numbering both food-yielding categories of fruit and vegetable plants which together accounted only 23.49%. 16.03% of the plant species are used for medicinal purposes. This is in contradiction to the general understanding that food plants are the most common species in most homegardens throughout the world. However, most of the plant species were of ornamental/commercial importance they may be due to increased incidence of urbanization among the rural households (Karyona, 1990; Drescher, 1996). Nevertheless, analysis of the socio-economic conditions of the homegarden owning families involved in the present study needed confirmation to this assumption. Thus, homegardens ensure plant genetic diversity conservation. Further, it is quite essential that these homegardens are properly maintained. Coastal region of Uttara Kannada shows rich species diversity (362 species) when compared to other parts of India.

KEYWORDS

- Apocynaceae
- *Areca catechu*
- Asteraceae
- *Cocos nucifera*
- Euphorbiaceae
- Fabaceae
- fence plants
- halakki
- havyaka
- *Hibiscus rosa-sinensis*
- homegardens
- *in situ* conservation
- India
- Kari-Vokkalu
- karivokkaliga
- Lamiaceae
- Malvaceae
- *Mangifera indica*
- *Musa paradisiaca*
- namadhary
- *Ocimum tenuiflorum*
- ornamental Plants
- Phyllanthaceae
- Uttara Kannada

REFERENCES

Bhat, K. G., (2003). *Flora of Udupi*. Indian Naturalists, Udupi, Karnataka, India (p. 913).

Bhat, S., & Rajanna, L., (2016). Plant diversity studies and role of Halakki community in agroforestry homegardens of Uttara Kannada district, Karnataka, India. *International Journal of Life Sciences, 5*(2), 54–62.

Black, G. M., Somnasang, P., Thamathawan, S., & Newman, J. M., (1996). Cultivating continuity and creating change: Women's homegarden practices in North-eastern Thailand – multicultural considerations from cropping to consumption. *Agriculture Human Values, 13,* 3–11.

Cooke, T., (1967). *Flora of Presidency of Bombay* (Vol. I-III). Botanical Survey of India, Calcutta.

Das, T., & Das, A. K., (2005). Inventorying plant biodiversity in home gardens – a case study in Barak Valley, Assam, North East India. *Curr. Sci., 89,* 155–163.

Drescher, A. W., (1996). Management strategies in African homegardens and the need for new extension approaches. In: Heidhues, F., & Fadani, A., (eds.), *Food Security and Innovations - Successes and Lessons Learned* (pp. 231–245). Peter Lang, Frankfurt.

Godbole, A., (1998). Home gardens: Traditional systems for the maintenance of biodiversity. In: Rastogi, A., Godbole, A., & Pei, S., (eds.), *Natural Resource Management - Traditional Home gardens* (pp. 9–12). International Centre for Integrated Mountain Development, Kathmandu, Nepal.

John, J., & Nair, M. A., (1999). Crop-tree inventory of the homegardens in southern Kerala. *J. Trop. Agric., 37,* 110–114.

Jose, D., & Shanmugaratnam, N., (1993). Traditional homegardens of Kerala: A sustainable human ecosystem. *Agroforest. Syst., 24,* 203–213.

Karyono, (1990). Homegardens in Java: Their structure and function. In: Landauer, K., & Brazil, M., (eds.), *Tropical Home Gardens* (pp. 138–146). United Nations University Press, Tokyo.

Kumar, B. M., George, S. J., & Chinnamani, S., (1994). Diversity, structure, and standing stock of wood in the homegardens of Kerala in peninsular India. *Agroforest. Syst., 25,* 243–262.

Nair, M. A., & Sreedharan, C., (1986). Agroforestry farming systems in the homesteads of Kerala, Southern India. *Agroforest. Syst., 4,* 339–363.

Nair, P. K. R., & Kumar, B. M., (2006). Introduction to homegardens. In: Kumar, B. M., & Nair, P. K. R., (eds.), *Tropical Homegardens: A Time-Tested Example of Sustainable Agroforestry* (pp. 1–10). Springer, Netherlands.

Pandey, C. B., Latha, K., Venkatesh, A., & Medhi, R. P., (2006). Diversity and species structure of homegardens in South Andaman. *Trop. Ecol., 47,* 251–258.

Pandey, C. B., Rai, R. B., Singh, L., & Singh, A. K., (2007). Homegardens of Andaman and Nicobar Islands. *Agroforest. Syst., 92,* 1–22.

Puskaran, K., (2002). Homegardens in Kerala as an efficient agroecosystem for conservation and sustainable management of biodiversity. In: Watson, J. B., & Eyzaguirre, P. B., (eds.), *Proceedings of the Second International Home Gardens Workshop*. IPGRI, Rome.

Shastri, C. M., Bhat, D. M., Nagaraja, B. C., Murali, K. S., & Ravindranath, N. H., (2002). Tree species diversity in a village ecosystem in Uttara Kannada district in Western Ghats, Karnataka. *Curr. Sci.*, *82*, 1080–1084.

Torquebiau, E., (1992). Are tropical agroforestry homegardens sustainable? *Agric. Ecosyst. Environ.*, *41*, 189–207.

CHAPTER 5

PRIMULA SPECIES AS INDICATORS OF FOREST HABITAT DIVERSITY IN GEORGIA, SOUTH CAUCASUS

NATALIA TOGONIDZE and MAIA AKHALKATSI

Department of Plant Genetic Resources, Institute of Botany, Ilia State University, Tbilisi, Georgia

5.1 INTRODUCTION

Global climate change is an existing reality, which will significantly affect the biological processes proceeding on the earth. Lately, the temperature has increased by 1.6–1.8 degrees, and at the end of the twenty-first century, the average temperature is likely to increase by 4–6 degrees in the result of global warming (Solomon et al., 2007). Climate change is the unity of numerous factors, such as increasing of average temperature, rising of CO_2 concentration in the atmosphere, change of amount and dynamics of precipitation (Menzel et al., 2006).

According to the present prognosis, the effect of global warming will be minimal in tropics, and maximal in highlands (Körner, 2012), i.e., it is expected that global warming and related change of precipitation and snow cover will, first of all, affect the vegetation of highlands (Guisan et al., 1995; Guisan, 1996; Guisan et al., 1998). Facts of climate change revealed in the European Alps in the 20th century coincide with the forecasted general tendencies of global warming. As it seems, the changes are most obvious in the West Europe and some regions of Asia (Diaz et al., 1997). Temperature minima rose in European Alps more than average indices, and the most sensitive habitats were revealed on the territories from the upper limit of forest to the nival belt. According to some considerations, the temperature change will primarily affect species adapted to the narrow temperature intervals (Körner, 2012).

Although the reasons of origin of global warming are debatable, results caused by it are clear. Investigations have shown that natural populations of living organisms already react on global warming (Parmesan, 2006). Direct effects of global warming are already recognizable, both in natural and in agricultural systems. These effects mainly cause the early starting of sexual reproduction, high or low reproductive ability and fertility in plants, that is dependent on peculiarities of region and species.

One of the most sensitive biological processes, on which global climate change would have influence, is the duration of the life cycle of a plant, or its phenological rhythm. Environmental factors conditioned by climate define favorable time interval which plant species have worked out in the result of historical adaptation for production, maturation, and spread of seed (Ladinig and Wagner, 2005; Körner, 2012). Thus, it is necessary to define tendencies of plant reproduction in a changeable environment.

It is very important to define the changes in phenological phases of plant and in the process of its seed production, which will be caused by variation of environmental factors in different habitats with individuals of one and the same species. It is particularly important for rare and endangered species, as well as for those plants which grow in climate-sensitive habitats, such as excessively humid territories, arid regions, and highlands.

5.2   INFLUENCE OF GLOBAL CLIMATE CHANGE ON PLANT REPRODUCTION

5.2.1   INFLUENCE OF GLOBAL CLIMATE CHANGE ON PHENOLOGICAL RHYTHM OF PLANTS

On the background of global climate change, phenology of plant flowering attracts special attention, as it appeared to be the important indicator for determining adaptation tendencies of the plant in connection with temperature change (Molou, 1989; Körner, 2012). Moreover, it is shown that rhythm of flowering phenology is directly connected with the structural peculiarities of reproductive organs; but, unfortunately, very little research has been carried out in this direction (Akhalkatsi and Wagner, 1996, 1997).

One of the manifestations of rising of temperature in the process of plant development is an early flowering, which affects the reproductive ability and generally, life cycle. Global warming could affect the duration of phenological stages of cultural plants, because increasing of temperature causes shortening of phases of development, and this is the reason of irregular productivity in agricultural systems. This topic is discussed by Hedhly et al. (2008), where the effect of temperature rising on the stages of plant development is considered. According to Hedhly et al. (2008), in some regions, mainly in northern latitudes, the temperature quickly changes in the period when the process of sowing of agricultural plants takes place that affects the rhythm of their germination and development and defines productivity. Main temperature changes were fixed in late winter and early spring.

The effect of high temperature on the duration of reproductive phases is more evident in those plant species which grow in sensitive habitats (Olesen et al., 2002). In this respect, the plants of highland and arctic zone are of particular importance, as the vegetative season is very short there and the plant has little time for the process of seed production. On the background of climate change this time could decrease up to critical phase, and the reproductive process could not be completed (Ladinig and Wagner, 2005; Milla et al., 2009).

The influence of climate change upon the forest species blooming in early spring, reproduction of which takes place before frondescence, is of the same great importance. It is expected, that vegetative season caused by climate change and started early or late could have an influence upon the successful reproduction of these species and cause certain shifting of phenological phases of flowering. This factor on its part would affect the quantitative and qualitative indices of seed production, which ultimately defines plant reproduction and survival of species. The researches have almost not been carried out in this direction, and that is why we have decided to study the phenology of forest plants blooming in early spring in different environmental conditions and determine the role of climate in the process of phenological phases.

5.2.2   INFLUENCE OF GLOBAL CLIMATE CHANGE UPON PLANT FERTILITY

Global climate change can cause a lessening of plant fertility and its ability to survive (Porter and Semenov, 2005). Transpiration and photosynthesis are those main processes upon which increasing of temperature and CO_2 have influence (Atkin and Tjoelker, 2003; King et al., 2006). It is shown (Long et al., 2004, 2006) that the attempts to increase plant fertility in agricultural systems cause the increasing of CO_2 concentration, which is in connection with other factors. Increasing of temperature is among them. The later can cause an increase of fertility in middle-aged plants of the temperate zone, while in tropics and semi-arid regions fertility will decrease (Tubiello et al., 2007). In the case, if the warming process continues it will cause a negative effect on fertility in all climate zones in the future. The studies have shown that effect of global warming causes decreasing of fertility by 17 % in cereals, for example in maize - *Zea mays*, and legumes, for example in soy - *Glycine max* (Lobell and Asner, 2003).

It is interesting that agrarian crops, as a rule, can adapt at a small range of temperature change, and thereby they can lessen flowering and seed production (Porter and Semenov, 2005; White et al., 2006). It is also shown that in the case of drought, initiation of flowers, i.e., lessening of the process of transformation of vegetative meristem into reproductive meristem, takes place in *Trogonella caerulea* and flowers are less in the period of drought than in control (Akhalkatsi and Lösch, 2005). Thus, environmental factors and particularly temperature and humidity affect the fertility of plant significantly.

5.2.3   INFLUENCE OF GLOBAL CLIMATE CHANGE UPON FUNCTIONING OF THE REPRODUCTIVE ORGANS OF PLANTS

Influence of temperature has a complementary effect upon the development of female and male sex organs of the plant. It accelerates or reduces the process of development. However, plants of different species differently adapt to changes of environmental conditions. Synchronization of formation of female and male organs is possible only in that case when both of them are in the same phase in the period of change of environmental factors (Wollenweber et al., 2003; Warner and Erwin, 2005). And the asynchronous formation of phases could reduce the ability of seed production (Prasad et al., 2002). It is known (Hedhly et al., 2008) that changes of environmental factors affect certain phases of reproduction, which can be divided into three main groups: (1) pregamic - formation of gametes; (2) prezygotic - the period from pollination to fertilization; and (3) postzygotic - formation of embryo.

Change of temperature has an influence as on male and female gametes, so on their interaction. Female sphere of the plant has not been paid much attention up today. The reaction of female gamete in pregamic phase to the increased temperature in the period of formation of the flower is explained as a genetically determined adaptive reaction (Warner and Ervin, 2005; Koti et al., 2005). It is shown that increasing of temperature causes fast development of ovules, while low-temperature conditions prolongation of the process of gametogenesis. The main works and observations are concentrated on male gamete and where it is shown (Aloni et al., 2001; Prasad et al., 2002) that increasing of temperature affects the amount of pollen grains, architecture, and morphology of their wall, as well as its chemical composition and metabolism. Reducing the number of pollen grains conditions reducing of seed production, that reacts upon the functioning of female gamete indirectly.

Change of temperature affects not only the amount of pollen grains, but reacts upon the character of its development, namely, on the quality of formed pollen (Erickson and Markhart, 2002; Prasad et al., 2002), the rate or growth of pollen tube and the ability of fertilization (Jóhannsson and Stephenson, 1998). Researches carried out in the phase of formation of pollen grain have shown that temperature stress is one of the selective factors (Jóhannsson and Stephenson, 1998). That means that in the process of pollen grain formation its genetic information can be changed under the influence of the temperature change that could become the ground of mutation in the following generations. In the result, two main effects of temperature influence could be singled out: acceleration, that is, the acceleration of growth of pollen tube, and potential influence of genetic changes of pollen grain upon selection (Sato et al., 2002; Hedhly et al., 2005).

Bukhholtz and Blackesley (1927) was the first to show the effect of temperature upon the rate of growing of pollen tube. Later this process was studied in the plant *Datura stramonium* (Lewis, 1942). After that, the acceleration and deceleration of growing of pollen tube in forest plants during the change of temperature was shown (Jefferies et al., 1982). In the result, change of rate of growing of pollen tube is considered as one of the preconditions of adaptation (Kakani et al., 2005). Changes in the process of formation of pollen grain are particularly important for those plants which are pollinated by insects.

If the plant responds on temperature change by the changing of reproductive processes, that means, that those processes are of great importance for adaptation. The plant reacts upon the changes of environmental factors by the selection of gametophyte and phenotype flexibility. However, this evolutionary change is not depended only on genetic variation. Environmental factors are also of great importance in this case (Hedhly et al., 2008).

It is known, that plants possess the ability of responding on climate changes with adaptation and migration (Davis and Shaw, 2001). After their prolonged study, Etherson and Ruth (2001) determined that numerous species had survived and kept their genetic structure in the period of many years of climate changes. The reason of it was that climate changes proceeded very slowly in the past and did not react upon the organism instantly.

It becomes clear from these works that plant migration is the fast reaction on global warming. This process takes place both in natural populations and in agricultural systems. In the result of it, plants can move to such territories, where they could not grow. Everything this creates a problem for those people who used these plants as food. However, various populations of grassy and garden plants were found, which were adapted to local environmental conditions, but despite it, they did not change their place on the background of climate changes. How it happened has not been explained yet (Hedhly et al., 2008). Supposedly, fragmentation of habitat could hinder potential migration of plants.

5.3   THE SKY EXPOSITION

Sky exposition is of great importance for the formation of the climate of microhabitat and, correspondingly, for the growing and development of plants. It is conditioned on the one hand by micro-topographic peculiarities of the landscape, and on the other hand by frequency of stories and cover of existing vegetation (Théry, 2001; Tabari, 2005). The sky exposition means not only lighting, which reaches plant, but many other environmental factors as well. We can roughly define what kind of humidity and temperature are in this or that microhabitat according to plant cover. But, certainly, the most important factor, which is defined by the sky exposition is lighting. This factor is able to react on the development and functioning of the plant both positively and negatively. But plants possess the ability of adaptation to destroying action of excessive light (Larcher, 2003). Light spectrum can directly affect the vital capacity of plants and animals, and its level could define diversity of plants (Théry, 2001).

Sustainability to light is different for every individual. Shade-tolerant and light-requiring plants are easily distinguished in the forest. Besides, the sprouts of both types and individuals in the juvenile stage require a wide spectrum of light (Larcher, 2003). It takes place till it is finally formed into light-requiring or shade-tolerant plant. All of them have nearly the same requirements towards the light at the early stage.

There are works in which it is clearly seen how sprouts of various plants adapt to different plant covers. Among them, Tabari (2005) on beech sprouts and their reaction to the size of cover and forest gap should be noted first of all. The investigation proceeded in Iran, in the regions near the Caspian Sea, where beech forests (*Fagus orientalis*) are spread on over 1500000 hectares. Oak, Alder, Zelkova, and others created a cover for the sprouts of this species. Exactly the character of this cover, because of the intensity of light and other environmental factors, affected the number of sprouts and the frequency of their size. The results of the investigation have shown that with lessening of the sky exposition, the ability of surviving of shoots decreases, and in the conditions of large opening of the leafage their mortality increases as well. The experiment revealed that there was a certain limit of cover, in the case of strong inclination from where sprouts lose their viability. Thus, beech shoots felt good and were growing intensively in the conditions of average lighting. On the grounds of this data conclusion of practical importance has been made, that is, shoots should be planted in the conditions of average lighting for the successful artificial regeneration of beech forests. In other conditions, i.e., under the strong and weak lighting, sprouts perish. That is, in order that shoots of *F. orientalis* can feel good, there is an optimal size of forest gaps, which conditions certain microclimate in the site of their sprouting (Tabari, 2005).

The sky exposition, on the background of global warming, will significantly affect the period of snow melting. One of the main manifestations of global warming will be melting of snow earlier than usual, and in the forest that will depend on plant cover and "forest gaps." First of all, this process will affect early spring flowers. Early melting of snow will have an especially negative effect in those regions, where there is a danger of drying of upper layers of soil and, correspondingly, there will be a deficiency of nutrient minerals in the middle of summer (Körner, 2012).

Moreover, in summer higher temperature will cause strengthening of metabolism and process of growing and, in the result of it, increasing of CO_2 excretion into the atmosphere in the belt of temperate forests. In the middle latitudes, the duration of the vegetative period will increase. In the case if deciduous plants retain present duration of period of frondescence (foliation), as their defoliation is connected more with photoperiodism, than with increasing of temperature (Larcher, 2003), vegetative period started early in plants of lower storey will affect the life cycle of plants blooming in early spring, which is adapted to finishing of reproductive period before forest frondescence (foliation). Thus, species blooming in early spring should be considered as an interesting object of study, since it is possible that expected changes of the vegetative period will react upon their reproductive productivity. That is why we have chosen species of the genus *Primula* as an object of our investigations, among which are forest species blooming in early spring, as well as species growing in other sensitive environments, namely, in alpine and humid meadows.

5.4   GENUS *PRIMULA*

The family Primulaceae unites about 30 genera and 1000 species, widely spread on the earth. A large part of them (about 500 species) grows in the temperate belt of the northern hemisphere (Whale, 1983, 1984). Genus *Primula* (Primrose) is one of the biggest and best-known genera in the family Primulaceae, which, according to present data, comprises 425 species. Most of them, about 300 species, are concentrated in the Himalayas and west China. Thirty-three species grow in Europe and only 20 in South America (Whale, 1983, 1984). The family Primulaceae is represented with 70 genera and 46 species in Georgia. Twenty-one species belong to the genus *Primula* (Gagnidze, 1985).

5.4.1 SPREAD AND DIVERSITY OF HABITATS OF PRIMROSE (PRIMULA) SPECIES

Species of Primrose occupy different habitats, among them sandy beaches of sea, deserts, forests, bushes, arid slopes, humid, and highland meadows. For example, *P. verticillata* is found on steep rocks in Arabian Desert, while *P. egaliksensis* grows in water-covered meadows in the Arctic, and *P. malacoides* is spread in grassy subalpine meadows as weeds (Whale, 1983, 1984). Our studied species are spread in Georgia, which (69.875 km²) is located in the south Caucasus region (Figure 5.1). The neighboring countries are Armenia, Azerbaijan, Russia, and Turkey (Figure 5.1). The study region of Kazbegi district (1081 km²) is situated northern from the Main Watershed Range of the Central Greater Caucasus, on the valley of R. Tergi (42°48'N; 44°39'E) until the border with Russia. This is morphologically most complex high-mountain region of the Central Greater Caucasus (Figure 5.1).

From the species spread in Georgia *P. juliae* (Caucasian endemic) grows on humid rocks in the forest belt. *P. abchasica* (Caucasian endemic), *P. komarovii*, *P. megaseifolia*, *P. saguramica* (Georgian endemic), *P. woronovii* (Caucasian endemic) grow in forests, outskirts of the forest and bushes, up to low or middle belt of mountain above the sea level. *P. macrocalyx*, *P. sibthorpii*, *P. vulgaris* are spread in forest zone and reach subalpine belt. *P. amoena* (Caucasian endemic), *P. pseudoelatior* (Caucasian endemic), *P. pallasii*, *P. ruprechtii* (Caucasian endemic), *P. luteola* (Caucasian endemic) grow in subalpine forest and alpine meadows. *P. algida*, *P. bayernii* (Caucasian endemic), *P. cordifolia* (Caucasian endemic), *P. darialica* (Caucasian endemic), *P. farinifolia* (Caucasian endemic), *P. kusnetzovii* (Caucasian endemic), *P. meyeri* (Caucasian endemic) are found only in subalpine and alpine meadows and debris. *P. auriculata* occupy quite different habitat - it grows on the banks of streams, on humid slopes of gorges, in subalpine and alpine marshy meadows.

There are works clearly showing how species of genus *Primula* adapt to some concrete habitat and what are their requirements for environmental factors. Among them, Whale (1983, 1984) are worth of noting, in which requirements for the environmental factors of three species of genus *Primula*, spread in Europe - *P. vulgaris*, *P. veris* and *P. elatior* are discussed. Our studied species are *P. woronowii* Losinsk., which grows in the most closed oak-hornbeam forest as the species of primrose spread in

Georgia; *P. macrocalyx* Bunge grows in beech forest; *P. amoena* M. Bieb. is
found both in subalpine birch forest and in subalpine and alpine meadows.
Two other species (*P. cordifoia* Rupr., *P. algida* Adams) are spread only in
subalpine and alpine meadows, and one of them (*P. auriculata* L.) is found
in humid meadows in subalpine and alpine zones (Figure 5.2).

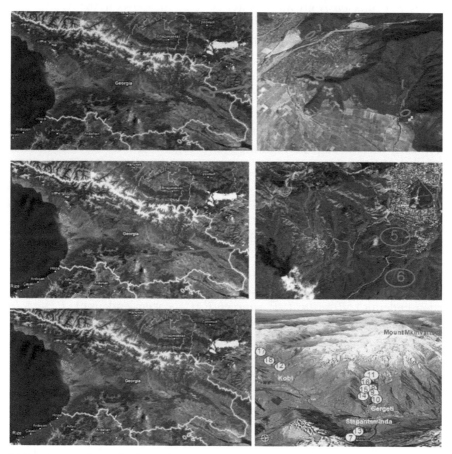

FIGURE 5.1   (See color insert.) Study areas: 1,2,3,4 - *Primula woronowii*, Oak-hornbeam
forest, Kvareli district, Shilda; 5,6 - *P. macrocalyx*, Bich forest, Tskhneti; 7–11 - *P. amoena*;
12 - *P. cordifolia*; 13–16 - *P. algida*; 17,18 - *P. cordifolia*, Stephantsminda district.

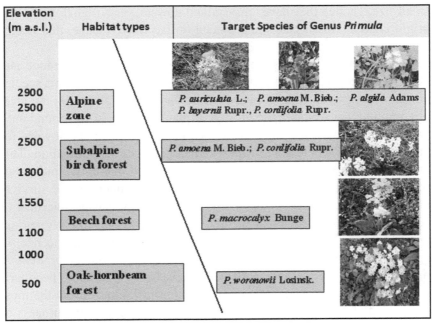

FIGURE 5.2   (See color insert.) Studied species of genus *Primula.*

5.4.2   REPRODUCTIVE BIOLOGY OF PRIMROSE

The flowers of primrose species are generally actinomorphic, bisexual, 5-merous, rarely, 4–7-merous, placed solitarily in axilla, or on long stalks growing out of axillae of radical leaves, or gathered in thyrsoid, racemose or umbelliform inflorescences. Stamens correspond to the number of teeth of crown and are placed opposite them. Ovary is upper, rarely semi-lower, consisting of 5 carpels, monothecal, ovules are numerous. Fruit is a poly-spermous capsule (Gagnidze, 1985).

Reproductive productivity of primrose, as well as of other plants, is dependent on numerous abiotic and biotic factors. Pollinators and their interaction with plant is one of the important biotic factors. Interaction of plant and animal has an important effect on the structure of community, dynamics of population and evolutionary processes of plant. On its part, structure of community and population conditions attraction of pollina-tors. The height of plant is an important factor to attract the pollinator. But individuals of *Primula* species have repent rosettes, and plants are of small size. Though they have long-stalked flowers, very often flowers with bright-colored crown petals and placed on tall flower scape, that makes them easily noticeable for pollinators.

It is expected, that vegetative season caused by climate change and started early or late could affect the successful reproduction of this species and cause certain shifting of phases of flowering phenology. On the level of community, it would cause growth of plant cover existing around until primrose finishes flowering, and thus hinder access to its flowers for pollinators, that would significantly reduce intensity of pollination. This factor, from its side, would affect quantitative and qualitative indices of seed production that ultimately defines plant reproduction and survival of species. That is why we have studied phenology and peculiarities of seed production of species of genus *Primula*, flowering in early spring, reproduction of which takes place in the forest before frondescence, in order to define sensitive peculiarities characteristic to their reproductive process. Moreover, we have considered necessary to compare peculiarities of habitat in places of spread of various species and reveal the most sensitive and changeable environment, where studying of the process of seed production and forecasting of expected changes would be important.

The investigation will enable us to define time duration, characteristic to species under study for seed production, so-called specific reproductive period, and determine what effect expected season shifting caused by global warming and vegetative season started early or late will have upon the character of seed production. In the case of primrose, we have observed the effect of frondescence in the forest, so-called spring index. It is also interesting if expected increase of snow cover and early snow melting causes early or late flowering of primrose.

5.5 SKY EXPOSITION IN HABITATS OF SPECIES OF *PRIMULA*

Average percentage index of the sky exposition of research places (18 places in total) strongly varies in alpine and forest species habitats. There is difference between the sky exposition indices of spring forest and summer forest as well, where index lessens nearly twice after leafing. Particularly low percentage index is observed in the case of *P. woronowii* in oak-hornbeam forest. In summer period the index is also low in subalpine birch forests. The percent of the sky exposition in alpine meadows exceeds nearly twice the index of spring forest and 6–8 times summer forest sky exposition.

The separate indices of the sky exposition for species show the same picture as average values. According to circular diagram, summer forest of *P. woronowii* having the lowest index is placed in the center, where measuring took place at four different sites of one population. But it should be noted that shadowing of habitat of this species in oak-hornbeam forest is less in spring than in the case of *P. macrocalyx*, growing in beech forest. The highest index of the sky exposition has been revealed for *P. algida*, *P. amoena* and *P. auriculata* (Figure 5.3).

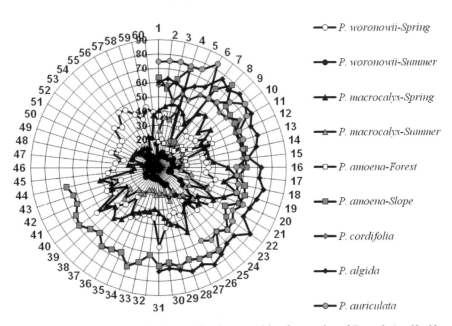

FIGURE 5.3   The separate indices of the sky exposition for species of *Primula* (n=60; 60; 40; 40; 20; 30; 16; 30; 25).

5.6   POLLINATION INDEX

One of the most important questions connected to early flowering of forest species is that which factor is significant for proceeding of life cycle by this scheme. According to data acquired by us, high correlative index has been found between the sky exposition and index of pollination of *P. woronowii* (r²=0.7357) (Figure 5.4A,B). It is known that pollinating agents such as insects and other living pollinators are found more in sunny places than in shady ones (Knight et al., 2005). According to our data, the index of pollination is greater in sunny places than in shady ones. That indicates that lighted place is more suitable for sexual reproduction of this species. Besides, it is known, that change of temperature not only affects number of pollen grains, but reacts on the character of its development as well, namely on the quality of formed pollen (Erickson and Markhart, 2002; Prasad et al., 2002), the rate of growth of pollen tube and the ability of fertilization (Jóhannsson and Stephenson, 1998).

Probably, flowering of this species in the forest before foliation is explained exactly by it, as the sky exposition considerably decreases in completely foliated forest and, correspondingly, the temperature drops on the stigma of flower where germination of pollen grain takes place. The temperature effect on the rate of growing of pollen tube which is considered as one of the preconditions of adaptation (Kakani et al., 2005) is well known (Buchholz and Blakeslee, 1927). And that indicates that temperature is an important factor for the successful accomplishment of the process of pollination, which is in direct connection with lighting.

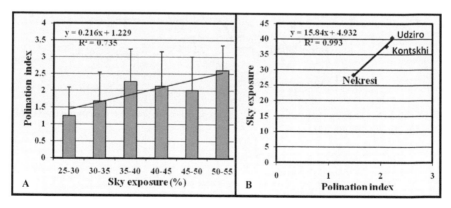

FIGURE 5.4 Relations between sky exposition percentage and pollination index for *P. woronowii*. A - The correlation between the average rate of study sites pollination index and sky exposition (n = 40); B - The correlation between pollination index and average index of separate groups of sky exposure (n = 120).

5.7 FLOWERS AND FRUITS

We have seen also an important correlation between the sky exposition and quantitative indices of reproductive organs, such as the number of flowers and fruit (Figure 5.5A,B). That is, the more the lighting, the more the number of flowers and, correspondingly, amount of fruit. Though such indices, as the number of anomalous fruit and ovules, did not correlate with the sky exposition that supposedly, was conditioned by maternal and genetic factors (Giovannoni, 2004). The number of flowers formed initially is dependent on the environment factors, for example, the process of transformation of vegetative meristem into reproductive meristem in *Trigonella coerulea* is connected to the conditions of drought (Akhalkatsi and Lösch, 2005). Humidity is also an important factor, influencing viability of *Primula* species (Whale, 1983, 1984). Although humidity is important in combination with other factors such as lighting and temperature.

The experiments conducted on European species have made it clear that the most favorable environment for *P. veris* is low humidity, high temperature, and excessive lighting, while *P. elatior* adapts well to the excessive humidity, average temperature and lighting, and *P. vulgaris* exists and develops best of all in the conditions of moderate humidity, temperature, and lighting. The object of our observation *P. woronowii* grows exactly in the habitats similar to those of *P. vulgaris*. It is widely spread in Georgia, and its habitat in the oak-hornbeam forest is moderately humid, with low lighting and temperature. Though this species is adapted to these factors, their maximal values are more suitable for reproductive production, than minimal ones.

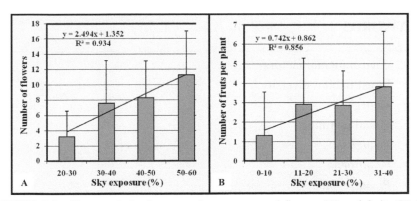

FIGURE 5.5 The correlation between sky exposure and flowers (A) and fruits (B) in three study sites of *P. woronowii*; mean and standard deviation (n=40).

5.8 ADAPTATIONS OF PRIMULA IN DIFFERENT HABITATS

The results of the research reveal a wide range of habitats to which *Primula* species adapt. For example, living environment of one of our objects, *P. amoena*, are subalpine birch tree, alpine meadows, stony slopes, snow-slides, rocky places, moraines, etc., that is, it grows in numerous habitats and feels well everywhere. There are same data for *P. nutans*, which also can exist and develop well in two different habitats considerably differing from each other by physical conditions of environment - on gentle slopes and crests. *P. nutans* was chosen for this experiment because this species is a wonderful marker of global warming in alpine ecosystems, as it is resistant to temperature evolution. The results of the study showed (Shen et al., 2006) that plants growing in different habitats differed by biomass, morphology, and size. The experiment has made it clear that *P. nutans* can adapt to different microhabitats as it possess an ability to acclimatize in various environments easily and manifest its high phenotypic flexibility. Variability of color of the flowers of *Primula* species is a good example of phenotypic variability. It is characteristic both for *P. woronovii* and for *P. amoena*. The color of their flowers changes from white to violet or purple. The same results are shown on *P. acaulis*, which is considered as a synonym of *P. woronowii* by some authors. It is shown in relation of this species (Shipunov and Butman, 2001), that there is a strong correlative relationship between the color of the crown petals and the height above the sea level. The color of the petals changed from blue to white with rising of height.

KEYWORDS

- **environmental factors**
- **forest habitat diversity**
- **Georgia**
- **global warming**
- ***P. algida***
- ***P. amoena***
- ***P. auriculata***
- ***P. nutans***

- *P. vulgaris*
- *P. woronovii*
- photoperiodism
- plant fertility
- pollination
- *Primula* species
- sky exposition

REFERENCES

Akhalkatsi, M., & Lösch, R., (2005). Water limitation effect on seed development and germination in *Trigonella coerulea* (Fabaceae). *Flora, 200*, 493–501.

Akhalkatsi, M., & Wagner, J., (1996). Reproductive phenology and seed development of *Gentianella caucasea* in different habitats in the Central Caucasus. *Flora, 191*, 161–168.

Akhalkatsi, M., & Wagner, J., (1997). Comparative embryology of three Gentianaceae species form the Central Caucasus and the European Alps. *Pl. Syst. Evol., 204*, 39–48.

Aloni, B., Peet, M., Pharr, M., & Karni, L., (2001). The effect of high temperature and high atmospheric CO_2 on carbohydrate changes in bell pepper (*Capsicum annum*) pollen in relation to its germination. *Plant Physiol., 112*, 505–512.

Atkin, O. K., & Tjoelker, M. G., (2003). Thermal acclimation and the dynamic response of plant respiration to temperature. *Trends Plant Sci., 8*, 343–351.

Buchholz, J. T., & Blakeslee, A. F., (1927). Pollen-tube growth at various temperatures. *Am. J. Bot., 14*, 358–369.

Davis, M. B., & Shaw, R. G., (2001). Range shifts and adaptive responses to quaternary climate change. *Science, 292*, 673.

Diaz, H. F., & Bradley, R., (1997). Temperature variations during the last century at high elevation sites. *Clim. Change, 36*, 253–279.

Erickson, A. N., & Markhart, A. H., (2002). Flower developmental stage and organ sensitivity of bell pepper (*Capsicum annum, L.,*) to elevated temperature. *Plant Cell Environ., 25*, 123–130.

Etterson, J. R., & Ruth, G. S., (2001). Constraint to adaptive evolution in response to global warming. *Science, 294*, 151.

Gagnidze, R., (1985). *Flora of Georgia* (Vol. 10, pp. 43–68). Metsniereba, Tbilisi.

Guisan, A., (1996). Alplandi: Évaluer la réponse des plantes alpines aux changements climatiques á travers la modélisation des distributions actuelles et future de leur habitat potentiel. *Bull Murith., 114*, 187–196.

Guistan, A., Holten, J. I., Spichiger, R., & Tessier, L., (1995). Potential ecological impacts of climate change in the Alps and Fennoscandian Mountains. *Publ. Hors-Série, 8*, 1–184.

Hedhly, A., Hormaza, J. I., & Herrero, M., (2005). Influence of genotype-temperature interaction on pollen performance. *J. Evol. Biol., 18*, 1494–1502.

Hedhly, A., Hormaza, J. I., & Herrero, M., (2008). Global warming and sexual plant reproduction. *Trends Plant Sci.*, *14*, 30–36.

Jefferies, C. J., Brain, P., Stott, K. J., & Belcher, A. R., (1982). Experimental systems and mathematical model for studying temperature effects on pollen-tube growth and fertilization in plum. *Plant Cell Env.*, *5*, 331–236.

Jóhannsson, M. N., & Stephenson, A. G., (1998). Effects of temperature during microsporogenesis on pollen performance is *Cucurbita pepo*, L., (Cucurbitaceae). *Int. J. Plant Sci.*, *159*, 616–626.

Kakani, V. G., Reddy, K. R., Koti, S., Vallace, T. P., Prasad, P. V. V., Reddy, V. R., & Zhao, D., (2005). Differences in *in-vitro* pollen germination and pollen tube growth of cotton cultivars in response to high temperature. *Ann. Bot.*, *96*, 59–67.

King, A. W., Gunderson, C. A., Carla, A., Post, W. M., Weston, D. J., & Wullschleger, S. D., (2006). Plant respiration in a warmer world. *Science*, *312*, 536–537.

Körner, C., (2012). *Alpine Treelines*. Springer, Basel (p. 220).

Koti, S., Reddy, K. R., Reddy, V. R., Kakani, V. G., & Zhao, D., (2005). Interactive effects of carbon dioxide, temperature, and ultraviolet-B radiation on soybean (*Glycine max*, L.,) flower and pollen morphology, pollen production, germination, and tube lengths. *J. Exp. Bot.*, *56*, 725–736.

Ladinig, U., & Wagner, J., (2005). Sexual reproduction of the high mountain plant *Saxifraga moschata* Wulfen at varying lengths of the growing season. *Flora*, *200*, 502–515.

Larcher, W., (2003). *Physiological Plant Ecology: Ecophysiology and Stress Physiology of Functional Groups*. Springer, Basel (p. 514).

Lewis, D., (1942). The physiology of incompatibility in plants. I. effects of temperature. *Proc. Royal Soc. London B.*, *131*, 13–26.

Lobell, D. B., & Asner, G. P., (2003). Climate and management contributions to recent trends in US agricultural yields. *Science*, *299*, 1032.

Long, S. P., Ainsworth, E. A., Leakey, A. D. B., Nösberger, J., & Ort, D. R., (2006). Food for thought: Lower-than-expected crop yield stimulation with rising CO_2 concentrations. *Science*, *312*, 1918–1921.

Menzel, A., Sparks, T. H., Estrella, N., Koch, E., Assa, A., Ahas, R., et al., (2006). European phenological response to climate change matches the warming pattern. *Glob Change Biol.*, *12*, 1–8.

Milla, R., Giménez-Benavides, L., Escudero, A., & Reich, P. B., (2009). Intra- and inter-specific performance in growth and reproduction increase with altitude: A case study with two *Saxifraga* species from northern Spain. *Funct. Ecol.*, *23*, 111–118.

Molau, U., (1993). Relationship between flowering phenology and life history strategies in tundra plants. *Arctic and Alpine Research*, *25*, 391–402.

Olesen, J. E., & Bindi, M., (2002). Consequences of climate change for European agricultural productivity, land use and policy. *Eur. J. Agron.*, *16*, 239–262.

Parmesan, C., (2006). Ecological and evolutionary responses to recent climate change. *Ann. Rev. Ecol. Evol. Syst.*, *37*, 637–669.

Porter, J. R., & Semenov, M. A., (2005). Crop response to climatic variation. *Philos. Trans. R. Soc. Lond. B. Biol. Sci.*, *360*, 2021–2035.

Prasad, P. V. V., Boote, K. J., Allen, L. H. Jr., & Thomas, J. M. G., (2002). Effects of elevated temperature and carbon dioxide on seed-set and yield of kidney bean (*Phaseolus vulgaris*, L.,). *Glob Change Biol*, *8*, 710–721.

Sato, S., Peet, M. M., & Thomas, J. F., (2002). Determining critical pre- and post-anthesis periods and physiological processes in *Lycopersicon esculentum* Mill. exposed to moderately elevated temperatures. *J. Exp. Bot.*, *53*, 1187–1195.

Shipunov, A. B., & Butman, P. S., (2001). Altitude and flower color in natural populations of *Primula acaulis* (L.) in Western Transcaucasia. *5th All-Russian Population Seminar* (pp. 111–113). Kazan.

Solomon, S., Qin, D., Manning, M., Chen, Z., Marquis, M., Avery, K. B., et al., (2007). *Climate Change 2007: The Physical Science Basis*. Cambridge University Press, Cambridge.

Tabari, M., Fayaz, P., Espahbodi, K., Staelens, J., & Nachtergale, L., (2005). Responses of oriental beech (*Fagus orientalis* Lipsky) seedlings to canopy gap size. *Forestry*, *78*, 443–450.

Théry, M., (2001). Forest light and its influence on habitat selection. *Plant Ecology*, *153*, 251–261.

Tubiello, F. N., Soussana, J. F., & Howden, S. M., (2007). Crop and pasture response to climate change. *Proc. Natl. Acad. Sci. USA*, *104*, 19686–19690.

Warner, R. M., & Erwin, J. E., (2005). Naturally occurring variation in high temperature induced floral bud abortion across *Arabidopsis thaliana* accessions. *Plant Cell Environ*, *28*, 1255–1266.

Whale, D. M., (1983). The response of *Primula* species to soil waterlogging and soil drought. *Oecologia*, *58*, 272–277.

Whale, D. M., (1984). Habitat requirements in *Primula* species. *New Phytol.*, *97*, 665–679.

White, M. A., Diffenbaugh, N. S., Jones, G. V., Pal, J. S., & Giorgi, F., (2006). Extreme heat reduces and shifts United States premium wine production in the 21st century. *Proc. Natl. Acad. Sci. USA*, *103*, 11217–11222.

Wollenweber, B., Porter, J. R., & Schellberg, J., (2003). Lack of interaction between extreme high-temperature events at vegetative and reproductive growth stages in wheat. *J. Agron. Crop Sci.*, *189*, 142–150.

Yakovlev, M. S., (1983). *Comparative Embryology of Flowering Plants* (pp. 241–243). Nauka, Leningrad.

CHAPTER 6

PROSPECTS OF COCONUT BIODIVERSITY FOR SAFE, HEALTHY, AND SUSTAINABLE EDIBLE OIL DISCOVERY

JOKO SULISTYO SOETIKNO

Faculty of Food Science and Nutrition, University Malaysia Sabah, Kota Kinabalu 88400, Sabah, Malaysia

6.1 INTRODUCTION

Coconut trees as valuable biodiversity resources have a role and become a source of food for humans since time eternity. Coconuts are also an integral part of the lives of the people of Malaysia. In traditional life, the coconut fruit is a source of nutrients that are filled with savory taste coconut milk, and oil. In many islands around the equator, coconut has enriched food population for decades, even hundreds of generations. The coconut tree is a tree of the most usefulness because almost every part of the tree can be utilized. Coconut also plays an important role in the local economy and culture of these countries. It is widely known as the 'tree of life' since it provides a large number of products that can be used to support the local economy. It is socially and culturally linked besides providing jobs and income to millions of people. In many islands of countries, it is the major source of revenue and is an integral part of the livelihood of the population. A wide variety made from raw coconut industry has evolved from traditional ones such as coconut oil and copra to the oil processing into chemical compounds that have high added value. In the processing and marketing subsystem, the downstream of coconut industry is still limited to the production of cooking oil, and primary products such as copra, grated coconut, or coconut milk. The processing technology of virgin coconut oil

(VCO), which is known to produce products with various health benefits, however, in fact, it has not yet provided any income to the farmers up to present (Rethinam, 2004; Lay and Pasang, 2012).

6.1.1 VALUE OF COCONUT AND COCONUT OIL

For thousands of years, tropical countries have used coconut from the tree *Cocos nucifera*, Family Aracaceae (*palm* family) as an integral part of their diet and livelihood. The coconut is known as "Tree of a Thousand Uses" in Malaysia. In the Philippines, it is commonly known as the "Tree of Life" for its important role in smallholders' livelihoods as a direct source of cash income, nutrition, and materials (Warner et al., 2007). All parts of the coconut palm are useful, with significant economic value (Five, 2004; Gervajio, 2005). In the Pacific, those benefits include shade for other crops, land stabilization, food, and drink, and material for construction, weaving, containers, fuel, and other uses.

The coconut is one of the commodities that have high economic value if could be managed properly. The coconut is a good source of vegetable protein and can be processed furthermore into various products that are beneficial to humans and can be used as raw material for cooking oil. In order to intensify the activities of the coconut farming system, a market share must be created with a certainty fix price to farmers eager to cultivate coconut plantation. One of the opportunities to maintain and protect the fixed price of coconut is through making a contract or agreement between coconut farmers and companies in the field of agro-industries that processing derivative products of coconut (Amin, 2000; Thantiyo, 2010).

The current competitiveness of coconut products are lies on downstream industry and no longer on the primary product, where the added value that can be created from its downstream products may provide multiply benefit compared to its primary product. The business of the downstream products is currently growing and has high feasibility to small, medium, and big enterprises. In turn, the downstream industry will be the locomotive of the upstream industry (Agency for Agricultural Research and Development, 2009). Coconut oil is the most valuable downstream product of coconut and is widely used as industrial raw material as well as cooking oil.

Coconut harvesting and primary processing are dominated by smallholders, as large coconut estates turn to more profitable crops. Papua New

Guinea, Fiji, Solomon Islands, Marshall Islands, Vanuatu, and Kiribati have substantial exports of copra and copra oil for further refining. In contrast, India, Indonesia, Malaysia, and the Philippines produce three-quarters of the world's coconuts, much of this in plantations, and fully process it locally. However, the greatest economic benefit to coconut producers has come from drying the coconuts into copra for further processing into copra oil. The coconut oil can be extracted from the coconut kernel or dried coconut kernel (copra). The oil content of copra is approximately 60–65%, while the coconut kernel is containing about 43% oil (Syafrini, 2013).

Copra and copra coconut oil are traded as commodities, with world prices driven by large Asian coconut processors, forcing Pacific producers to be price takers in the marketplace. Declining and fluctuating copra and coconut oil prices and rising market standards around quality and consumer safety have severely affected the viability and competitiveness of all Pacific Island copra producers, and many coconuts are not being harvested. The current situation provides little or no incentive for further investment or replanting (Warner et al., 2007). Coconut oil or Copra oil is edible oil extracted from the kernel of mature coconuts of the coconut palm. In recent years this oil has attained superstardom in the health food world. Celebrities are adopting its use, nutritionists advocating it, and patients acclaiming its many virtues. Yet, despite the growing popularity, some people are skeptical. Its many health benefits sound too good to be true (Warner et al., 2007).

In principle, the making of coconut oil is generally done by using dry and wet method. The dry method is done by pressing a copra which is obtained by drying the coconut kernel. This method is implemented at the coconut oil processing plant due to its requirement cost and complicated equipment. Whilst, the wet is conducted by extraction of coconut milk derived from coconut kernel prior to heating separated oil from its pretreated emulsion. Coconut oil produced by the wet manner requires a long heating process, and thus it requires a lot of energy. This method is less efficient due to long processing time and high cost for energy (Hasbullah, 2001).

The quite latest technology today in manufacturing coconut oil is by application of fermentation and enzymatic processing technology. This method is expected to be the one of alternative to overcome some obstacles dealing with traditional method, thus it may be implemented to improve the quality, equipment constraints and system used for processing control to optimize the production both qualitatively and quantity, as well as in reducing cost and energy (Arsa et al., 2004).

The success of the fermentation process depends on the exact type of microbial starter culture according to the product and the ingredients used. In this study, starter culture was used for enzymatic bioprocesses of coconut milk due to the culture characteristic which has potential capability to produce hydrolytic enzymes which may work on carbohydrates, proteins, cellulose, hemicellulose, and pectin which bind to the fat globules in coconut milk. Whereas, the starter culture may activate to break down the coconut emulsion to separate out fats and oils, while the carbohydrates are digested as energy providers and carbon sources for biosynthesis of oils and mineral salts and other growth factors are optimally utilized to synthesize fats (Rindengan, 2005). The way of enzymatic bioprocesses has several main advantages are the effectiveness of reaction, saving energy due to the relatively short processing time, low production cost and no complicated equipment is involved, high quality of end product which is characterized in high nutritional value (Sukmadi et al., 2002).

Producing coconut oil is an important post-harvest action dealing with coconut production, where coconut oil is the most valuable part of the coconut. Coconut oil is often used as industrial raw materials and cooking oil production. In addition, good coconut oil is used to improve one's health. Hence, undoubtedly the coconut oil or as known well as virgin coconut oil (VCO) has become target product of most people. The good coconut oil manufacturing techniques can improve and thus maintain the quality and quantity of the oil production (Rindengan et al., 2005). Given the need for coconut oil as a health food commodity continues to increase, it is necessary to carry out many ways to produce coconut oil in sufficient quantities. One of the efforts that need to be taken is to diversify production technology through its processing. Many ways of coconut oil processing that have been known well having their respective advantages and disadvantages.

6.1.2 REVITALIZING COCONUT INDUSTRY IN MALAYSIA

In Malaysia, coconut is the fourth important industrial crop after oil palm, rubber, and paddy in terms of total planted area. It is also one of the oldest agro-based industries. As an industry, coconut contributes very little to the overall economy of Malaysia. However, coconut still plays an important role in the socio-economic position of Malaysian rural population that involves 80,000 households. About 63% of coconut production is for domestic consumption, and 37% is for export and industrial processing. The domestic demand for coconut products takes in the form of fresh coconut, tender coconut, coconut oil, and cream powders. In terms of exports, the country has seen an increase in the export of end-products of coconut such as desiccated coconut, coconut milk powder and activated carbon (Sivapragasam, 2008).

Reports have indicated that there is an increasing demand for raw material for the local coconut based processing industries for local consumption and for the expanding demand and market for coconut products worldwide. Thus, it is very interesting to note that the Government of Malaysia has put up proactive policies and concrete investments to revitalize its coconut industry and increase coconut productivity with the existing land devoted to the crop. In spite of the decreasing land planted to coconuts, the Government of Malaysia is determined to increase coconut farm productivity and the income of the smallholders through appropriate policies, investments, and implementation strategies. For its national investment for the coconut sector, the national Government of Malaysia has approved a funding allocation totaling RM 71,5 million under its 9th Malaysia Plan, Mid-Term Review Fund and Economic Stimulus Package. Under its Coconut Industrial Road Map, it aims to replant 10,200 hectares and rehabilitate 51,200 hectares of coconut land involving small landholders up to 2015. To ensure higher productivity, improved coconut varieties like the MATAG hybrids and others will be used in the Replanting Program. The Coconut Rehabilitation Program will include provisions for fertilizer inputs, and agro-know-how on coconut based farming systems like intercropping and livestock integration.

The expected impact of these programs has been identified in terms of increase in coconut production from 3,500 nuts/ha/year to 20,000 nuts/ha/year in replanted farms with hybrids. Under the rehabilitation programs, farms with existing Malayan Tall varieties are expected to double its productivity from 3,500 nuts/ha/year to 7,000 nuts/ha/year. The income of the coconut smallholders are likewise expected to increase from RM 350/month to RM 1000/month. The increase in coconut production is expected to provide adequate raw materials for the coconut-based processing industries and for home consumption. With the strong determination of the Government of Malaysia and with the support and collaboration of the private sector, there is no doubt that the proactive policies and the implementing strategies that have been formulated will indeed revitalize the Malaysian coconut industry in the years to come (http://www.coconut-protectors.com).

6.1.3 MARKETING AND TRADE

There are two major markets for coconuts - copra and fresh, the latter commanding a higher price than the former. The market for copra and oil is worldwide. Most large/medium coconut producers crush the copra themselves and have oil mills. As a consequence of this, only about 4% of the copra produced is exported. The majority is exported as oil. Coconut oil exports have been increasing over the last decade mainly because of the greater global need for the essential characteristics of coconut oil. In 2008, just over 2 million tonnes coconut oil was traded on the world market. The Philippines was the largest exporter of coconut oil in 2008, with 42% of world exports. The main destination markets for the oil were USA and Europe accounting for 24% and 25% of imports, respectively. Supplies of coconut oil on the global market being adversely affected by problems affecting production are pests and disease, aging plantations and harvesting problems. It is also affected by competition for the fresh nuts for coconut water. Organizations such as the EU provide assistance in the form of preferential tariffs as well as price support to imports from the Pacific Islands. Many manufacturing companies showed how coconut oil has a good prospect. However, the oil processing industry is inseparable from the problems and weakness technical or non-technical, among other things, the lack of availability of raw materials quality assurance, product is not yet optimal and uniform, yet satisfy the tastes of the market design, the low quality of human resources, waste management poor, poor business management, and market information that has not been steady (http://www.unctad.info/en/Infocomm/AACP-Products).

6.2 HEALTH ISSUES

Factors of aflatoxin appeared in the eighties and again become a threat. In 1989, the European Union set a standard of 200 ppb aflatoxin levels, in 1991 the EU countries set a new standard of 20 ppb. Currently, the Codex Alimentarius Commission of the FAO is in negotiations to reduce levels of aflatoxin B_1 in copra to 5 ppb. By way of making existing copra in South East Asian countries, it is very difficult to achieve the level of 5 ppb aflatoxin. In addition to the Aflatoxin, some countries in the EU have started to ban smoke-dried copra as a polycyclic aromatic hydrocarbon (PAH) standard stipulation in the oil. PAH is a cancer-causing element that goes into copra because of the smoke.

6.2.1 COCONUT OIL AND COOKING OIL

In general, the market has to offer two kinds of cooking oil that are derived from plants and that is derived from animals, known as tallow and lard. Vegetable cooking oils are including palm oil, coconut oil, corn oil, soybean oil, and olive oil. In Malaysia, vegetable oil that is widely circulated and consumed by the public at large is palm oil. Malaysia is the largest country in the world as a palm oil exporter. Many advantages of this kind of oil compared with others, such as beta-carotene and saturated fat that is pretty much contained inside so that when it is heated, it will not be easily damaged. Cooking oil, in addition to functioning as a medium of heat, give texture, appearance, and good taste, also have health function. Thus, in choosing a cooking oil, should pay attention to the list of ingredients contained in the packaging, standard quality, and "Halal" guarantee.

Basically all cooking oils contain the same chemical that fatty acids (FAs) and glycerol. Saturated fatty acids (SAFA) such as palmitic and stearic has the advantage of resistant or stable against heating, but has the disadvantage that it is a solid at room temperature and the suspected bad effect on health. Unsaturated fatty acids such as oleic, linoleic, linolenic, and arachidonic has the advantage that it can prevent the occurrence of deep vein thrombosis, but the drawbacks are not resistant to heating to high temperatures even in unsaturated fatty acids are easily damaged and conformational changes of geometric shapes that had *cis* into a *trans*. These conformational changes then can be caused by high temperatures can also be caused by the hydrogenation process. Hydrogenation process carried out to convert unsaturated fatty acids into saturated, for example in the manufacture of butter. In the United States has enacted, each product must include the oil and fat-free label *trans*. *Trans* oil is not good for health, thus choosing oil or fat must be in accordance with the needs, for frying or eaten directly.

Oils and fats can be damaged. Damaged oil is characterized by the emergence of odor, rancid, the colors are not clear even blackish brown, and foamy. The cause of this damage is due to the absorption of odors, hydrolysis, and oxidation. Absorption of odors can be caused by packaging or the surrounding environment. Hydrolysis caused by the presence of water and oxidation triggered by oxygen (air), repeated heating at high temperatures, the presence of iron and copper metal and light. A process of oxidation forms oil-chain compounds such as hydrocarbons, aldehydes, acids, alcohols, and ketones.

Damage oils and fats cannot be prevented, but it can be slowed. Damage of cooking oil can be slowed down by utilizing correctly. Although many studies have found about dangers of cooking, however still many do. Thus, it is necessary a way to recycle oil simply to reduce further damage. Ways to recycle used cooking oil such as through the use of coconut shell charcoal, rice flour, noni, *Aloe vera*, red onion, made into soap and biodiesel. How to recycle used cooking oil using activated charcoal which relies on the absorption of active charcoal to very large dirt. Activated charcoal has a surface area that is very much so effective in filtering out impurities that exist in the vicinity either on the water or in the air. Recycling with rice flour relying on the surface area of the powder is very broad. Flour is able to bind to particles around it along the bottom layer of sediment. Color white flour does not contaminate the oil itself can even provide the more whitening effect. Noni or *Morinda citrifolia* which are high in the content of vitamin C can act as antioxidants so that it can prevent the occurrence of oxidation to stabilize the free radicals that are nearby. *Aloe vera* is a type of plant has been known for thousands of years ago and used as hair growth, wound healing, skin care and even to recycle cooking oil. *Aloe vera* can absorb impurities contained in cooking.

Shallots are rich in antioxidants, so it functions as an antioxidant, which is expected to provide a positive influence on cooking. Shallots are believed to be efficacious for the health of the body. Shallots are rich in substances like essential oils, cycloalyin, methylalyin, dihydroalyin, flavon glycoside, quercetin, saponin, peptides, phytohormones, vitamins, starch, and other components that are beneficial to health. The content of this essential oil spread the aroma when we fry onions. Another way to recycle used cooking oil is to be used as a soap product. The fat content is pretty much used cooking oil. The reaction between fats with NaOH (caustic soda) formed soap. To obtain the desired aroma who lived added essential oils or extracts of flowers, fruit, and spices according to taste. Along with fuel oil reserves the used cooking oil can be recycled into biodiesel. Another advantage of the basic ingredients of cooking oil is that the rest of the cooking oil can also be used, thereby reducing the environmental burden due to the garbage. Compared with diesel oil, cooking oil can reduce pollution due to lower sulfur content, and less noisy. The use of cooking oil for fuel does not increase the number of gaseous carbon dioxide, because the oils derived from plants. Shortage of cooking oil is more viscous, so it is easy to clog the channel, especially if mixed with solid fraction, used oil or if the weather is cold.

Cooking oil is an oil derived from plant or animal fat which is purified and liquid at room temperature, and is usually used for frying food (Sitepoe, 2008). Cooking oil from plant materials is usually produced from crops such as coconut, seeds, nuts, corn, soybean, and canola (Sitepoe, 2008). Cooking oil usually can be used up to 3–4 times a frying process (Tomskaya et al., 2008). If it is used repeatedly, then the oil will have color change (Sitepoe, 2008). When frying is done, there is a double bond in unsaturated fatty acids that will be broken up and forming a saturated fatty acid (Tomskaya et al., 2008). Good oils are oils that contain unsaturated fatty acids more than the saturated fatty acid content (Sitepoe, 2008). After frying for many times, fatty acids contained in the oil will be more saturated (Tomskaya et al., 2008). Thus, the oil can be said to have been damaged or can be referred to as waste cooking oil (Hou et al., 2007). The use of oil for many times of frying will make the double bond of oil is oxidized and to form a peroxide group and cyclic monomers. Oils are like this can be said to have been damaged and harmful to health (Herlina et al., 2002). The higher the temperature and the longer the heating of frying may lead the levels of saturated fatty acids will further increase (Tomskaya et al., 2008). Vegetable oil with saturated fatty acid levels are too high will result in foods that are processed are harmful to health (Sitepoe, 2008). Moreover, not only because of the way of repeatedly frying, but the oil also can be damaged due to incorrect storage within a certain period, so that the bond of triglycerides may break down into glycerol and free fatty acids (FFA) (Hou et al., 2007). Some factors that can influence the occurrence of oil damage are: (a) Oxygen and oil double bond - the more double bonds, and oxygen contained, then the oil will be more quickly oxidized (Herlina et al., 2002); (b) Temperature - higher temperatures will accelerate oxidation process (Hou et al., 2007); (c) Light and metal ions - act as a catalyst that accelerates oxidation process (Pikiran Rakyat, 2002), and (d) Antioxidants - makes oil more resistant to oxidation (Pikiran Rakyat, 2002).

World Health Report (2002) cites cardiovascular diseases (CVDs) as the largest cause of death and disability in India by 2020, where fats are an essential part of a healthy balanced diet, there is an evidence to show that limiting saturated and *trans* fat intake is important, while it is prudent to have fat in your meals; it helps the body to absorb vitamins like vitamin A, D, E, and K. Studies have shown that it is not a single fat source, but a combination of various fats having fatty acid compositions and additional minor components like tocotrienols, oryzanol, and a good balance of saturated fatty acid (SFA), monounsaturated fatty acid (MUFA) and polyunsaturated fatty acid (PUFA) that will bring about favorable serum lipid profiles which help in guarding against suffering or mortality from coronary heart diseases (CHD). It has been consistently shown in all studies that the possibility that the proportional increase in plasma HDL concentration produced by saturated fat somewhat compensates for its adverse effect on LDL level. In metabolic studies, different classes of SFAs have different effects on plasma lipid and lipoprotein levels2 specifically, SFAs with 12–16 carbon atoms tend to increase plasma total and LDL cholesterol levels, whereas stearic acid (18:0) does not have a cholesterol-raising effect in comparison with oleic acid (18:1). Among the cholesterol-raising SFAs, myristic acid (14:0) appears to be more potent than lauric acid (12:0) or palmitic acid (16:0), but the data are not entirely consistent. In contrast, intakes of longer-chain SFAs (12:0–18:0) were each separately associated with a small increase in risk. Saturated fats are consumed as clarified butter and coconut as fresh, dry, and as coconut oil along the coastal and southern parts of India (Keys, 1980; Temme et al., 1996; Kris-Etherton and Yu, 1997; Hu, 1999). American Heart Association and National Cholesterol Education Program recommend that the fat which should be predominant in our diets should be monounsaturated. Dietary fat plays an important role in cardiovascular health. Numerous studies have been carried out to identify the type of fat that correlates with the CVD. Excessive intake of saturated fat raises total and LDL cholesterol levels. It also has a negative effect on blood pressure and arrhythmias. TFAs generated during partial hydrogenation of fat can elevate LDL cholesterol and reduce HDL cholesterol, whereas mono-unsaturated fat is neutral when substituted with carbohydrates. The greatest CVD risk reduction is associated with PUFA intake whereas a lesser risk reduction is associated with monounsaturated fat. However, substitution of MUFA to PUFA results in the formation of LDL which are less susceptible to oxidation (Hu, 1994; Temme et al., 1996).

Milk products such as condensed milk (khoya), cream, full-fat paneer are other sources of saturated fats. Cheese and also whole milk and its various products like milkshakes, ice creams, etc. form a major source of saturated fats in urban India. Coconut oil can be used for frying, because the structure of the oil does not have a double bond, so that the type of oil is a stable saturated fat. In addition, the coconut oil contains essential fatty acids which cannot be synthesized by the human body. The fatty acids are including palmitic acid, stearic, oleic, and linoleic. In addition, the coconut oil, are essential fatty acids, which cannot be synthesized by the body (Sitepoe, 2008).

6.2.2 PALM OIL VERSUS COCONUT OIL

Wright (2012) mentioned that one tablespoon of coconut oil contains 117 calories, 14 g total fat and 12 g of saturated fat. Meanwhile, unsalted butter only contains 102 calories, 12g total fat, and 7 g saturated fat. A research study says that coconut oil contains a lauric fatty acid, an essential fatty acid that can increase the amount of good cholesterol (HDL). However, these fatty acids can also increase the amount of bad cholesterol (LDL). Coconut oil is a little healthy compared to palm oil which can be bad for health. However, palm oil consumption is also not good for health. In this case, it is suggested to replace coconut oil and butter with olive or canola oil. Every serving olive oil contains 119 calories, 13.5 g total fat and 1 gram of saturated fat. While canola oil contains 124 calories, 14 g total fat and 1 g of saturated fat.

Palm oil or coconut oil, naturally processed, and did not undergo hydrogenation. This causes the cooking oil made from palm oil and coconut oil, contains no *trans* fatty acids (TFA) which adversely affects health. TFA can increase blood cholesterol levels. The presumption on consumption of coconut oil with the material in the daily menu in accordance with the needs of no proven adverse health effects instead fosters a healthy heart and blood vessels. It is also supported by the results of a study of Sri Lankan society shows the ratio of LDL/HDL increases after replacing the consumption of coconut oil with corn oil. The most significant evidence was prior research on Polynesian inhabitants are turning to modern oil have increased levels of total cholesterol. Cholesterol content in coconut oil is in the range of 0 to 14 ppm, where the value is the lowest value compared with cholesterol levels in other vegetable oils despite the dangers of cholesterol levels much lower than animal fats and dairy products. Frying foods with palm oil also make food more durable and not easily rancid. Coconut oil contains 0.02% FFA. Acid content is too high, above 0.5% (Prior et al., 1981).

6.2.3   BEST OF EDIBLE OILS

This one question is increasingly becoming important. What kind of oil is the best? Animal oil or vegetable oil? Among vegetable oils, which is the best among of palm oil, coconut oil, corn oil, sunflower, olive oil, sesame, avocado, macadame oil, and others. Malaysia as one of the largest producers of palm oil, always in an irritable mood when palm oil was accused of being healthy oils and high cholesterol levels. First of all, that should be clarified is that we are talking about edible oils, so instead of non-edible oils commonly used for dressing or delicious maker for snacks or salads. Non-edible oils have many good product collections. In general, when oil made its way in the form of cold pressing, then the oil is entirely good. An example is an olive oil, coconut oil, avocado oil, and some other types. On the product label, it is described as cold-pressed oil or oil being milked in cold conditions (without heating or boiling to remove oil). Nanji et al. (1995) report that a diet enriched in saturated but not unsaturated fatty acids reversed alcoholic liver injury in their animals, which was caused by dietary linoleic acid. These researchers concluded that this effect might be explained by the down-regulation of lipid peroxidation. This is another example of the need for adequate saturated fat in the diet. Cha and Sachan (1994) studied the effects of saturated fatty acid and unsaturated fatty acid diets on ethanol pharmacokinetics. The hepatic enzyme alcohol dehydrogenase and plasma carnitines were also evaluated. The researchers concluded that dietary saturated fatty acids protect the liver from alcohol injury by retarding ethanol metabolism, and that carnitine may be involved.

Nanji et al. (1995) postulated that they would find that diets rich in linoleic acid would also cause acute liver injury after acetaminophen injection. In the first experiment, two levels of fat (15 g/100 g protein and 20 g/100 g protein) were fed using corn oil or beef tallow. Liver enzymes indicating damage were significantly elevated in all the animals except for those animals fed the higher level of beef tallow. These researchers concluded that diets with high linoleic acid might promote acetaminophen-induced liver injury compared to diets with more saturated and monounsaturated fatty acids. For edible oil, it turns out the best product is coconut oil. Coconut oil can still be rivaled and surpassed by olive oil. As of cooking oil, it has the advantage, which at high temperature, this oil has properties stable even in high temperatures. It does not quickly transform into fatty acids or *trans*-fat, because the more it contains what is referred to as medium-chain fatty acid (MCFA). Coconut oil has so many advantages for health. One of the advantages that are not possessed by some other oils is lauric acid that is called as a deterrent against viruses, bacteria, fungi, and mycobacteria.

Olive oil is superior to the non-edible oils, turns when used as cooking oil, will be denatured at high temperatures. The worst oil for cooking oil is derived from margarine is made from palm oil as raw material. Margarine or the like was indeed made by making the palm oil becomes saturated (solidified). The trick is to heat the oil until it reaches the temperature was high, and the oil changed the nature of their liquid into a solid. Margarine is one of the unhealthiest foods in the world. It is a little ironic, considering that many bakers or cake, replacing the recipe of butter into margarine, with a view to avoiding animal fats. In fact, margarine is much worse than butter. In a certain dose, butter was still quite healthy.

When cooking oil is applied to a high heat, the oils are properly used for cooking should have high stability and will not undergo oxidation or go rancid easily. By means that when oils undergo oxidation, they react with oxygen to form free radicals and harmful compounds that may be definitely high risk to be consuming. The most important factor in determining an oil's resistance to oxidation and rancidification, both at high and low heat, is the relative degree of saturation of the fatty acids in it. Saturated fats have only single bonds in the fatty acid molecules, monounsaturated fats have one double bond, and polyunsaturated fats have two or more. It is these double bonds that are chemically reactive and sensitive to heat. Saturated fats and monounsaturated fats are pretty resistant to heating, but oils that are high in polyunsaturated fats should be avoided for cooking (Grootveld et al., 2001). When it comes to high heat cooking, coconut oil is your best choice. Over 90% of the fatty acids in it are saturated, which makes it very resistant to heat. This oil is semi-solid at room temperature, and it can last for months and years without going rancid. Coconut oil also has powerful health benefits. It is particularly rich in a fatty acid called lauric acid, which can improve cholesterol and help kill bacteria and other pathogens (Nevin and Rajamohan, 2004; Marina et al., 2009; Assunção et al., 2009). The fats in coconut oil can also boost metabolism slightly and increase feelings of fullness compared to other fats. It is the only cooking oil that made it to my list of superfoods (St-Onge and Jones, 2002; Assunção et al., 2009).

6.3   THE COCONUT OIL GREATNESS

Coconut oil has a long reputation and is highly respected in many cultures around the world, not only as a high-value food, but also as a good medicine. It is used throughout the tropics in many systems of traditional medicine. In India, coconut oil is an essential ingredient of Ayurvedic potions. A treatment has been practiced in India for thousands of years and is still used as a form of primary treatment by millions of people. In Central America, people were known to drink from a glass of coconut oil to help them cope with the pain. They have learned from generation to generation that consume coconut oil recovery from illness. It is considered as a health tonic that is good for the heart. Among the Polynesians, coconut oil is very valuable to exceed all other crops in terms of nutrition and health properties. The healing properties using this oil have long been known in the development of this plant. But only recently, the benefits are increasingly recognized around the world. Unlike other fats, coconut oil protect against heart disease, cancer, diabetes, and other degenerative diseases. This oil supports and strengthens the immune system that helps the body flush infection and disease attack. This oil is unique and different from other types of oil as it accelerates weight loss, which made him known as the only low-calorie fats in the world.

6.3.1 PRIVILEGED VIRGIN COCONUT OIL

VCO is oil that is processed without purification (bleaching, deodorizing, and refinery), without heating or with slightly heating. Coconut oil contains high lauric acid (45–55%) and also other acids. Lauric acid is a medium chain saturated fat which is known as medium chain triglycerides (MCT). Since VCO containing MCT is easily digested in the gastrointestinal tract, it will be directly absorbed through the intestinal wall without undergoing a process through enzymatic hydrolysis, and is directly supplied into the bloodstream and go through liver to be processed into energy to improve function of all endocrine glands, organs, and tissues of the body. While other vegetable oils, such as sunflower oil, soybean oil, corn oil which are having a larger size of fatty acid molecules need to be processed first in the digestive tract through hydrolysis and emulsion with the help of bile and pancreatic enzymes gland. Decomposition units of free fatty acids are reassembled and packaged into lipoprotein, then is supplied to the liver for metabolism and its products are distributed to all endocrine glands, organs, and tissues of the body in the form of energy, while the rest of the cholesterol and fat deposited into body tissues.

Triglycerides of coconut oil in the body will be split into simple diglycerides, monoglycerides (MGs) and free fatty acids. Monoglycerides and free fatty acids are having antimicrobial properties. Among monoglycerides that have bioactive properties are lauric acid and capric acid. Lauric acid and capric acid also has antiviral properties, thus to penetrate the outer layer of lipids of the virus, hence, later on, they were developed as antiviral agents of HIV and Hepatitis B and C. There is another aspect to the coronary heart disease picture. This is related to the initiation of the atheromas that are reported to be blocking arteries. Recent research is suggestive that there is a causative role for the herpes virus and cytomegalovirus (CMV) in the initial formation of atherosclerotic plaques and the recloging of arteries after angioplasty (New York Times, 1991). What is so interesting is that the herpes virus and CMV are both inhibited by the antimicrobial lipid monolaurin, but monolaurin is not formed in the body unless there is a source of lauric acid in the diet. Thus, ironically enough, one could consider the recommendations to avoid coconut and other lauric oils as contributing to the increased incidence of coronary heart disease. Perhaps more important than any effect of coconut oil on serum cholesterol is the additional effect of coconut oil on the disease fighting capability of the animal or person consuming the coconut oil. Coconut oil and certain coconut products contain approximately 50% lauric acid and approximately 6–7% capric acid. The question frequently arises whether ingestion of lauric acid can lead to significant endogenous production of monolaurin. There do not appear to be any clear data concerning how much monolaurin is actually formed from lauric acid in the human body. Nevertheless, there is evidence that some are formed. Lauric acid is the main antiviral and antibacterial substance found in human breast milk.

Kabara (1978) and others have reported that certain fatty acids (medium-chain saturates) and their derivatives (monoglycerides) can have adverse effects on various microorganisms: those microorganisms that are inactivated include bacteria, yeast, fungi, and enveloped viruses. The medium-chain saturated fatty acids and their derivatives act by disrupting the lipid membranes of the organisms (Isaacs and Thormar, 1991; Isaacs et al., 1992). In particular, enveloped viruses are inactivated in both human and bovine milk by added FAs and monoglycerides as well as by endogenous FAs and MGs (Isaacs and Thormar, 1986; Thormar et al., 1987; Isaacs et al., 1990, 1992). All three monoesters of lauric acid are shown to be active antimicrobials, i.e., alpha-, alpha'-, and beta-MG. Additionally, it is reported that the antimicrobial effects of the FAs and MGs are additive and total concentration is critical for inactivating viruses (Isaacs and Thormar, 1990). The properties that determine the anti-infective activities of lipids are related to their structure; e.g., monoglycerides, free fatty acids. Comparatively speaking, lauric acid (C12) has a greater antiviral and antibacterial activity than other medium-chain triglycerides such as caprylic acid (C8), capric acid (C10), or myristic acid (C14). Monolaurin is many times more biologically active than lauric acid in killing viruses and bacteria, leading to an interesting question concerning the conversion rate in the human body. Unlike these medium-chain fatty acids, diglycerides, and triglycerides are inactive against microorganisms (Kabara and Vrable, 1977). Research has suggested that monolaurin exerts virucidal and bactericidal effects by solubilizing the lipids and phospholipids in the envelope of the pathogen causing the disintegration of its envelope. Recent evidence has also indicated that the antimicrobial effect is related to its interference with signal transduction in cell replication. The action attributed to monolaurin is that of solubilizing the lipids and phospholipids in the envelope of the virus causing the disintegration of the virus envelope. In effect, it is reported that the fatty acids and monoglycerides produce their killing/inactivating effect by lysing the plasma membrane. However, there is evidence from recent studies that one antimicrobial effect is related to its interference with signal transduction (Projan et al., 1994).

Some of the viruses inactivated by these lipids, in addition to HIV, are the measles virus, herpes simplex virus–1, vesicular stomatitis virus, visna virus, and CMV. Many of the pathogenic organisms reported to be inactivated by these antimicrobial lipids are those known to be responsible for opportunistic infections in HIV-positive individuals. For example, concurrent infection with CMV is recognized as a serious complication for HIV+ individuals (Macallan et al., 1993). Thus, it would appear to be important to investigate the practical aspects and the potential benefit of an adjunct nutritional support regimen for HIV-infected individuals, which will utilize those dietary fats that are sources of known anti-viral, antimicrobial, and anti-protozoal monoglycerides and fatty acids such as monolaurin and its precursor lauric acid. No one in the mainstream nutrition community seems to have recognized the added potential of antimicrobial lipids in the treatment of HIV-infected or AIDS patients. These antimicrobial fatty acids and their derivatives are essentially non-toxic to man; they are produced *in vivo* by humans when they ingest those commonly available foods that contain adequate levels of medium-chain fatty acids such as lauric acid. According to the published research, lauric acid is one of the best "inactivating" fatty acids, and its monoglyceride is even more effective than the fatty acid alone (Kabara, 1978; Sands et al., 1978; Fletcher et al., 1985; Kabara, 1985). The lipid coated (envelop) viruses are dependent on host lipids for their lipid constituents. The variability of fatty acids in the foods of individual accounts for the variability of fatty acids in the virus envelop and explains the variability of glycoprotein expression. Monolaurin also inhibits production of staphylococcal toxin–1 effectively (Projan et al., 1994) including protein A, alpha-hemolysin, B-lactamase, and the induction of vancomycin resistance (Preus et al., 2005; Ruzin and Novick, 1998).

6.3.2 COCONUT KERNEL

VCO is obtained from the fresh and mature kernel of the coconut (*Cocos nucifera* L.) by mechanical or natural means with or without the application of heat, which does not lead to alteration of the nature of the oil. VCO has not undergone chemical refining, bleaching or deodorizing. It can be consumed in its natural state without the need for further processing. Virgin coconut oil consists mainly of medium chain triglycerides which are resistant to peroxidation. The fatty acids in virgin coconut oil are distinct from animal fats which contain mainly of long-chain saturated fatty acids. Virgin coconut oil is colorless, free of sediment with a natural fresh coconut scent. It is free from rancid odor or taste.

VCO is the naturally processed, chemically free and additive-free product from fresh coconut meat or its derivative, which has not undergone any chemical processing after extraction. It is the purest form of coconut oil, water white in color, contains natural vitamin E with very low free fatty acid content and peroxide value. It has a mild to intense fresh coconut scent depending on the type of process used for production. Virgin coconut oil can only be achieved by using fresh coconut meat or what is called non-copra and is produced using a variety of methods. Generally, good quality virgin oil is produced at the lowest temperature possible and will vary depending on the method used. Whichever method is used to produce the oil, the moisture content of the resulting oil needs to be 0.1% or less. Otherwise, the oil can become rancid. Disadvantages of VCO take place when the body is used to a low-fat diet regimen. Since lauric acid is antibacterial and anti-viral, there could also be "die-off" effects from the VCO as these organisms are eliminated from the body. The most common side effect is diarrhea. Like any food, some people could possibly have allergic reactions to VCO as well, although it has traditionally nourished millions if not billions of people throughout Asia for thousands of years.

Fats in coconut meat contains 90% saturated fatty acids and 10% unsaturated fatty acids. Although it is classified as saturated oil, coconut oil is categorized as medium chain fatty acids (MCFA). The excellence of MCFA over long-chain fatty acids (LCFA) is a chain of fatty acids that are more easily digested and absorbed. The MCFA when consumed can be directly digested in the intestines without hydrolysis, and enzymatic processes directly supplied to the bloodstream and transported to the liver to be metabolized energy.

The oil has LCFA in the digestive be processed first before it is absorbed through the intestinal wall several long processes to get to the heart. Another advantage of MCFA in the body is that it is not converted into fat or cholesterol and does not affect blood cholesterol. The MCFA especially lauric acid has specific abilities such as antiviral, antifungal, antiprotozoal, and antibacterial. Lauric acid in the human body and the animal will be transformed into monolaurin. Monolaurin have health effects similar to human breast milk, which can boost the immune system in infants from viral infections, bacteria, and protozoa. Therefore, monolaurin likely to be developed as a medicine for a severe acute respiratory syndrome or severe acute respiratory syndrome (SARS). Lauric acid has also been shown to be virucidal and bactericidal although monolaurin has even greater activity. Enteral feedings have been reformulated without coconut oil and formula such as Ensure and Nutren are made with hydrogenated oils. Both humans and animals can metabolize some monolaurin from lauric acid (Enig, 1998). However, it is worth repeating that the level of this metabolizing has not been quantified. Dr. Kabara has postulated that, if monolaurin is formed from lauric acid in coconut fat, the level is no greater than 3% (Kabara, 2005). Mothers' milk, a rich source of lauric acid, may also provide a lipase that converts the triglycerides to monoglycerides by the infant. Even small amounts of monolaurin converted from lauric acid in coconut fat, or mothers' milk and lauric acid are still virucidal and bactericidal. In the past, infant formula were good sources of lauric acid, because a greater amount of coconut oil was used. However, they have been reformulated with little or no coconut oil, thus losing their similarity to mothers' milk, which provides antimicrobial and antiviral fats. Human milk provides approximately 3.5% of calories as lauric acid for the human infant. Mature human milk has been noted to have up to 12% of the total fat as lauric acid. Some of the viruses inactivated to some extent by monolaurin include HIV, measles, *Herpes simplex*–1, vesicular stomatitis, visna virus, and CMV (Enig, 1998). Coconut oil does not have a deleterious effect on cholesterol or other blood lipids. In fact, it may raise high-density lipoprotein cholesterol. Coconut oil is rich in "good" saturated fatty acids that conserve the elongated omega–3 fatty acids. Animal studies have shown that some omega–3 fatty acids can be formed as a result of ingestion of coconut oil. Coconut oil does not contain TFA that have deleterious effects on blood lipids and insulin binding (Enig, 1998).

6.3.3   COCONUT OIL CHARACTERISTICS

Coconut meat contains 10 kinds of essential amino acids that can be categorized as a food ingredient with a high-quality protein. High-quality protein is a protein that can provide essential amino acids in a ratio equaling human needs. Generally, high-quality protein is derived from animal materials such as meat, eggs, and milk. For a population with relatively low levels of income, of course, very difficult to meet the needs of amino acids by eating animal protein because the price is not affordable. Therefore, consuming food made from coconut is one of the solutions to meet the needs of essential amino acids.

The use of coconut oil as edible oil needs to glance back to the community, especially the coconut farmers. Today most of the people tend to eat a meal that is processed oil from other vegetable oils, because of cooking oil on the market is palm oil, soybean oil, corn oil, and sunflower oil. Increasing awareness of consumers to consume coconut oil will encourage the development of fresh coconut oil processing. This condition is an opportunity for farmers or farmer groups in these areas to re-process the coconut oil from coconut meal. Furthermore, consumption of food products made from coconut meat, such as cooking oil, dried grated coconut, coconut milk, VCO, coconut candy, coconut butter, coconut ice, and coconut tart unwittingly have made use of medium chain fatty acids is high enough which is very useful for boosting the immune system against viral infections, bacteria, and protozoa, as well as a number of essential amino acids needed by the body. Consuming VCO as much as 3.5 tablespoons/day is equivalent to eating 7 ounces of fresh coconut meat or dried grated coconut bowl 2.5 or 10 ounces of coconut milk. Coconut oil is important for the metabolism of the body because it contains vitamins which are fat soluble, namely vitamins A, D, E, and K and pro-vitamin A (carotene). In addition, coconut oil contains a number of saturated fatty acids and unsaturated fatty acids. Composition of fatty acids were analyzed from varieties of copra is 36.12 to 38.28% lauric acid, 13.42–15.90% myristic acid, 8.78 to 11.10% caprylic acid, 6.38 to 8.08% capric acid, 6.48–7.95% palmitic acid, 4.27 to 5.26% oleic acid, 1.76 to 2.54% stearic acid and 1.44 to 1.66% linoleic acid. VCO analysis results obtained from four varieties average fatty acid chain is 56–57% with 43% lauric acid levels. Other MCFA that have health benefits are capric acid, oleic acid (omega–9) and linoleic acid (omega–6).

6.3.4 BENEFITS OF COCONUT OIL

A study was conducted on 15 patients with HIV in San Lazaro Hospital, Manila. Patients are given purely in the form of capsules monolaurin and coconut oil and are divided into three groups. Patients in the first group were given a high dose of monolaurin, in the form of capsules containing 7.2 g of monolaurin three times a day, or 21.6 g/day. Patients in the second group were given a low dose monolaurin, which were given capsules containing 2.4 g of monolaurin three times a day or 7.2 g/day. Patients in the third group were given a coconut oil 15 ml three times daily or 45 ml/day. After 6 months of treatment, 9 of the 15 patients experienced improvement against HIV, which is characterized by decreasing the amount of HIV virus. Of the 9 patients, 2 capsules containing 7.2 g monolaurin, 4 people eating capsules containing 2.4 g monolaurin, and 3 people consuming coconut oil. A study of people living in one of the islands in the Pacific are eating comes from coconut oil showed total cholesterol and good cholesterol (high-density lipoprotein, HDL cholesterol) increased and bad cholesterol (low-density lipoprotein, LDL cholesterol) decreases. On the other groups who migrated to New Zealand and rarely consume coconut oil, was total cholesterol and increases good cholesterol while decreasing bad cholesterol. A research conducted in India showed that cardiovascular attacks in Nicobar Island are very low, because of the people are living on the island eating coconuts. Similarly, the population on the island who eat coconut meat and coconut oil as edible oil, it is very low heart disease cases. It is well known that the use of coconut oil as a baby food formula can help increase calcium absorption.

6.4 COCONUT OIL PROCESSING METHOD

Coconut oil is the most valuable part of coconuts. Oil content on old coconut meat is 34.7%. Palm oil is used as an industrial raw material or as cooking oil. Coconut oil can be extracted from fresh coconut meat, or extracted from dried coconut (copra). Limited to small industrial capital capability, it is advisable to extract oil from fresh coconut meat. This method is easy to do and not a lot of costs. The disadvantage is the lower yield obtained. Producing coconut oil begins with the manufacture of coconut milk which is a liquid extracted from the grated coconut with water. When milk silenced, it will happen slowly separating the rich with oil (called cream) with a poor section with oil (called skim). Since it is lighter than the skim so that the cream is at the top and the skim is below.

Coconut oil can be prepared in various ways; one way is called 'wet' relatively simple way. Extraction is a way to get oil from oil-containing materials. There are three oil processing process generally carried out, namely: method of extraction, which is a wet process, dry process pressure, and process by using a solvent. Wet process is characterized by the addition of water, while a dry process without adding water. Coconut oil extraction process generally takes two forms of energy, namely mechanical energy and thermal energy. Mechanical energy is used to break the cell wall, while thermal energy in addition to damaging the cell wall separately also to reduce the viscosity of the oil and adjust the water content. The third method specifically using heat, which aims to an agglomerate protein on the cell walls and to break the cell walls so easily penetrated by the oil contained therein. There are several ways to extract coconut oil, which is through the process of physical, chemical, and fermentation. The traditional process through physical means heating produces low quality because of high water content and causes rancidity. While the extraction of oil by chemical means will cause a decrease in the quality of some essential nutrients, such as lauric acid and tocopherol and causes high peroxide. VCO which is processed by fermentation, many of them have the advantage of having shelf life for longer, not rancid quickly, contain antioxidants (vitamin E, tocopherol) and lauric acid content in high amounts, equivalent to the levels of lauric acid in breast milk (43–53%) (Sulistyo, 2012). Different types of coconut oil for edible purposes are available, viz, virgin coconut oil from wet coconuts (unrefined grade); coconut oil from dry coconuts (unrefined grade); and coconut oil by solvent extraction method (refined from coconut expeller cake). Virgin coconut oil is claimed to have more health benefits compared to coconut oil extracted from copra.

6.4.1   DRY METHOD

A method of making palm oil is the dry method; first coconut meat is made in the form of copra. To be made in the form of copra, the coconut meat is made into dry by sun drying or dried through a drying oven. Drying coconut meat with drying is highly dependent on weather conditions, so the drying will be better when it is in the summer. And if the drying is done in the rainy season, the drying process may take longer. A long time in the drying process will greatly interfere with the quality of copra produced due to the biological process. For the drying process by using the oven will be faster compared to by drying in the sun. Drying using oven will take greater operational costs. The steps of making coconut oil by the dry method are as follows: (a) Copra minced, then mashed into a coarse powder; (b) Copra powder is heated, and then pressed so to remove the oil and the resulting pulp still contains oil - pulp finely ground, then heated and pressed to remove the oil, and, (c) The collected oil is precipitated and filtered. Oil-treated following screening results: (1) The addition of an alkaline compound (KOH or NaOH) to neutralize (eliminate free fatty acids); (2) The addition of absorbent materials (absorbent) color, usually using activated charcoal in order to produce the oil is clear and transparent, (3) Jetting steam into the oil to evaporate and eliminate compounds that cause undesirable odor. The oil that has clean, clear, and odorless, packed in tin boxes, plastic bottles or glass bottles. Physicochemical properties and fatty acid compositions were evaluated. Sensory studies are also carried out and found that virgin coconut oil is superior to the commercial sample. Scale-up runs were carried out for the production of virgin coconut oil.

6.4.2 WET METHOD

6.4.2.1 TRADITIONAL WET METHOD

The wet coconuts are subjected to pressing to ooze the oil out along with coconut milk. This is processed afterward without employing heat, shear, chemicals, refining, and is known as virgin coconut oil. Shredded pieces of meat then add water and squeezed, so that secrete milk. After the separation of oil from coconut milk by heating, coconut milk is heated so that the water is evaporated and the solids stay to clot. Oil is separated from protein waste material by filtration. The protein waste material mass still contains a lot of oil so that the oil can still be taken by way of squeezed.

6.4.2.2   CENTRIFUGAL WET METHOD

Separation of oil from coconut milk can be done by centrifugation. The coconut milk treated centrifugation at a speed of 3000–3500 rpm resulting in the separation of the cream from the skim. Furthermore, the cream is acidified, then treated again centrifuged to separate the oil from the oil instead. Oil separation can also be done with a combination of heating and centrifugation. Coconut milk is centrifuged to separate the cream. After, the cream is heated to agglomerate solids instead of oil. The oil is separated from the non-oil by centrifugation. The oil obtained was filtered to obtain clean oil.

6.4.2.3   WET FERMENTATION METHOD

Coconut oil is extracted from coconut paste by a new enzymatic process, and the method used less energy than the conventional processes. How to wet fermentation is somewhat different from the traditional wet method. In the wet method of fermentation, allowed to stand for separating skim milk from the cream. Furthermore, the cream is fermented to facilitate clotting parts instead of oil (especially proteins) of heating oil at a time. Microbes that develop during fermentation especially are acid-producing microbes. The resulting acid causes the milk protein clots and easily separated upon heating. Prepare tubes of mineral water, then grate some coconuts and squeezed into coconut milk. After the milk is diluted with water and allowed to stand for 8 hours until there is a separation between the "milk primed" with water. Prime coconut milk (condensed), the acidity is lowered to acidic conditions (pH 4.2) using vinegar. This mixture is then left 12–24 hours. In this process, milk is separated into three parts primed. The bottom layer is water, then solid blob and on top of virgin oil. Filtered oil with a tissue or gauze smooth, through taps that had been prepared beforehand. The collected oil, then heated for 15 minutes plus antioxidants. Oil produced can be packaged and consumed. This process can be continued to the next oil making process. The coconut milk residue (solid blob) which is not only rich in protein and fat, it contains as well active probiotic microorganism for manufacturing of cooking oil. Oil mass liquid sprayed onto the entire surface of the gallon and allowed to dry. Then enter the liquid milk and left for 12 hours in a warm room temperature. After that period, the oil will separate themselves. The next process, as the first is heated to heat the nail for 15 minutes, and the resulting oil ready to be packed.

6.4.2.4   WET METHOD "LAVA PROCESS AND KRAUSS MAFFEI PROCESS"

Lava wet method process somewhat similar to the way wet fermentation. On this way, the milk is treated so that there is a separation skim centrifugation of cream. Furthermore, the cream is acidified by adding acetic acid, citric, or HCl to pH 4. After the milk is heated and treated as a traditional wet method or wet method of fermentation. Skim milk is processed into protein concentrates in the form of granules or powder. In the wet method, the milk is treated centrifugation, resulting in the separation of skim from the cream. Furthermore, the cream is heated to agglomerate the solids. After it was treated centrifugation so that the oil can be separated from clumps of solids. The solid is separated from the oil proceeds centrifugation and pressed to remove residual oil. Furthermore, the oil is filtered to remove impurities and solids. Skim milk is processed into coconut flour and coconut honey. After fermentation, the cream is processed like a traditional wet processing method.

6.4.2.5   ACIDIFICATION METHOD

Destruction of proteins or protein denaturation to be able to get coconut oil can be done by acidification. In principle, this acidification technique is a method of protein denaturation due to zwitterion formation in iso-electronic condition. Zwiter ion formed because the molecules have the opposite charge in the respective ends. In the protein itself actually contains more NH_2 groups had a slight positive charge and negatively charged carboxylic groups. In order to achieve this iso-electronic condition, then the coconut milk made in acidic conditions. Usually, the pH adjustment to obtain iso-electrically condition that at pH 4.5 is done with the addition of acetic acid or vinegar. By way of this acidification is formed of three layers as well, which is the top layer of oil, then the middle layer and bottom layer protein is water. The oil obtained from acidizing color will be crystal clear.

6.4.2.6   ENZYMATIC TECHNIQUE

An enzymatic technique is a method to denature the protein with the aid of enzymes. Several types of enzymes that can be used in this process are papain, bromelain, galacturonase, amylase, protease or pectinase. Stages of making coconut oil by means of the enzymatic process are the manufacture of coconut milk produced by the addition of coconut water. The purpose of the use of coconut water is to accelerate the clotting process. Coconut milk further coupled with an enzyme to be used for the fermentation process by way of settling for one night. The next day separation between oil phase with a protein or biomass phase will be performed (Figure 6.1). In the processing of VCO through fermentation or enzymatic technique may produce as well coconut milk residue as by product of incubation process. It is considered useful to be reprocessed into a good and healthy cooking oil. Thus it is expected to increase a value added in the processing of VCO (Table 6.1). The main disadvantages of this process are low oil recovery and fermented odor, which masks the characteristic coconut flavor of the oil.

FIGURE 6.1   (See color insert.) Cooking coconut oil obtained by heating of coconut milk residue.

TABLE 6.1 Quality of Coconut Cooking Oil Derived from Byproduct of VCO

Fat and Vitamin	Fatty Acids	Content (%)
Saturated fatty acids	Caproic	0.59
	Caprilic	4.55
	Capric	4.50
	Lauric	37.29
	Miristic	20.13
	Palmitic	21.32
	Stearic	11.62
Unsaturated fatty acids	Oleic (ω–3)	0.00
	Linoleic (ω–6)	0.00
	Linolenic (ω–9)	0.00
Vitamin	Vitamin E	113 ppm

6.4.2.7 INDUCING METHOD

How to fishing in the manufacture of coconut oil is coconut milk emulsion splitting system by regulating the surface tension increase. To be able to lure the oil out of the emulsion system used bait in the form of oil as well. The use of bait will greatly affect the outcome of the quality of the oil. If the bait used is oil in good quality, you will get good quality oil anyway, but on the contrary, if the oil is used as bait in quality is not good, the result was also obtained oil quality is not good.

6.4.2.8 COOLING TECHNIQUE

Cooling method is based on the difference between the freezing point of water and the freezing point of the oil. The freezing point of the oil is in the range of 15°C while the water has a freezing point at 0°C. Therefore the use of this cooling technique will freeze first oil than water. With other, rich oil will clot early and then be separated by a water component.

6.4.2.9 MECHANICAL TECHNIQUE

A mechanical technique is done with the intention of damaging proteins and water that surrounds the drops of oil. The trick is to incorporate milk into the mixer or stirring occurs. With the continuous stirring of water molecules and proteins can be damaged, molecules that eventually drops of oil can come out.

6.4.2.10   MICROWAVE TECHNIQUE

The use of microwaves in the manufacture of coconut oil intended to damage the structure of proteins because of the combination of the orientation of polar molecules (protein and water) thermal emulsion constituent. Because of the damage, the oil component will be out of the emulsion system.

6.4.2.11   SOLVENT EXTRACTION METHOD

This method uses a solvent which can dissolve the oil. The solvent used low boiling, volatile, does not interact chemically with the oil and non-toxic residue. Although this method is quite simple, but it is rarely used because the cost is relatively expensive. To make the oil by solvent extraction, coconut meat is also prepared as copra. The principle of this method is to use a solvent that can dissolve the oil. The characteristics of the solvent used for the extraction of coconut oil should be volatile and not chemically interact with the oil and non-toxic residue.

6.5   THE PLAIN TRUTH ABOUT COCONUT OIL

In 2003, the Jakarta Post published an article entitled "the simple truth about cholesterol" written by Melissa Southern-Garcia. While the article tries to enlighten readers about cholesterol, it is sad to note that the author gave inaccurate and misleading information about coconut oil. The author implied that since coconut oil belongs to saturated fat, it has a negative effect on our health by increasing our blood level of harmful LDL cholesterol. This article is meant to give a scientifically proven fact about coconut oil that may be missed out by common people and professionals. A study done in two groups of community living in New Zealand who consume a large number of coconut oil has proved that they have rare incidents of hypercholesterolemia and heart attack. Two groups of Polynesians from the Cook Islands derive 35% and 27% of their calories from coconut oil, but their mean cholesterol values are low, i.e., 153 mg% and 195 mg% bodyweight, respectively. Prevalence of heart attacks also is low in these groups compared to the usual New Zealand population (http://www.thejakartapost.com).

About 70% Sri Lankans are consuming coconut oil for over 1000 years, but the epidemic of hypercholesterolemia and heart disease is of recent origin. Before 1950, heart attacks were not common in Sri Lanka. The hospital admission rate for heart attacks was 57.3 in 1970 to 182 in 1992. On the other hand, the Central Bank of Sri Lanka figures out that the coconut consumption has gone down from 132 nuts per person per year in 1952 to 90 per person per year in 1991. It indicates that the increase in heart attacks incidents in Sri Lanka is not due to the increased consumption of coconut (http://www.apccsec.org/truth.html). In the study, 10 medical students tested diets consisting of different levels of animal fat and coconut oil. When the ratio of animal fat and coconut oil at a ratio of 1:1, 1:2, 1:3 no significant change in cholesterol but when animal fat level increased total calories reached 40% and blood cholesterol increased. This study indicated that not only did coconut have no effect on cholesterol levels, it even reduced the cholesterol elevating effect of animal fat. Enig (1996) demonstrated that coconut oil was not a "bad oil" when they compared essential fatty acid-rich safflower oil (SFO) with an equal mixture of safflower oil and coconut oil on 10 hyper-cholesterolemic males, 8 of whom were survivors of myocardial infarction. They showed that both SFO and SFO-coconut oil (SFO-CNO) caused marked decrease in the serum cholesterol and that the effect was obtained regardless of whether it was fed before or after the safflower oil (Kaunitz and Dayrit, 1992; Enig, 1996).

There were experiments which concluded that coconut oil caused hyper-cholesterolemics, but these experiments turned out to be unacceptable due to some reasons. First, these experiments used hydrogenated coconut oil in which the coconut oil became more saturated and its essential fatty acid, linoleic fatty acid, got destroyed. Second, most of the research work has been done using animals, and the number of animals used is very small. In some experiments, only four animals were used. Third, the rabbit model used for most of the research work cannot be compared to man. It has been found that when corn oil is administered to a man, it will make his serum cholesterol level come down, while in the rabbit model, corn oil increases serum cholesterol level (Rethinam and Muhartoyo, 2003).

The fact that coconut oil belongs to saturated oil cannot be automatically justified to be the cause of increasing LDL cholesterol as coconut oil has its own unique properties. Moreover, people may not know what saturated oil means. Chemically, oil is made up of chains of carbon, hydrogen, and oxygen called fatty acid. All fatty acids consist of a chain of carbon atoms with varying amounts of hydrogen atoms attached to them. A molecule that has two hydrogen atoms attached to each carbon is said to be "saturated" with hydrogen because it is holding all the hydrogen atoms it possibly can. A fatty acid that is missing a pair of hydrogen atoms on one of its carbons is called monounsaturated fat. If more than two hydrogen atoms are missing, it is called polyunsaturated fat (Fife, 2004). It must be noted that there are different groups of fatty acids contained in major oils and fat. Generally, they are grouped into MCFA and LCFA. The two fatty acids have different behavior and health effect on a human being. Those who equate coconut oil with other saturated fats are not conscious of the existence of subgroups within the broad category of saturated fatty acids. The medium fatty acids have a lower melting point, smaller molecular size and greater solubility in water and biological fluids compared with those of the long chain fatty acids (Thampan, 1998).

Coconut oils are grouped into MCFA as 57% its fatty acids consisting of C8 (capric acid) and C12 (lauric acid). A number of noted scientists have revealed the superiority of MCFA. Coconut oil has approximately 50% lauric acid. Lauric acid has the additional beneficial function of being formed into monolaurin in the human or animal body. Monolaurin is the antiviral, antibacterial, and antiprotozoal monoglyceride used by the human or animal to destroy lipid-coated viruses such as HIV, herpes, CMV, influenza, various pathogenic bacteria, including *Listeria monocytogenes* and *Helicobacter pylori*, and protozoa such as *Giardia lamblia*. Some studies have also shown antimicrobial effects of the free lauric acid (Enig, 1999; http://www.thejakartapost.com).

Coconut oil has also approximately 6–7% capric acid. Capric acid has a similar beneficial function when it is formed into monocaprin in the human or animal body. Monocaprin has also been shown to have antiviral effects against HIV and is being tested for antiviral effects against herpes simplex and antibacterial effects against chlamydia and other sexually transmitted bacteria (Enig, 1999).

Garcia who stated in her article that vegetable oils did not have cholesterol is not accurate. The latest research finding concludes that cholesterol can also be found in vegetable oils. John et al. (2002) indicate that vegetable oils contain cholesterol although in small amounts. It is further stated that coconut oil has the lowest cholesterol amounts (5–24 parts per million) compared to palm kernel oil, sunflower oil, palm oil, soy oil, cottonseed oil, rapeseed oil, and corn oil. The leading coconut producer nations are (in order) the Philippines, Indonesia, India, Sri Lanka, Thailand, and Malaysia. While production from nations such as Fiji, Papua New Guinea, Solomon Islands, Samoa, Tonga, and Vanuatu is relatively small, the coconut trade is a major source of export revenue for these countries. In some years, copra comprises more than 50% of Vanuatu's export income (http://aciar.gov.au).

The coconut industry is the highest net foreign exchange earner of agricultural exports in the Philippines, accounting for about 1.5% of GNP. It employs, directly or indirectly, some 20 million people and earns more than US$510 million annually. However, the industry's ability to meet demand and expand may be jeopardized by declining share of coconut oil in the world's oils and fats market, a proposed levy on vegetable oil imports to the European Community (EC) and a campaign against coconut oil in the US, more stringent aflatoxin regulations imposed in the international copra market, erosion of the European desiccated coconut market, lack of a market development and expansion program, and low incomes for coconut farmers. Aware that failure to respond to changing patterns in world trade in coconut products could have adverse effects on employment and revenue, the Coconut Authority and Department of Agriculture requested ACIAR assistance. As a result, this project was set up: to define the factors that drive international coconut product markets, assess the factors that determine demand in the major consumer nations of the world, and identify threats to the industry's viability posed by US and EC trade restrictions, competition from other oils, and mycotoxin contamination (Retinam, 2004; 2005; Retinam and Amrizal, 2005).

6.6   THE PROSPECT OF VIRGIN COCONUT OIL

VCO is obtained from the fresh and mature kernel of the coconut (*Cocos nucifera* L.) by mechanical or natural means with or without the application of heat, which does not lead to alteration of the nature of the oil. VCO has not undergone chemical refining, bleaching or deodorizing. It can be consumed in its natural state without the need for further processing. Virgin coconut oil consists mainly of medium chain triglycerides, which are resistant to peroxidation. The fatty acids in VCO are distinct from animal fats which contain mainly of long-chain saturated fatty acids. Virgin coconut oil is colorless, free of sediment with a natural fresh coconut scent. It is free from rancid odor or taste.

The market is then developed in the field of pure coconut oil or coconut oil fresh or VCO has created opportunities for the export of oil-producing countries. Excess VCO for production on a small scale in order to maintain its quality, opportunities for coconut farmers to gain more profits and increase revenue from coconut crops. Incessant publicity in various media and promotion of many parties in Indonesia about VCO, has created a great entrepreneurial spirit, so that the business of making VCO popping up everywhere. And because the VCO can be made in several ways, it appears a variety of process technologies. Each technology this process suggests that the best technology. This situation would require the attention of the parties concerned because of the quality and standardization into things that determine the quality of business continuity.

Buyers and users VCO come from a variety of industries, which can be divided into two categories; buyers using VCO as drugs, vitamins, food oils and ointments; and buyers with the aim to be used as raw material food, skin, and hair care industry, artificial beauty industry or the pharmaceutical industry. As the first product in the category of users, has led to the VCO potential to be great, so it requires a serious effort from all sides to prove that the VCO has the potential to be commercialized patents, which raised the initiative being proven scientifically, clinically, and commercially (Rindengan and Karaow, 2003).

While the market for a second user category (as raw material) requires chemical-free raw materials. With soaring petroleum prices as a source of petrochemical raw materials, the competitiveness of VCO becomes stronger and can be an option to use, however, the issue probably faced is offering VCO as the raw material of choice. In addition, the problem also lies in the availability, marketing considering the VCO to the needs of industrial raw materials more demanding, especially in industrialized countries. This means trade VCO must adhere to the rules of international trade with all the obstacles. Examples of marketing success VCO international scale and boost incomes of coconut farmers, is in Southern Luzon, Philippines, and Fiji Islands, South Pacific (http://www.unctad.info/en/Infocomm/AACP-Products). According to The Coconut Industry Market

Intelligence Report (Singh et al., 2007) which was commissioned as part of the CARICOM Regional Transformation Programme for Agriculture, the high price of petroleum-based fuel has brought about a renewed interest in the use of coconut oil for conversion into bio-fuel with the main focus being on the on the commercial products from the regional Industry. Coconut oil prices increased to USD 674/tonne in 2004 from a low of USD 324/tonne in 2001. The Caribbean produced an average of 512,000 tonnes of coconuts annually over the period 1990/2002 with CARICOM accounting for 301,000 tonnes or 59%. Jamaica was the major producer with 170,000 tonnes (56%) followed by Guyana with 59,000 tonnes (20.6%) and Trinidad and Tobago with 23,000 tonnes (8%). Despite this, CARICOM remains a net importer of oil for table consumption and industrial use. There is potential for supplies to both the regional and export markets in the USA, Canada, and European Union markets which are major destinations for coconut oil and coconut products.

The industry is however faced with some constraints which have resulted in the loss of market share. Principal among these is the suspected adverse health and nutrition effects on humans, but studies, such as that conducted by Spade and Dietchy (1988) and Singh et al. (2007) have shown that coconut oil prevents the formation of hepatic cholesterol esters. In addition to this, the lauric acid found in coconut oil provides the disease-fighting fatty acid monolaurin which boosts the immune system. Although this controversy was not sustained, supplies of coconut oil in the global market remained at low levels due to the effect of pests and diseases on production (http://www.agricarib.org/primary-dropdown/coconut).

VCO is simply coconut oil that has not gone through the intermediary phase of copra but instead is extracted directly from fresh coconut meat. It has appealed so far to health a conscious niche market which unfortunately does not result in a substantial premium at the production end. In turn, this tends to encourage compromise solutions to the extraction of virgin oil. The quality of virgin oil depends almost entirely on the extraction technology used. There is great and increasing ambiguity as to what is VCO. All qualities are being produced and marketed as virgin oil, but sometimes they are simply a good quality version of copra oil. Unless there is greater care, we will simply erode the market for virgin oil and yet, there is no one to supervise what the oil is called. Consumers have to become more careful.

There appear to be two broad categories of approach depending on whether heat is applied or is not. Heat may be applied before pressure as in the DME process, after separation to speed up separation from cream or boiling the meat to skim off the oil. The cold alternative is to allow settling and/or use of centrifuges. The differences in quality may be marginal in chemical composition and have not been measured to any great degree as yet, and differences in flavor and fragrance and cell structures will have to be studied further. Virgin oil is cleaner than copra based oil and has a more natural aroma and flavor. Virgin oil, of a crude quality, has always been produced on a cottage or village basis in coconut producing countries. There is a more advanced way of doing it that is widely being used now known as immersion drying. Since frying is involved at high temperatures, there is a consequent change in the quality of the resulting oil.

VCO has not become a major commodity but rather remains a niche health food product. The reasons for this lack of success are many, probably a weakness in marketing and market information is the most important but also the lack of differentiation between qualities from the different techniques is important. In any case, the only driver of the process is the niche market for healthier foods which has motivated small scaled relatively crude production of a relatively crude product. Those who claim to value virgin oil and help arrange small-scale production on isolated islands are not doing a favor to the poor farmers since this small scaled production at low values discourages a significant breakthrough. Much more can be done to develop production and consumption of VCO, but there is nothing underway that offers to do that. Relying on a verbal association with olive oil and a trend that favors natural products is not enough. After all, traditional copra oil is equally organic compared to the VCO being offered. Those are promoting sales in health food outlets make some marginal headway by using the argument. There has been no consideration of using the advantages of characteristics of VCO beyond edible oils, and there are applications that could be developed for a high-quality product (http://www.ruraldevelopment.info). VCO is a pure coconut oil, or coconut oil is obtained through a process that does not involve high heat and without the addition of chemicals. VCO has been widely known, has many good benefits for health, fitness, beauty, and so on. Making VCO that do not involve high heating makes the existing content in the VCO still intact, and nothing is lost during processing. Therefore, the VCO quality benefits because it contains many constituent compounds are still intact.

The emergence of the VCO is rich in benefits to encourage the emergence of many manufacturers that produce and sell various brands. Their processing methods are diverse; therefore the quality of the oil produced is also different. As a consumer, we should be careful in selecting the VCO to be used. Here are tips on choosing a qualified VCO so that it can provide optimal efficacy according to the quality standards of the Asian Pacific Coconut Community (APCC) (Table 6.2).

TABLE 6.2 Coconut Oil Specification According to Asia Pacific Coconut Community (APCC)

Essential Composition and quality	APCC Standard	Obtaining VCO
Identity Characteristics		
Relative density	0.915–0.920	0.92
Reactive Index at 40°C	1.4480–1.4492	1.45
Moister % wt. Max	0.1–0.5	0.12
Insoluble impurities % by mass max	0.005	0.04
Saponification Value	250–260 min	252.00
Iodine Value	4.1–11.0	4.90
Unsaponifiable matter % by mass max	0.2–0.5	0.32
Specific gravity at 30°C	0.915–0.920	0.92
Acid Value max	0.5	0.42
Polenske Value min.	13	14.00
GLC Ranges of FA Composition (%)		
C 6: 0 (Caproic)	0.4–0.6	0.51
C 8: 0 (Caprylic)	5.0–10.0	6.01
C 10: 0 (Capric)	4.5–8.0	7.5
C 12: 0 (Lauric)	43.0–53.0	48.02
C 14: 0 (Myristic)	16.0–21.0	17.02
C 16: 0 (Palmitic)	7.5–10.0	7.21
C 18: 0 (Stearic)	2.0–4.0	3.11
C 18: 1 (Oleic)	5.0–10.0	5.41
C 18: 2 (Linoleic)	1.0–2.5	1.3
C 18: 3-C 24: 1 (Linolenic)	> 0.5	0.00
Quality Characteristics		
Color	Water Clean	Clean
Free Fatty Acid	≤ 0.5 %	0.31
Peroxide Value	°3 meg/Kg oil	2.00
Total Plate Count	< 10 cfu	3.00
Odor and Taste	Free	Free
Contaminants		
Matter volatile at 105°C	0.20%	0.17
Iron (Fe)	5 mg/Kg	3.00
Copper	0.4 mg/Kg	0.19
Lead	0.1 mg/Kg	0.05
Arsenic	0.1 mg/Kg	0.00

6.6.1 SOURCE OF GOOD FAT

Coconut oil is included into the category of saturated fat (no double bonds). Fall within the category of saturated fat that fatty acids with MCFA (consisting of 8–12 carbons), and LCFA (consisting of 14 or more carbon). Based on this classification, the VCO is a group of oil containing approximately 64% MCFA consists of about 50% lauric acid (C12), 6–7% caproic acid (C10), and 8% caprylic acid (C8). The content of VCO is generally composed of saturated fatty acids (92% of total content), 6% MUFA and 2% PUFA. With the very least because of PUFA and saturated, the VCO is very stable and oxidation resistant making it difficult to become rancid. Because already saturated, hydrogenated VCO does not need anymore. Hence the VCO is absolutely not contained trans fatty acids, a type of fat is harmful to the body resulting from hydrogenation. VCO is difficult to oxidize, and it will not release free radicals that harm the body. In contrast, PUFA oils are very easy to produce free radicals that damage the body's cells. VCO contains large amounts of natural vitamins that act as antioxidants (particularly vitamin E). This is the antioxidant that helps prevent the VCO becomes rancid. This VCO is outstanding that the country's anti-palm oil began quietly producing VCO, as they continue to campaign on the issues so bad about coconut oil consumers shunned, thus that the soybean oil and corn can keep control of the world market. A statement that vegetable oils do not have cholesterol is not accurate. The latest research finding concludes that cholesterol can also be found in vegetable oils. INFORM (2002) published by American Oil Chemistry Society indicates that vegetable oils contain cholesterol although in small amounts. It is further stated that coconut oil has the lowest cholesterol amounts (5–24 ppm) compared to palm kernel (9–40 ppm), sunflower (8–44 ppm), palm (13–19 ppm), soy (20–35 ppm), cottonseed (28–108 ppm), rapeseed (25–80 ppm), corn (18–95 ppm), beef tallow (800–1400 ppm), butter (2200–4100 ppm) and lard (3000–4000 ppm). Hence, the coconut oil is edible oil that containing the lowest cholesterol (Retinam and Muhartoyo, 2003).

6.6.2 DIFFERENCES BETWEEN SATURATED FATTY ACIDS AND TRANS FATTY ACIDS

SAFA will increase HDL cholesterol, the good cholesterol that protects the heart and reduces the prevalence of hypertension, whereas TFA lowers

HDL cholesterol. Hence, it is clear that the VCO to lower LDL and raise HDL (Mensink and Katan, 1990; Mesink et al., 2003; Judd et al., 1994). The results also showed that the SAFA lower level component causes hardening of the arteries, which lipoprotein (a) or Lp(a), hence the prevalence of stroke and myocardial infarction decreased significantly, whereas SAFA actually increase Lp(a) (Khosla and Hayes, 1996; Hornstra et al., 1991; Clevidence et al., 1997). SAFA keep extra omega–3 acids, whereas TFA is precisely causing the loss of omega–3. From these results, it appears that the VCO role in the mechanism of the increase in HDL cholesterol in the body tissues (Sugano and Ikeda, 1996; Gerster, 1998). SAFA-VCO does not increase insulin binding, whereas TFA actually causes insulin binding. SAFA role in the cells of the body's energy supply VCO is easily absorbed because of the cell without the help of enzymes and insulin. The research proved that VCO increases insulin secretion and glucose consumption, causing a decrease in blood sugar levels were significantly (Garfinkel, 1992).

SAFA of VCO has the capability of lowering levels of total cholesterol, triglycerides, phospholipids, and LDL and increase HDL in the blood serum and tissue, reducing the formation of carbonyl and oxidative stress than other oils (Nevin and Rajmohan, 2004). Provision of VCO is almost similar to the effect of statins in patients with hypercholesterolemia with the aim of lowering cholesterol levels. The difference is, with the cholesterol-lowering statin drug administration led to a decrease in testosterone levels due to a deficiency of cholesterol in the cell so that it will cause a decrease in libido. In contrast to the decrease in cholesterol, VCO will increase the libido (de Graff et al., 2004). Diets enriched SAFA able to maintain the integrity of liver cells resulting from the use of alcohol and oxidative stress caused by linoleic acid, with a lower fat peroxidase, so that does not happen too high ROS that damage and causes endothelial dysfunction. Clear that the VCO can reduce oxidative stress and reduce the occurrence of endothelial dysfunction that underlies the onset of diabetes complications (Nanji et al., 1995). In another study, it was shown that monolaurin as a result of the metabolism of lauric acid contained in breast milk could effectively kill the virus by lysis cell walls are composed of double-layered phospholipids. This wall will dissolve in fats derived from the VCO, while the effect of monolaurin alone will disrupt signal transduction in the virus. Another effect of lauric acid is a disturbing development and maturation of the virus. MCFA-VCO ability in disrupting the membrane of the fat is also shown by the MCFA virus present in breast milk.

6.6.3   ABILITY TO KILL BACTERIA AND VIRUSES

The ability of VCO in killing bacteria and viruses is based on the levels of lauric acid and capric acid, which is transformed into monolaurin in the body monocaprin. MCFA and its derivatives have the effect of killing a wide range of microbes, including bacteria, fungi, yeasts, and viruses. The ability to kill the virus mainly derived from lauric acid, caprylic acid (C–8), capric acid (C–10) and myristic acid (C–14). The ability to kill the virus began popularized through Hierholzer and Kabara (1982) in collaboration with the Center for Disease Control of US Public Health Service. Viruses can be inactivated by fatty acid VCO is the HIV virus, Measles virus, Herpes Simplex Virus–1, Vesicular stomatitis virus, Visna virus, and CMV. Many of these pathogens are opportunistic infections in HIV patients. Therefore, many physicians provide extra nutrition in the form of a VCO to aid durability of HIV patients (Enig, 1998; Macallan et al., 1993). Even hydrogel (a type of jelly ointment) contains monokaprin able to inactivate through *in vitro* experiment sexually transmitted disease Herpes Simplex Virius–2 (HSV–2), HIV, and *Neisseria gonorrhoeae* (Isaacs et al., 1990).

Monolaurin, a monoester formed from lauric acid (medium chain fatty acids), has profound antiviral and antibacterial activity. Generally, mono-laurin is more effective against gram-positive bacteria such as *Staphylococcus* and *Streptococcus*. In a study conducted by Preuss et al. (2005) monolaurin was shown to be bactericidal to *S. aureus* and *Mycobacterium terrae* but not *Escherichia coli* and *Klebsiella pneumoniae*, which are both gram-negative, confirming prior work on monolaurin, and was shown to be static to a variant of the virulent anthrax pathogen, *Bacillus anthracis* (Preuss, et al., 2005). Monolaurin also inhibits production of staphylococcal toxic shock toxin–1 effectively (Projan et al., 1994). The monoester is effective against CMV and the expression of virulence factors including protein A, alpha-hemolysin, B-lactamase, and the induction of vanco-mycin resistance in *Enterococcus faecalis* (Ruzin, 2000; Preuss, 2005). Monolaurin has also been shown to inactivate *Listeria monocytogenes, Streptococcus agalactiae,* and Groups A, F, and G *streptococci* (Boddie and

Nickerson, 1992; Issacs et al., 1992; Ruzin, 2000). It says that monolaurin 5000 times more potent than ethanol in inhibiting bacterial growth of *L. monocytogenes* and *S. aureus* (Oh and Marshall, 1993). A very interesting is the ability of monolaurin in killing the bacteria *Helicobacter pylori* in gastric heartburn sufferers. Bacteria will be inactivated by the MCFA, and almost no resistance arises against the fatty acid (Petschow, 1996).

Various bacteria, including *H. pylori*, *C. pneumoniae*, gram (-) bacteria, are bacteria that contributed to a hardening of the arteries. These bacteria will provoke the onset of the inflammatory process that results in the oxidation of lipoproteins, together with the induction of cytokines and production of proteolytic enzymes that arise later, as a phenomenon typical of atherosclerosis. Giving VCO with an antibacterial effect can reduce the occurrence of atherosclerosis which is the basis of the occurrence of degenerative diseases such as hypertension, heart blockages, and other blood vessel disorders (Saikku, 1997). In addition to bacteria, *Candida albicans* and various kinds of yeasts, fungi, and protozoa (parasites), including *Giardia lamblia* worms are also inactivated by monolaurin (Bergsson et al., 2001).

A study showed that administration of the VCO enriched source of fat of thermogenic diet that was able to reduce the accumulation of white fat. Another mechanism underlying the decrease of obesity is that VCO as a natural source of MCFA, after being absorbed directly into the portal vein to the liver, and directly generates energy and is not deposited as fat. Instead, LCFA be esterified in the gut, entered through the lymphatic system of the intestine, and then into the bloodstream edges, and broken down into energy by the cells through the enzymatic breakdown of VLDL. The remaining portion will be deposited as fat. Thus the body will increase the rate of metabolism, thus preventing the accumulation of fat and even reduce weight (Portillo et al., 1998; Thampan, 1998).

Currently, MCFA of the VCO has been added to beverages athletes to produce energy quickly. MCFA energy is 6.8 kcal/g, while other fats will produce higher energy of about 8–9 kcal/g. The ability of the VCO in improving the health of the body originates from the same mechanism with the ability to improve the health of infant breast milk. Thus the children formula enriched with VCO can protect babies from infection and boost the immune system, especially for infants born prematurely and patients who suffer from digestive disorders (Berger et al., 2004).

Research with laboratory animals fed VCO showed remarkable results. Total cholesterol, triglycerides, LDL, and VLDL, decreased significantly, while HDL in serum and tissue were significantly increased as well. Besides, the VCO is also able to prevent the oxidation of LDL, thereby reducing the formation of carbonyl compounds that are harmful to the body. VCO is also able to reduce the inflammatory reaction that occurs due to *E. coli* endotoxin (Nevin and Rajmohan, 2004). This happens because of the VCO ability to inhibit leukotriene production and interleukin–1 (IL–1), which is thought to decrease through prostaglandins (mediators of inflammation) (Wan and Grimble, 1987). Another study on tumor necrosis factor alpha (TNF-α) and endotoxin-induced by inflammatory mediators prostaglandin E 2 (PGE–2) in the test animals were given VCO does not show an increase in PGE–2 or do not show inflammation. Most likely this effect is modulated through the reduction of the arachidonic acid content of phospholipids (Bibby and Grimble, 1990).

Other amazing effects of VCO are its ability in reducing the formation of toxic glutamate adverse reactions that exist in nerve tissue culture (Dave et al., 1997). Recent research suggests that monolaurin is able to induce the proliferation of T-lymphocytes (Witcher et al., 1996). Another effect particularly impressive is the ability of the VCO to treat some skin diseases. If used as a moisturizer, the VCO will be able to repair the skin disease in the form of xerosis (Agero and Velarro-Rowell, 2004). Similarly, if the VCO is used as hair oil was able to prevent hair damage as a hair fertilizer (Rele and Mohile, 2003).

The National Committee on Coconut Oil for Health Research of Philippines formed in 2004 and began to sharpen the focus of the research *in-vitro* anti-microbial capability of the VCO to (a) a common pathogenic bacteria present in the room baby in neonatal intensive care units; (b) the prevention of sepsis (infection that penetrated throughout the body via the bloodstream) in adult patients in the ICU; and (c) the ability to kill parasites. Even the United States and Canada have far-away days researching the VCO. The two countries have even had the VCO producers under Omega Nutrition and Carotec Inc. This plant has been producing VCO for half a decade ago. Without adequate research then it is likely that Malaysia will import the VCO from the US or neighboring countries.

Research on VCO in Indonesian Institute of Sciences began when researchers from Research Center for Biology-LIPI, Indonesia conducted a survey in order to study microbial biodiversity in palm plantations area in 1996 according to information obtained from local inhabitants living around the plantation. The information was about discovering the occurrence of coconut which had long been fallen from their tree, however, they did not undergo rotten. The coconut fruits had naturally fermented hence to form oil layer inside the fruits and fragrant. The natural coconut oil was believed having medicinal properties and hence furthermore used by inhabitants to cure some kinds of skin diseases suffered by many inhabitants. Through investigation on the samples of coconut oil found that it contained and fermented by several types of microbial strains including yeast, bacteria, lactic acid bacteria, and fungi. The cultures were then grown onto growth media and isolated to be applied as an enzymatic starter for processing fermented virgin coconut oil.

6.7 PRECLINICAL TEST OF FERMENTED VCO

Our VCO is fermented using a microbial starter of *Lactobacillus plantarum* culture that acts as probiotics, and the extraction process is conducted without heating or addition of synthetic chemical solvents. Hence it meets the quality of VCO according to specification by APCC. The effectiveness of VCO as antibacterial has been tested *in vitro* and *in vivo* at laboratorium (Figure 6.2). Parameters were observed and measured in this study were body weight, blood sugar, cholesterol, HDL, LDL, and triglyceride, pathology, and histopathological changes. Synergistic effect of organic acid produced by *L. plantarum* bacterial cultures was used as a starter for the enzymatic fermentation VCO with lauric acid contained in the resulting VCO; allegedly the two compounds have similar properties as an antimicrobial. The nature of the fatty acid that does not dissociate VCO, having ester groups, while the antimicrobial properties of the organic acids contained in the starter enzymatically dissociated due to the nature of which can lower the pH of the cytoplasm. Organic acids that act as the antimicrobial metabolites produced by LAB during logarithmic phase cultures in the growth medium (Sulistyo, 2012). Assay on antimicrobial properties of VCO was tested on agar medium through diffusion method. The test results on the test culture of *Listeria monocytogenes*, *Bacillus cereus*, *Escherichia coli*, and *Salmonella typhimurium* as potentially pathogenic microbial infectious diseases, indicating that the VCO mixed with culture extract of *L. plantarum* that was used for producing starter culture of VCO could inhibit the growth of *L. monocytogenes, B. cereus,* and *S. typhimurium.* The presence of fatty acids and organic acids can increase the antibacterial activity of VCO, since both of monoglyceride of VCO and organic acid of BAL could work synergistically to inhibit the growth of tested bacteria after diffused into a medium which had been overgrown by tested bacterial strains (Figure 6.3).

FIGURE 6.2   (See color insert.) Coconut milk as raw material for fermentation using starter culture of *Lactobacillus plantarum* (A); crude coconut oil (*top layer*) obtained after overnight fermentation (B); separation and extraction of obtaining oil layer using glass funnel (C); virgin coconut oil obtained after filtration and separation oil from water and protein-rich coconut milk residue (D).

The mechanism of microbial cell damage caused by the reaction mixture of VCO and metabolites from LAB starter culture is due to differences in the nature, structure, and chemical composition of the cell wall and membrane. Cell damage can be observed from the changes in cell shape due to antimicrobial activity. In normal cells, they should have formed slightly rounded and elongated cells with a diameter and length of 1.2 mμ and 2 mμ, respectively. Effect of antimicrobial metabolites caused normal bacterial cells appeared elongated up to 3 μm with a diameter of 1.4 μm and microscopically some cell cavities visibly found on the surface of its cells (Figure 6.4).

To determine the effectiveness of the VCO to control blood chemical, an *in vivo* testing using DDY mice as an animal model was carried out. Parameters measured were weight gain of mice and blood sugar levels, cholesterol, LDL, HDL, and triglycerides in the serum of mice. In addition, pathological, and histopathological examination on liver, kidney, spleen, and intestine of mice were also conducted to determine the effect of VCO towards the organs.

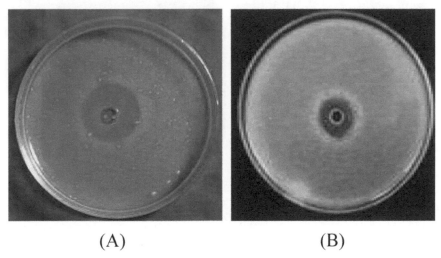

(A)   (B)

FIGURE 6.3   (See color insert.) Antimicrobial activity of starter culture of VCO against pathogenic microbial growth of *Listeria monocytogenes* (A) and *Salmonella typhimurium* (B) (courtesy by Dr. Asriani, examined by Dr. Joko Sulistyo).

A pre-clinical trials using 160 of 2–3 months mice with average weight of 20 gram were monitored for 28 days observation on 8 groups which consist of 20 mice according to protocol of treatment: Group (I), control mice; Group (II), mice administered with dose of 50 μl VCO; (III), mice administered with dose of 500 μl VCO; (IV), mice administered with dose of 500 μl sunflower oil; (V), mice administered with dose of 50 μl VCO prior to infection with *E. coli*; (VI), mice infected with *E. coli* prior to

administer with dose of 50 μl VCO; (VII), mice administered with dose of
500 μl VCO prior to infection with *E. coli*, and, (VIII), mice infected with
E. coli prior to administer with dose of 500 μl VCO. Table 6.3 showed that
up to day–14, body weight of mice in groups V and VII went to decline,
however, the group VII which were provided more doses of VCO (500 μl)
weight loss could be moderately retained compared to the group V which
were provided with a lower VCO dose (50 μl) (Sulistyo et al., 2009).
Figure 6.4 and 6.5 exhibited that the body weight of mice of groups II, III,
and IV showed an increase in the body weight following the treatment.
This was due to the administration with doses 50–500μl of VCO, and with
doses, 500μl of sunflower oil could increase the body weight of mice.

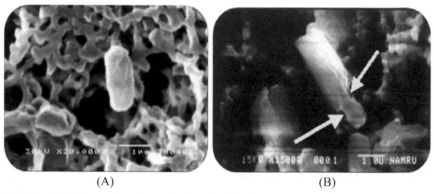

(A)   (B)

FIGURE 6.4   (See color insert.) Normal cell of *L. monocytogenes* (A, 20,000X) and
effect of an extract of a starter culture of VCO against cells of *S. typhimurium* (B, 15000X)
(courtesy by Dr. Asriani and examined by Dr. Joko Sulistyo).

Oral administration with dose 500μl of sunflower oil resulted in
increasing body weight higher than administration with doses 500μl
of VCO. This indicated that sunflower oil is a kind of hydrogenated oil
which could increase the body weight rapidly. Mice of group V and VII
showed higher weight loss compared to the control group. This was due
to mice infected with *E. coli* prior to administration with VCO underwent
to be infected. Providing VCO hereafter being infected could not be able
to increase the body weight, since providing VCO thereafter resulting in
VCO to work extra hard to improve metabolism and helps the body to
protect the cells in the body from damage. Mice groups VI and VIII which
were infected with *E. coli* and given VCO at the same time could gain body

weight. This was due to VCO which has entered into the body instantly could be used to help the body to defend against the occurrence of infection. Thus oral administration with doses 500µl of VCO to the group VIII resulted in gaining body weight higher than group VII which was given dose 50 µl of VCO. Based on these results demonstrated that administration of the VCO prior to infection with *E. coli* provided the difference in weight gain higher compared to administration of VCO thereafter infection with *E. coli* was occurred. Consequently, the oral administration of VCO prior to the occurrence of infection enabled the body to overcome the infections resulted in obtaining the weight gain was higher (Sulistyo et al., 2009). Analysis of blood sugar of mice in groups II and III those were administered with doses 50µl of VCO and 500µl of VCO decreased by 5.33 mg/dl and 0.33 mg/dl, respectively on the day–14, and decreased by 8.5 mg/dl and 1,2 mg/dl on the day–28, respectively. This suggested that the decrease in blood sugar levels required a longer time. Serum was taken on the day–28 which was an administration with VCO could produce blood sugar lower than serum was taken on the day–14 (Sulistyo et al., 2009) (Table 6.4).

TABLE 6.3 Observations on Body Weight Changes of Mice Before and After Treatment With VCO and *E. coli* Infection on Days 0–28

Group	Average of body weight (gram) on observation days					
	0	7	14	18	21	28
I	19.48	20.28	20.74	22.34	21.29	22.16
II	19.95	21.18	22.08	24.35	21.92	22.52
III	19.60	21.49	23.76	25.60	23.45	23.75
IV	17.59	20.14	22.34	27.43	21.79	22.89
V	23.71	24.30	21.90	23.24	21.31	22.33
VI	21.10	22.17	24.22	23.57	22.24	22.53
VII	24.49	25.07	25.51	25.74	23.17	25.77
VIII	22.99	25.01	27.2	29.12	26.12	26.94

(I) Control; (II) Oral administration with VCO 50 µl; (III) Oral administration with VCO 500 µl; (IV) Oral administration with sunflower oil 500 µl; (V) Infected with *E. coli* 3 days and oral administration with VCO 50 µl 5 days; (VI) Infected with *E. coli* plus VCO 50 µl 3 days and oral administration with VCO 5 days; (VII) Infected with *E. coli* 3 days and oral administration with VCO 500 µl 5 days; (VIII) Infected with *E. coli* plus VCO 500 µl 3 days and oral administration with VCO 5 days.

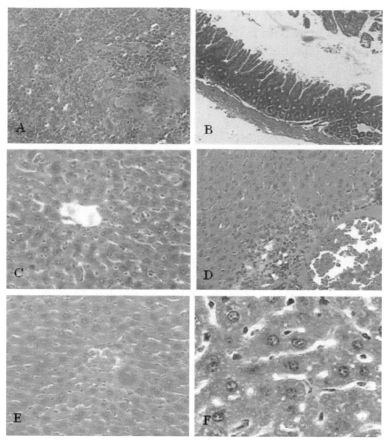

FIGURE 6.5   (See color insert.) Microscopic observation on preparation of mice spleen (A), intestinal (B), liver of control group (C) and liver of mice group after infection with *E. coli* continued with VCO oral administration (D), liver of mice group treated with single treatment of VCO (E) and sunflower oil (F), at 400× magnification.

TABLE 6.4   Observation on Blood Serum of Mice on Day–0 and Days–28

Blood chemistry (mg/dl)	Day–0	Day–28			
	Average	Group I	Group II	Group III	Group IV
Glucose	20	17.5	9	16.3	11.67
Cholesterol	160	133	124	104.67	121.3
HDL cholesterol	76.70	95.33	100	108	96.33
LDL cholesterol	156.3	20.5	32	12.3	18.3
Triglyceride	164.3	150.5	98	107.67	109.3

Analysis of blood cholesterol levels of mice in the group II which were administered with doses 50 μl of VCO showed decreasing at day–28 by 9 mg/dl. Whereas the cholesterol in mice of group III were administered with doses 500 μl of VCO showed decreasing to 3.6 mg/dl on day–14 and 28.33 mg/dl on day–28, respectively, while mice of group IV were administered with doses 500 μl of sunflower oil showed decreasing in blood cholesterol started from the day–14 (3.3 mg/dl) to the day–28 (11.7 mg/dl). HDL cholesterol of mice of groups II and III were administered with doses 50 μl and 500 μl of VCO and 500 μl of sunflower oil showed increasing on the day–14 (4.67 mg/dl, 12.67 mg/dl, and 1.0 mg/dl, respectively), however, showed decreasing on the day–28 (10.2 mg/dl, 11.5 mg/dl, and 1.5 mg/dl, respectively) (Table 6.4). Although the administration of VCO and sunflower oil could increase HDL levels, however, administration with the dose 500 μl of VCO exhibited more effective compared to the administration with doses 50 μl of VCO as well as administration with dose 500 μl of sunflower oil (Sulistyo et al., 2009). Analysis of LDL cholesterol of mice in the group II were administered with doses 50 μl of VCO showed decreasing on the day–14 (5.6 mg/dl), however, then increasing on the day–28 (11.5 mg/dl). The mice of groups III administered with doses VCO of 500 μl showed decreasing on levels of LDL cholesterol was 8.2 mg/dl. Comparing to the group IV which were administered with sunflower oil (2.2 mg/dl), the group of mice that were subjected to the VCO 500 μl showed lower LDL cholesterol level in their blood serum. Oral administration with doses 50 μl of VCO could lower triglycerides on the day–14 (18.6 mg/dl) until the day–28 (52.5 mg/dl). With a dose of 50μl of VCO showed more effective in lowering triglycerides than the dose 500 μl of VCO either with a dose of 500 μl of sunflower oil. These results indicated that administration with a lower dose of VCO (50 μl) took a long time to give the best result while within higher dose (500 μl) required a shorter time to give the good results and was more effectively used rather than the use of sunflower oil. Figure 6.5 showed microscopic observations on preparations of a tissue sample of mice spleen. Observation on the liver tissue of all groups of mice was carried out on the day–14, 17, 23 and 28. Based on histopathological examination of the tissues of spleen and intestine, showed no change at day–13 until the day–28, whereas from the sample of liver and kidney tissues revealed the occurrence of changes.

Histopathological examination on the spleen and intestinal tissue of mice (Figure 6.5A and 6.5B) showed no change. Likewise, to the group II which was administered with doses 50µl of VCO and the group III was administered with doses 500µl of VCO, it did not show any changes on their sample of liver tissue compared to sample of liver tissue of mice was administered with sunflower oil. Microscopic observation showed that the liver of mice which had been infected with *E. coli* and simultaneously administered with VCO underwent infection resulted in the occurrence of focal necrosis due to pathogenic infection of *E. coli*. Oral administration with doses 500µl of VCO could cause fatty liver occurrence, while administration with doses 50µl of VCO indicated that the occurrence of fatty liver was no longer appeared. Hence the use of VCO could be suggested to administer at appropriate doses (Figure 6.5C and 6.5D). The presence of vacuoles of fat in the liver of the group of mice was administered with sunflower indicating the occurrence of the fatty liver due to administration with high fatty oil. A different result was exhibited through the samples of the liver of mice was administered with VCO at the same doses which did not cause fatty liver since VCO containing saturated fatty acid does not cause fat deposits into the liver organ. While in the case of sunflower oil, since the percentage of unsaturated fatty acids is high, and consequently, it will be converted into triglycerides into the blood (Figure 6.5E and 6.5F). Histopathological changes in kidneys were observed such as swelling of the glomerulus, and renal tubule epithelial necrosis appeared in the organ of some mice at different groups of treatment (Sulistyo et al., 2009). However, the histopathologic changes in the kidneys did not find in a sample of mice in the control group. Administration with dose 50µl of VCO did not cause changes to the kidneys of mice since day–13 until the end of the study on day–28 compared to the sample of mice were administrated with 500 µl of VCO which indicated a change in the form of swelling on glomeruli and necrosis of renal tubule epithelial cells (Table 6.5). Oral administration with a dose of 500 µl of VCO resulting in a burden of kidneys to work harder and hence affected kidney's function which was characterized by swelling on the renal glomerulus. Groups of mice infected with *E. coli* and simultaneously administered with VCO exhibited swelling on glomerular and necrosis on epithelial cells of kidney's tubules. Infection by *E. coli* might lead to sepsis. Thus bacteria could penetrate the kidneys resulted in nephritis of the kidney which was characterized by inflammation and necrosis of the glomerular epithelial cells of renal tubule (Sulistyo et al., 2009).

TABLE 6.5 Histopathology Observation on Samples of Mice Kidney

Group	Day–14	Day–18	Day–21	Day–28
I	NC	NC	NC	NC
II	NC	NC	NC	NC
III	NC	NC	NC	Glomerular swelling and necrosis of epithelial tubule
IV	NC	NC	Glomerular swelling and necrosis of epithelial tubule	Glomerular swelling and necrosis of epithelial tubule
V	NC	NC	NC	NC
VI	NC	NC	Glomerular swelling and necrosis of epithelial tubule	NC
VII	NC	Glomerular swelling and necrosis of epithelial tubule	Glomerular swelling and necrosis of epithelial tubule	Glomerular swelling and necrosis of epithelial tubule
VIII	NC	Glomerular swelling and necrosis of epithelial tubule	Glomerular swelling and necrosis of epithelial tubule	Glomerular swelling and necrosis of epithelial tubule

6.8 CONCLUSION AND FUTURE PERSPECTIVES

Coconut oil is nature's richest source of MCFA. Not only do MCFA raise the body's metabolism leading to weight loss, but they have special health-giving properties as well. The most predominant MCFA in coconut oil is lauric acid, a monoglyceride in coconut oil that is similar to fats in mother's milk and has similar nutraceutical effects. These health effects were recognized centuries ago in Ayurvedic medicine. Modern research has now found a common link between these two natural health products. The MCFA and monoglycerides found primarily in coconut oil and mother's milk have miraculous healing power. Outside of a human mother's breast milk, coconut oil is nature's most abundant source of lauric acid and MCFA. Eating MCFA of coconut oil, especially VCO is similar to filling a vehicle with a high-octane fuel since MCFA are easily absorbed into the cells, thus increasing metabolism, and the energy produced by metabolism will produce stimulating effects throughout the body. In addition to increasing the energy level of the body, there are some other benefits along with the increased metabolism, such as an increase in resistance against disease and healing from disease. The cells of the body work more efficient in forming new cells and replace damaged cells quickly. Thus the body does not only have more energy but also has a shelf life longer. VCO will not go rancid even at room temperatures in for a couple of years. Conversely, refined oils are very unstable and turn rancid (oxidation) quickly, and oxidized oils are very toxic to the body and they can cause widespread free-radical damage. Hence, it is said that VCO is a kind of edible oil that is the safest and most healthful oil in the world.

KEYWORDS

- *Cocos nucifera*
- DDY mice
- edible cooking oil
- enzymatic fermentation
- histopathology
- *Lactobacillus plantarum*
- lauric acid
- medium chain fatty acid
- medium chain triglycerides
- monocaprin
- monoglyceride
- monolaurin
- pre-clinical test
- saturated fatty acid
- trans fatty acid
- virgin coconut oil

REFERENCES

Agero, A. L., & Verallo-Rowell, V. M. A., (2004). Randomized double-blind controlled trial comparing extra virgin coconut oil with mineral oil as a moisturizer for mild to moderate xerosis. *Dermatitis, 15*(3), 109–116.

APCC, (2003). *Standard for Virgin Coconut Oil.* Asian and Pacific Coconut Community, Jakarta, Indonesia, www.apccsec.org.

Arsa, M., Putra, A. A. B., Sahara, E., Asih, I. A. R. A., Bogoriani, N. W., Bawa, I. G. A. G., & Simpen, I. N., (2004). Making coconut oil using fermentation method. *Udayana Mengabdi, 3*(1), 21–26.

Assunção, M. L., Ferreira, H. S., Dos Santos, A. F., Cabral, Jr., C. R., & Florêncio, T. M. M. T., (2009). Effects of dietary coconut oil on the biochemical and anthropometric profiles of women presenting abdominal obesity. *Lipids, 44*(7), 593–601.

Berger, A., Jones, P. J. H., & Abumweis, S. S., (2004). Plant sterols: Factors affecting their efficacy and safety as functional food ingredients. *Lipids in Health and Disease, 3*(5), 1–19.

Bergsson, G., Arnfinnsson, J., Steingrímsson, O., & Thormar, H., (2001). *In vitro* killing of *Candida albicans* by fatty acids and monoglycerides. *Antimicrob Agents Chemother., 45*(11), 3209–3212.

Bibby, D. C., & Grimble, R. F., (1990). Tumor necrosis factor-alpha and endotoxin induce less prostaglandin E2 production from hypothalami of rats fed coconut oil than from hypothalami of rats fed maize oil. *Clinical Science (Colch), 79,* 657–662.

Boddie, R. L., & Nickerson, S. C., (1992). Evaluation of post-milking treat germicides containing lauricidin, saturated fatty acids, and lactic acid. *Journal of Dairy Science, 75,* 1725–1730.

Ceriello, A., Mercuri, F., Quagliaro, L., Assaloni, R., Motz, E., Tonutti, L., & Taboga, C., (2001). Detection of nitrotyrosine in the diabetic plasma: Evidence of oxidative stress. *Diabetologia, 44,* 834–838.

Cha, Y. S., & Sachan, D. S., (1994). Opposite effects of dietary saturated and unsaturated fatty acids on ethanol-pharmacokinetics, triglycerides, and carnitines. *Journal of the American College of Nutrition, 13,* 338–343.

Clevidence, B. A., Judd, J. T., Schaefer, E. J., Jenner, J. L., Lichtenstein, A. H., Muesing, R. A., Wittes, J., & Sunkin, M. E., (1997). Plasma lipoprotein (a) levels in men and women consuming diets enriched in saturated, *cis-*, or *trans-*monounsaturated fatty acids. *Arterioscler Thromb. Vasc. Biol., 17,* 1657–1661.

Dave, J. R., Koenig, M. L., Tortella, F. C., Pieringer, R. A., Doctor, B. P., & Ved, H. S., (1997). Dodecylglycerol provides partial protection against glutamate toxicity in neuronal cultures derived from different regions of embryonic rat brain. *Molecular Chemistry and Neuropathology, 30,* 1–13.

De Graaf, L., Brouwer, A. H., & Diemont, W. L., (2004). Is decreased libido associated with the use of HMG-CoA-reductase inhibitors? *Br. J. Clin. Pharmacol., 58*(3), 326–328.

Enig, M. G., (1996). Health and nutritional benefits from coconut oil: An important functional food for the 21st century. *AVOC (ASEAN Vegetable Oils Club) Lauric Oils Symposium* (pp. 1–12). Ho Chi Min City, Vietnam.

Enig, M. G., (1998). Lauric oils as antimicrobial agents: Theory of effect, scientific rationale, and dietary applications as adjunct nutritional support for HIV-infected individuals. In: Watson, R. R., (ed.), *Nutrients and Foods in AIDS* (pp. 81–97). CRC Press, Boca Raton.

Enig, M. G., (1999). *Coconut: In Support of Good Health in the 21st Century.* Paper presented at the 36th Meeting of APCC, Singapore.

Fife, B., (2004). *The Coconut Oil Miracle* (pp. 1–7). Avery, USA.

Fletcher, R. D., Albers, A. C., Albertson, J. N., & Kabara, J. J., (1985). Effects of monoglycerides on *Mycoplasma pneumoniae* growth. In: Kabara, J. J., (ed.), *The Pharmacological Effect of Lipids* (pp. 59–63). American Oil Chemists Society, Champaign, IL.

Garfinkel, M., Lee, S., Opara, E. C., & Akwari, O. E., (1992). Insulinotropic potency of lauric acid: A metabolic rationale for medium chain fatty acids in TPN formulation. *Journal of Surgical Research, 52,* 328.

Gerster, H., (1998). Can adults adequately convert alpha-linolenic acid (18: 3n–3) to eicosapentaenoic acid (20: 5n–3) and docosahexaenoic acid (22: 6n–3)? *International Journal for Vitamin and Nutrition Research, 68*(3), 159–173.

Gervajio, G. C., (2005). *Fatty Acids and Derivatives From Coconut Oil* (6th edn., pp. 1–56), John Wiley and Sons, USA.

Grootveld, M., Silwood, C. J. L., Addis, P., Claxson, A., Serra, B. B., & Viana, M., (2001). Health effects of oxidized heated oils. *Journal of Food Service, 13*(1), 41–55.

Hasbullah, (2001). In: Sawedi, E., (ed.), *Appropriate Technology for Small Agroindustry in West Sumatra.* Dewan Ilmu Pengetahuan, Teknologi Dan Industri, Sumatera Barat.

Herlina, N., Ginting, M., & Hendra, S., (2002). *Fats and Oils* (pp. 1–8). North Sumatera University, Indonesia.

Hierholzer, J. C., & Kabara, J. J., (1982). *In vitro* effects of monolaurin compounds on enveloped RNA and DNA viruses. *Journal of Food Safety, 4,* 1–12.

Hornstra, G., Van Houwelingen, A. C., Kester, A. D., & Sundram, K. A., (1991). Palm oil-enriched diet lowers serum lipoprotein in normocholesterolemic volunteers. *Atherosclerosis, 90,* 91–93.

Hou, X., Qi, Y., Qiao, X., Wang, G., Qin, Z., & Wang, J. L., (2007). Acid-catalyzed transesterification and esterification of high free fatty acid oil in subcritical methanol. *Kor. J. Chem. Eng., 24*(2), 311–313.

Hu, F. B., Stampfer, M. J., & Manson, J. E., (1999). Dietary saturated fat and their food sources in relation to the risk of coronary heart disease in women. *Am. J. Clin. Nutr., 70*(6), 1001–1008.

Isaacs, C. E., & Thormar, H., (1986). Membrane-disruptive effect of human milk: Inactivation of enveloped viruses. *Journal of Infectious Diseases, 154,* 966–971.

Isaacs, C. E., & Thormar, H., (1990). Human milk lipids inactivated enveloped viruses. In: Atkinson, S. A., Hanson, L. A., & Chandra, R. K., (eds.), *Breastfeeding, Nutrition, Infection, and Infant Growth in Developed and Emerging Countries* (pp. 167–174). Arts Biomedical Publishers and Distributors, St. John's, N. F., Canada.

Isaacs, C. E., & Thormar, H., (1991). The role of milk-derived antimicrobial lipids as antiviral and antibacterial agents. In: Mestecki, J., & Ogra, P. L., (eds.), *Immunology of Milk and the Neonate* (pp. 159–165). Plenum Press, New York.

Isaacs, C. E., Kashyap, S., Heird, W. C., & Thormar, H., (1990). Antiviral and antibacterial lipids in human milk and infant formula feeds. *Archives of Disease in Childhood, 65,* 861–864.

Isaacs, C. E., Litov, R. E., Marie, P., & Thormar, H., (1992). Addition of lipases to infant formulae produces antiviral and antibacterial activity. *Journal of Nutritional Biochemistry, 3,* 304–308.

John, J., Bhattacharya, M., & Turner, R. B., (2002). Characterization of polyurethane foams from soybean oil. *Journal of Applied Polymer Science, 86*(12), 3097–3107.

Judd, J. T., Clevidence, B. A., Muesing, R. A., Wittes, J., Sunkin, M. E., & Podczasy, J. J., (1994). Dietary trans fatty acids: Effects on plasma lipids and lipoproteins of healthy men and women. *American Journal of Clinical Nutrition, 59,* 861–868.

Kabara, J. J., & Vrable, R., (1977). Antimicrobial lipids: Natural and synthetic acids and monoglycerides. *Lipids, 12,* 753–759.

Kabara, J. J., (1978). In: Kabara, J. J., (ed.), *The Pharmacological Effects of Lipids* (pp. 15–24). The American Oil Chemists Society, Champaign IL.

Kabara, J. J., (1985). In: Kabara, J. J., (ed.), *Inhibition of Staphylococcus aureus in the Pharmacological Effect of Lipids* (pp. 71–75). American Oil Chemists Society, Champaign IL.

Kabara, J. J., (2005). Pharmacological effects of coconut oil vs. monoglycerides. *Inform June,* 286.

Kaunitz, H., & Dayrit, C. S., (1992). Coconut oil consumption and coronary heart disease. *Philippine Journal of Internal Medicine, 30,* 165–171.

Keys, A. B., (1980). *Seven Countries: A Multivariate Analysis of Death and Coronary Heart Disease.* Harvard University Press, Cambridge, MA.

Khosla, P., & Hayes, K. C., (1996). Dietary trans-monounsaturated fatty acids negatively impact plasma lipids in humans: Critical review of the evidence. *Journal of the American College of Nutrition, 15,* 325–339.

Kris-Etherton, P. M., & Yu, S., (1997). Individual fatty acids on plasma lipids and lipoproteins: Human studies. *Am. J. Clin. Nutr., 65*(5), 1628S–1644S.

Lay, A., & Pasang, P. M., (2012). Strategy and implementation of future coconut product development. *Perspektif, 11*(1), 1–22.

Macallan, D. C., Noble, C., Baldwin, C., Foskett, M., McManus, T., & Griffin, G. E., (1993). Prospective analysis of patterns of weight change in stage IV human immunodeficiency virus infection. *American Journal of Clinical Nutrition, 58,* 417–424.

Marina, A. M., Che Man, Y. B., & Amin, I., (2009). Virgin coconut oil: Emerging functional food oil. *Trends in Food Science and Technology, 20,* 481–487.

Mensink, R. P., & Katan, M. B., (1990). Effect of dietary trans fatty acids on high-density and low-density lipoprotein cholesterol levels in healthy subjects. *New Engl. J. Med., 323,* 439–445.

Mensink, R. P., Zock, P. L., Kester, A. D., & Katan, M. B., (2003). Effects of dietary fatty acids and carbohydrates on the ratio of serum total to HDL cholesterol and on serum lipids and apolipoproteins: A meta-analysis of 60 controlled trials. *Am. J. Clin. Nutr., 77*(5), 1146–1155.

Nanji, A. A., Sadrzadeh, S. M., Yang, E. K., Fogt, F., Maydani, M., & Dannenberg, A. J., (1995). Dietary saturated fatty acids: A novel treatment for alcoholic liver disease. *Gastroenterology, 109,* 547–554.

Nevin, K. G., & Rajamohan, T., (2004). Beneficial effects of virgin coconut oil on lipid parameters and *in vitro* LDL oxidation. *Clinical Biochemistry, 37,* 830–835.

Oh, D. H., & Marshall, D. L., (1993). Antimicrobial activity of ethanol, glycerol monolaurate or lactic acid against *Listeria monocytogenes*. *International Journal of Food Microbiology, 20,* 239–246.

Petschow, B. W., Batema, R. P., & Ford, L. L., (1996). Susceptibility of *Helicobacter pylori* to bactericidal properties of medium-chain monoglycerides and free fatty acids. *Antimicrobial Agents and Chemotherapy, 40,* 302–306.

Portillo, M. P., Serra, F., Simon, E., Del Barrio, A. S., & Palou, A., (1998). Energy restriction with high-fat diet enriched with coconut oil gives higher UCP1 and lower white fat in rats. *International Journal of Obesity and Related Metabolic Disorders, 22,* 974–979.

Pramita, Y., (2002). *Behind the Scum of Used Cooking Oil, Stimulates Colon Cancer.* Pikiran Rakyat, (in Indonesian).

Preuss, H. G., Echard, B., & Zonosi, R. R., (2005). The potential for developing natural antibiotics: Examining oregano and monolaurin. *Original Internist, 12,* 119–124.

Prior, I. A., Davidson, F., Salmond, C. E., & Czochanska, Z., (1981). Cholesterol, coconuts, and diet on Polynesian atolls: A natural experiment: The Pukapuka and Tokelau Island studies. *American Journal of Clinical Nutrition, 34,* 1552–1561.

Projan, S. J., Brown-Skrobot, S., Schlievert, P. M., Vandenesch, F., & Novick, R. P., (1994). Glycerol monolaurate inhibits the production of beta-lactamase, toxic shock toxin–1 and other staphylococcal exoproteins by interfering with signal transduction. *Journal of Bacteriology, 176,* 4204–4209.

Ravnskov, U., (1998). The questionable role of saturated and polyunsaturated fatty acids in cardiovascular disease. *Journal of Clinical Epidemiology, 51*(6), 443–460.

Rele, A. S., & Mohile, R. B., (2003). Effect of mineral oil, sunflower oil, and coconut oil on prevention of hair damage. *J. Cosmet. Sci., 54*(2), 175–192.

Rethinam, P., & Amrizal, I., (2005). Changes, challenges, and opportunities for coconut and its products of the Asia and Pacific countries. *Asia–Pacific Business Forum,* Bangkok, Thailand.

Rethinam, P., (2003). Muhartoyo. *The Plain Truth About Coconut Oil.* Asian and Pacific Coconut Community, Jakarta.

Rethinam, P., (2004). World coconut industries: Past, present, and future. *Coconut World Meeting,* Bali, Indonesia.

Rethinam, P., (2005). Global coconut production, challenges, and solutions of coconut industry with special emphasis to Sri Lankan coconut industry. *Coconut Cultivation Board,* Colombo, Sri Lanka.

Rindengan, B., & Karaow, S., (2003). Opportunity in development of pure coconut oil. *Proceedings of 5th National Conference on Coconut* (pp. 146–153). Tembilahan, Riau Province.

Rindengan, B., & Novarianto, H., (2005). *Making and Utilizing Virgin Coconut Oil.* Swadaya Publishers, Jakarta.

Ruzin, A., & Novick, R. P., (1998). Glycerol monolaurate inhibits induction of vancomycin resistance in *Enterococcus faecalis. J. Bacteriol., 180,* 182–185.

Ruzin, A., & Novick, R. P., (2000). Equivalence of lauric acid and glycerol monolaurate as inhibitors of signal transduction in *Staphylococcus aureus. J. Bacteriol., 182,* 2668–2671.

Saikku, P., (1997). Chlamydia pneumonia and atherosclerosis - an update. *Scandinavian Journal of Infectious Diseases, S104,* 53–56.

Sands, J. A., Auperin, D. D., Landin, P. D., Reinhardt, A., & Cadden, S. P., (1978). Antiviral effects of fatty acids and derivatives: Lipid-containing bacteriophages as a model system.

In: Kabara, J. J., (ed.), *The Pharmacological Effect of Lipids* (pp. 75–95). American Oil Chemists Society, Champaign IL.

Singh, R. B., Mori, H., Chen, J., Mendis, S., Moshiri, M., Zhu, S., et al., (1996). Recommendations for the prevention of coronary artery disease in Asians: A scientific statement of the International College of Nutrition. *J. Cardiovasc. Risk, 3*(6), 489–494.

Singh, R. H., Seepersaud, G., & Rankine, L, B., (2007). The regional coconut industry global market intelligence. *CARICOM Regional Transformation Programme for Agriculture, The Regional Coconut Industry: Global Market Intelligence, 1–56.*

Sitepoe, M., (2008). *Doodles of the Village Boy, Double Profession* (pp. 15–18). Gramedia Popular Literature, Jakarta (in Indonesian).

Sivapragasam, A., (2008). Coconut in Malaysia, current developments and potential for revitalization. *Second International Plantation Industry Conference and Exhibition* (pp. 1–9, 18–21,). Shah Alam, Malaysia.

St-Onge, P. M., & Jones, P. J. H., (2002). Physiological effects of medium-chain triglycerides: Potential agents in the prevention of obesity. *Recent Advances in Nutritional Sciences, American Society for Nutritional Sciences, 329–332.*

Sugano, M., & Ikeda, I., (1996). Metabolic interactions between essential and trans-fatty acids. *Current Opinions in Lipidology, 7*, 38–42.

Sukmadi, B., & Nugroho, N. B., (2002). Study of inoculum use in coconut oil production by fermentation. *Journal Biosains dan Bioteknologi Indonesia, 2*(1), 12–17.

Sulistyo, J., (2012). Fermentation process in extraction of coconut oil. *Food Review Indonesia, 7*(9), 36–40.

Sulistyo, J., Rahayu, R. D., & Poeloengan, M., (2009). Enzymatic extraction of virgin coconut oil and pre-clinical test analysis using white mice. *Berkala Penelitian Hayati, S3A,* 101–106.

Sulistyo, J., Rahayu, R. D., & Poeloengan, M., (2009). Enzymatic extraction of virgin coconut oil and pre-clinical test using mice-DDY. *Proceeding of the 20th Seminar on Biology and the 16th Congress of PBI* (pp. 24, 25). Malang, East Java, Indonesia.

Sundram, K., Hayes, K. C., & Siru, O. J., (1994). Dietary palmitic acid results in lower serum cholesterol than does a lauric-myristic acid combination in normolipidemic humans. *Asia Pac. J. of Clin. Nutr., 59,* 841846.

Temme, E. H., Mensink, R. P., & Hornstra, G., (1996). Comparison of the effects of diets enriched in lauric, palmitic, or oleic acids on serum lipids and lipoproteins in healthy women and men. *Am. J. Clin. Nutr., 63*(6), 897–903.

Thampan, P. K., (1998). *Facts and Fallacies About Coconut Oil* (pp. 1–39). Asian and Pacific Coconut Community, Jakarta.

Thormar, H., Isaacs, E. C., Brown, H. R., Barshatzky, M. R., & Pessolano, T., (1987). Inactivation of enveloped viruses and killing of cells by fatty acids and monoglycerides. *Antimicrobial Agents and Chemotherapy, 31,* 27–31.

Tomskaya, L. A., Makarova, N. P., & Ryabov, V. D., (2008). Determination of the hydrocarbon composition of crude oils. *Chem. Tech. Fuel. Oil, 44,* 280–283.

Wan, J. M., & Grimble, R. F., (1987). Effect of dietary linoleate content on the metabolic response of rats to *Escherichia coli* endotoxin. *Clinical Science (Colch), 72,* 383–385.

Wang, L. L., & Johnson, E. A., (1992). Inhibition of *Listeria monocytogenes* by fatty acids and monoglycerides. *Applied and Environmental Microbiology, 58,* 624–629.

Warner, B., Quirke, D., & Longmore, C., (2007). *A Review of the Future Prospects for the World Coconut Industry and Past Research in Coconut Production and Product. Australian Centre for International Agricultural Research,* 1–78.

Witcher, K. J., Novick, R. P., & Schlievert, P. M., (1996). Modulation of immune cell proliferation by glycerol monolaurate. *Clinical and Diagnostic Laboratory Immunology, 3,* 10–13.

Wright, B., (2012). *Is Coconut Oil Healthier Than Butter?* Eating Well, Inc.

FIGURE 4.1 Overview of some of the traditional homegardens in the study area.

FIGURE 4.2 Vegetables, fruits, and beans maintained in the homegardens of coastal taluks of Uttara Kannada.

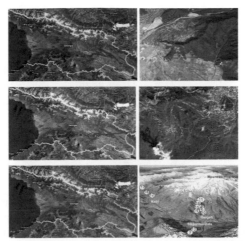

FIGURE 5.1   Study areas: 1,2,3,4 - *Primula woronowii*, Oak-hornbeam forest, Kvareli district, Shilda; 5,6 - *P. macrocalyx*, Bich forest, Tskhneti; 7–11 - *P. amoena*; 12 - *P. cordifolia*; 13–16 - *P. algida*; 17,18 - *P. cordifolia*, Stephantsminda district.

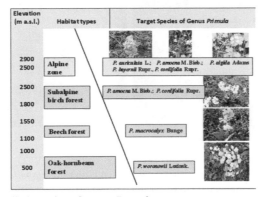

FIGURE 5.2   Studied species of genus *Primula*.

FIGURE 6.1   Cooking coconut oil obtained by heating of coconut milk residue.

FIGURE 6.2 Coconut milk as raw material for fermentation using starter culture of *Lactobacillus plantarum* (A); crude coconut oil (*top layer*) obtained after overnight fermentation (B); separation and extraction of obtaining oil layer using glass funnel (C); virgin coconut oil obtained after filtration and separation oil from water and protein-rich coconut milk residue (D).

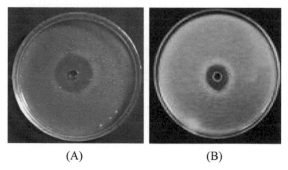

(A)   (B)

FIGURE 6.3 Antimicrobial activity of starter culture of VCO against pathogenic microbial growth of *Listeria monocytogenes* (A) and *Salmonella typhimurium* (B) (courtesy by Dr. Asriani, examined by Dr. Joko Sulistyo).

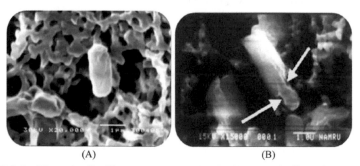

(A)   (B)

FIGURE 6.4 Normal cell of *L. monocytogenes* (A, 20,000X) and effect of an extract of a starter culture of VCO against cells of *S. typhimurium* (B, 15000×) (courtesy by Dr. Asriani and examined by Dr. Joko Sulistyo).

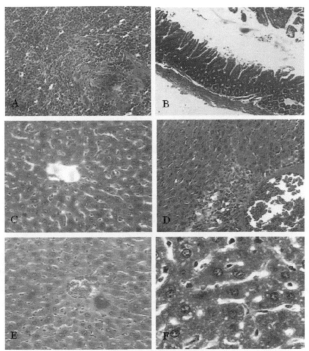

FIGURE 6.5 Microscopic observation on preparation of mice spleen (A), intestinal (B), liver of control group (C) and liver of mice group after infection with *E. coli* continued with VCO oral administration (D), liver of mice group treated with single treatment of VCO (E) and sunflower oil (F), at 400× magnification.

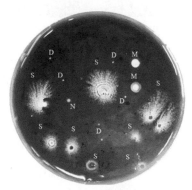

FIGURE 7.1 Actinomycete colonies grown on HV agar supplemented with cyloheximide (50 mg/l) and nalidixic acid (25 mg/l). The air-dried soil was heated at 110°C for 1 h and was diluted with sterile basic lauryl sulfate solution [SDS 1 g; KH_2PO_4 1.75 g; K_2HPO_4 3.5 g; 1000 ml distilled water, pH 7.0], and 1 ml of each resultant was spread on the agar plate. The plate was incubated at 30°C for 21 days (S, *Streptomyces*; M, *Microbispora*; D, *Dactylosporangium*; N, *Nocardia*).

FIGURE 12.1 Local people being interviewed.

FIGURE 12.2 Night view of the floating restaurant.

FIGURE 12.3 Indigenous people village and tourists are visiting the indigenous village.

FIGURE 13.1   (a) and (b): Entrance of Pandanus Trail and Bukit Tupai Trail.

FIGURE 13.2   Mountain Torq office in Kinabalu Park.

FIGURE 13.3   (a–c): Poring Hot Spring Visitor Centre, Butterfly Farm Exhibition Gallery and Rafflesia blooming.

FIGURE 13.4 (a–c): Mesilau Nature Resort, Mesilau Nature Center, and map of Mesilau complex.

FIGURE 13.5 (a), (b) and (c): Interpretive guiding by Sabah Parks' staff, visitors at the Exhibition Hall and Mountain guiding activity in Kinabalu Park.

CHAPTER 7

DIVERSITY AND METABOLITES OF ACTINOMYCETES FROM PEAT SWAMP FOREST SOILS

WONGSAKORN PHONGSOPITANUN and
SOMBOON TANASUPAWAT

Department of Biochemistry and Microbiology, Faculty of Pharmaceutical Sciences, Chulalongkorn University, Bangkok 10330, Thailand

7.1 INTRODUCTION

Actinomycetes, belonging to phylum *Actinobacteria*, class *Actinobacteria*, are Gram-stain-positive filamentous bacteria with high guanine and cytosine contents of their genomic DNA. According to the Bergey's Manual of Systematic Bacteriology (Volume 5), the *Actinobacteria* can be classified into six classes as *Actinobacteria*, *Acidimicrobiia*, *Coriobacteriia*, *Nitriliruptoria*, *Rubrobacteria*, and *Thermoleophilia*. Among them, the *Streptomyces* belongs to the family *Streptomycetaceae*, the largest genus of phylum *Actinobacteria* are well known for the bioactive compound producers (antibiotics, antiviral, antitumor, immune suppressant agents, herbicide, insecticide, phytotoxins, plant growth regulator and antiparasitic agents) (Berdy, 2005). Many compounds produced from actinomycetes are clinically useful antibiotics including actinomycin D (Waksman and Woodruff, 1940), amphotericin B (Steinberg et al., 1956), neomycin (Lechevalier, 1975), avermectin (Burg et al., 1979), vancomycin (Levine, 2006) and platensimycin (Wang et al., 2006).

Actinomycetes are widely distributed in environments especially in terrestrial soils. However, in the last decade, many of actinomycete species are discovered from others habitats, including insects (Currie et al., 1999), plant materials (Qin et al., 2012); deep marine sediment (Pathom-aree et al., 2006), marine sponges as well as animals such as pufferfish (Wu et al., 2005). Although terrestrial soils are the major habitat of actinomycetes, peat swamp forest soils show the unique habitat and the actinomycete biodiversity remains poorly understood. Peat swamp forests are the world's wetlands that provide a variety of benefits in the form of forestry and fisheries products, energy, flood mitigation, water supply and groundwater recharge (UNDP, 2006). In Southeast Asia, Indonesia, Malaysia, Brunei, and Thailand have tropical peat swamp forests that are the unique ecosystem under enormous threat from logging, fire, and land conversion (Posa et al., 2011). This chapter describes the isolation, taxonomic criteria for identification, diversity, secondary metabolites and biological activities of actinomycetes from peat swamp forest soils.

7.2 ISOLATION AND CULTIVATION METHODS

The population of actinomycetes in soils is intermediate between bacteria and fungi. Their colonies are interfered by other microorganisms when they were cultivated on certain media. Before the genomic era, the first step to study actinomycetes is "how to obtain them from the environments?" The traditional technique for the isolation is the standard plate dilution technique. Several attempts have been described to overcome the isolation including the use of selective media (Table 7.1), pretreatment of soil samples before isolation and the supplement of the selective inhibitors or antibiotics, to the medium (Williams and Davies, 1965). The appropriate medium should be closely similar to the nature of the samples (pH, salinity, trace element) and enrich the growth of actinomycetes. In nature, approximately 97% of actinomycetes are *Streptomyces* strains while other 3% are the rare actinomycetes. Although the antibiotics are supplemented, the other bacteria and fungi are reduced. However, only *Streptomyces* strains were redundantly obtained from normal air-dried sample treatment. Because of the isolation strategies of rare actinomycetes typically started with "how to eliminate the *Streptomyces*" and then "what method will stimulate the growth of rare species?." The pretreatments of soil sample before isolation of actinomycetes are the powerful methods to obtain those rare actinomycetes.

TABLE 7.1   Selective Media for Isolation of Actinomycetes From Peat Swamp Forest Soils

Medium	Formula (g/L)	References
Humic acid vitamin agar	Humic acid 1.0 g*; Na_2HPO_4 0.5 g; $MgSO_4.7H_2O$ 0.05 g; $FeSO_4.7H_2O$ 0.01 g; $CaCO_3$ 0.02 g B-vitamin**; agar 18 g; water 1 L	Hayakawa and Nonomura, 1987
Arginine glycerol salt	Arginine monohydrochloride 1.0 g; glycerol 12.5 g; K_2HPO_4 1g; $Fe_2(SO_4)_3.6H_2O$, 0.010 g; NaCl 1.0 g; $MgSO_4.7H_2O$ 0.5 g; $CuSO_4.5H_2O$, 0.001 g; $ZnSO_4.7H_2O$, 0.001 g; $MnSO_4$ H_2O, 0.001 g; and agar, 15.0 g; water 1 L	El-Nakeeb and Lechevalier, 1962
Starch casein nitrate	Glycerol (or starch) 10g; casein 0.3 g; KNO_3 2g; NaCl 2 g; K_2HPO_4 2 g; $MgSO_4.7H_2O$ 0.05 g; $CaCO_3$ 0.02 g; $FeSO_4.7H_2O$ 0.01 g; Agar 18 g; distilled water 1 L, pH 7.0–7.2	Küster and Williams, 1964
Chitin agar	Colloidal chitin 2.5 g; K_2HPO_4 0.7 g; KH_2PO_4 0.3 g; $MgSO_4.7H_2O$ 0.5 g; $FeSO_4.7H_2O$ 0.01 g; $ZnSO_4.7H_2O$ 0.001 g; agar 20 g	
Water proline agar	Proline 10 g; agar 1.5 g; Tap water 1000 ml, pH 7.0	Inahashi et al., 2015
Humic acid-MOPS	Nitrohumic acid 0.5 g; $CaCl_2$ 3mM; MOPS (morpholinepropanesulfonic acid) 5 mM; $FeSO_4.7H_2O$ 1 mg; $MnCl_2.4H_2O$ 1 mg; $ZnSO_4.7H_2O$ 1 mg; $NiSO_4.6H_2O$ 1 mg; Gellan gum 7 g Distilled water 1 l, pH 7.2	Suzuki, 2001
LSV-SE	Lignin 1.0 g; Soil extract 100 ml; Soybean flour 0.2 g; $CaCO_3$ 0.02 g; NaH_2PO_4 0.5 g; KCl 1.7 g; $MgSO_4.7H_2O$ 0.5 g; $FeSO_4.7H_2O$ 0.01 g; B-vitamin**; distilled water 1 l; agar 18 g; pH 7.2	Hayakawa et al., 1996

The media are supplemented with selective antibiotics to suppress the fungi, and Gram-negative as well as others unicellular bacteria. The most frequently used antifungal is cycloheximide 100 mg/l, nystatin 25–50 mg/l, amphotericin B 30 mg/ml. Novobiocin 25 mg/l, nalidixic acid 10–25 mg/l are anti-bacterial. *Dissolve in 10 ml of 0.2N NaOH**0.5 mg each of thiamine-HCl, riboflavin, niacin, pyridoxine-HCl, inositol, Ca-pantothenate, aminobenzoic acid, and 0.25 mg of biotin. Vitamins were filter-sterilized.

Physical pretreatment, especially dried heat of soil sample (100–120°C, 1 h), is one of the easy methods to isolate the rare species (Nonomura and Ohara, 1969a,b; 1971a,b,c,d). Dried heat method can reduce the other unwanted bacteria as well as some *Streptomyces* species. Furthermore, this can be stimulated by the germination of actinomycetes spores. The ability of spores to resist the deleterious agent leads to the development of chemical treatment method. Hayakawa et al. (1991a, 1997) proposed the usage of 1.5% phenol treatment for isolation *Micromonospora* species and chloramine-T to isolate *Herbidospora*, *Microbispora*, *Microtetraspora*, *Nonomuraea*, and *Streptosporangium*. Other than a single method, the combination between physical pretreatment and chemical treatment was used to isolate *Actinomadura viridis* (Hayakawa et al., 1995), *Microbispora* (Hayakawa et al., 1991a), *Streptosporangium*, *Dactylosporangium* (Hayakawa et al., 1991b) and *Microtetraspora* (Hayakawa et al., 1996).

The incorporating method between flooding the air-dried soil and centrifugation method lead to the selective isolation of zoospore-producing actinomycetes including *Dactylosporangium*, *Actinoplanes*, *Catenuloplanes*, and *Kineospora* (Hayakawaet al., 2000). Hayakawa et al. (1991b) proposed the method for isolation of rare actinomycetes in the genera *Streptosporangium* and *Dactylosporangium* using the dry heating and treatment with benzethonium chloride. As the *Streptosporangium* and *Dactylosporangium* produced the sporangiospores and globose bodies (aleuriospore), respectively, they resisted to this method while eliminated other bacteria and *Streptomyces*. To isolate *Streptosporangium* species, the air-dried sample was heated at 120°C, 1 h. Then the water suspension of the heated sample was treated with 0.01% benzethonium chloride. To isolate *Dactylosporangium* species, 0.03% of benzethonium chloride was used. The resultant of serial dilution was placed on HV agar.

Hayakawa et al. (1991a) established the method to isolate members of the genera *Micromonospora* and *Microbispora* using the chemical treatment, and found that the spores of both genera could resist to deleterious agents. To isolate *Micromonospora* species, the air-dried soil is treated with 1.5% phenol (30 min) and then diluted using water and then cultured on HV agar. To isolate *Microbispora* species, the air-dried soil is treated with 1.5% phenol and 0.03% chlorhexidine gluconate for 30 min and then diluted with water, the resultant of each dilution is plated on HV agar.

Hayakawa et al. (1996) proposed that the water suspension of a soil sample dry heat (110°C, 1h) which was subjected to a treatment with

0.05% benzethonium chloride and cultivated on the LSB-Se agar (supplemented with kanamycin, norfloxacin, and nalidixic acid) was found to be the suitable method for isolation of *Microtetraspora* species. Chemotaxis, γ-collidine and vanillin, and baiting, with *Pinus* pollen grains, were useful to isolate *Actinoplanes*, *Catenuloplanes*, *Dactylosporangium*, and *Virgisporangium* (Hayakawa et al., 1991a,b, 1995). The rehydration of air-dried soil with10 mM phosphate buffer containing 10% soil extract at 30°C (90 min) provided the motility and releasing of zoospore, followed by the centrifugation at 1,500 × g, 20 min. The supernatant, containing the zoospore, was placed on the HV medium supplemented with nalidixic acid and trimethoprim and incubated at 30°C for 4 weeks. This rehydration and centrifugation method proposed by Hayakawa et al. (2000) was useful to obtain those species.

Suzuki (2001) combined the heat treatment of soil sample and centrifugation method (flooding with a suitable buffer to stimulate the motility of zoospore, then centrifuge, wait for releasing the zoospore and collect the supernatant containing zoospore for further isolation) to obtain the above genera. The combination of gellan gum and 2 mM $CaCl_2$ is necessary for the growth of *Sporichthya* strains. To isolate the *Planobispora* and *Planomonospora* species, the trace salt ($FeSO_4.7H_2O$, $MnCl_2.4H_2O$, $ZnSO_4.7H_2O$, and $NiSO_4.6H_2O$) and alkaline isolation medium (pH 9.0) were found to stimulate sporangium formation. To stimulate the formation of sporangia, the humic acid vitamin agar modified with trace salt solution and gellan gum was used instead of normal humic acid vitamin agar.

In addition, the cellulose ester membrane filter method based on the ability of the actinomycete to propagate and pass through the pore of the filter was used by Hirsch and Christensen (1983). According to this method, agar medium was overlaid with 0.22–0.45 μm-pore cellulose ester membrane filter and the filter was inoculated. After incubation, the mycelia of actinomycetes penetrate through the filter pores to the agar medium while other-bacteria are trapped on the filter surface. After removing the membrane, the agar medium should be further incubated to obtain a clear colony of actinomycetes. This approach was adapted by Gavrish et al. (2008) to obtain the rare actinomycetes.

In Thailand, there are 25 peat swamp forests where most of them are located in the southern part of the country. Several isolation methods as mentioned above were found to be successful to isolate actinomycetes from peat swamp forest soil samples. In 2004, Thawai isolated many

Micromonospora species from peat swamp forest soils collected in Trang, Pattaloong, Yala, and Narathiwat provinces using standard dilution technique plating on starch-casein nitrate agar supplemented with nystatin, novobiocin, and tetracycline. Niemhom et al. (2012) isolated rare actinomycetes from peat swamp forest soils collected from Phu Sang National park (pH 6.5–8.0), Phayao province and Doi Inthanond National park (pH 3.5–6.5), Chiangmai Province, Thailand using air-dried, heat, 1.5% phenol treatment and flooding treatment. Phongsopitanun et al. (2015a,b) isolated actinomycetes from peat swamp forest soils collected from Rayong Province using the standard dilution plating method. Briefly, one gram of the soil sample was suspended in the basic lauryl-sulfate solution [0.1 g SDS, 1.75 g KH_2PO_4, 3.5 g K_2HPO_4 in 1000 ml distilled water (pH 7.0)] and was diluted to 10^{-4}. The resulting suspension was spread on humic-acid vitamin agar (Hayakawa and Nonomura, 1987) supplemented with nalidixic acid 25 mg/l and cycloheximide 50 mg/l. After 14 days of incubation at 30°C, a single colony of isolates was isolated and the pure isolate was cultivated on ISP 2 medium (Figure 7.1).

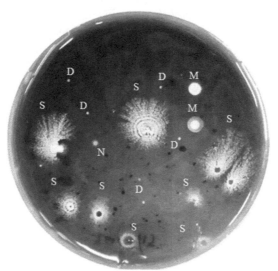

FIGURE 7.1   (See color insert.) Actinomycete colonies grown on HV agar supplemented with cyloheximide (50 mg/l) and nalidixic acid (25 mg/l). The air-dried soil was heated at 110°C for 1 h and was diluted with sterile basic lauryl sulfate solution [SDS 1 g; KH_2PO_4 1.75 g; K_2HPO_4 3.5 g; 1000 ml distilled water, pH 7.0], and 1 ml of each resultant was spread on the agar plate. The plate was incubated at 30°C for 21 days (S, *Streptomyces*; M, *Microbispora*; D, *Dactylosporangium*; N, *Nocardia*).

Jeffery et al. (2011) used the multi-stage Dispersion and Differential Centrifugation (DDC) method to isolate actinomycetes from soil samples collected from Malaysian Agricultural Research and Development Institute (MARDI) peatland research station at Sessang, Sarawak. After centrifugation, serial dilutions (10^{-1}–10^{-3}) were done for all fractions and spread plate onto Starch Casein Agar pH 7.0, supplemented with cycloheximide (50 µg/ml). The plates were incubated at $28 \pm 2°C$ for 14 days. Niyasom et al. (2015) isolated acidophilic actinomycetes from acidic soil samples using starch casein agar, soil extract agar, and humic vitamin agar, pH 4.5, containing nystatin and nalidixic acid. The total acidophilic actinomycete counts ranged from 3.9–8.2×10^3 CFU/g of the soil sample.

7.3   IDENTIFICATION OF ACTINOMYCETES

The polyphasic approach including the morphology, physiological properties, chemotaxonomic, and genotypic characteristics are frequency useful for the classification and identification of actinomycetes. The actinomycetes showed high variety of morphology. The presence and absence of aerial mycelia, the shape, size, texture, position, and the orientation of spores, as well as the presence of sclerotia, globose bodies, pycnidial-, sporangia, and synnemata-like structure, are the key characteristics to classify the actinomycetes in the genus level. In addition, cultural characteristics, the color of spore mass, the color of vegetative mycelia including soluble pigment and melanin formation, are the key factor that differentiated between actinomycetes strains. The standard method for characterization of *Streptomyces* species using several media, ISP (International *Streptomyces* project) was used for comparison of the morphological and cultural characteristics (Shirling and Gottlieb, 1966).

The chemotaxonomic characteristics including the isomers of diaminopimelic acids, whole-cell sugars, menaquinones, cellular fatty acids, and phospholipids composition as well as DNA guanine + cytosine content (mol%) are useful characteristics for the identification of actinomycetes. Actinomycetes contained diaminopimelic acid (DAP) at the third position of the peptide residues, which bound through the *N*-terminal to the carboxylic residue of muramic acid in cell wall peptidoglycan. *Streptomyces* strains contained LL-DAP while rare actinomycetes had *meso*-DAP or the combination of *meso*-DAP and LL-DAP or 3OH-*meso*-DAP (Table 7.2).

TABLE 7.2 Morphological and Chemotaxonomic Characteristics for Differentiating of the Genera of Actinomycetes[*]

Genus	Morphology	Isomer of diaminopimilic acid	Diagnostic sugar	Major menaquinone	Major phospholipids[γ]	Fatty acid pattern[α]	G+C content (mol%)
Family *Streptomycetaceae*							
Streptomyces	Spiral spore on aerial mycelia	LL-	None	MK–9(H$_{6,8}$)	PII	2c	66–78
Kitasatospora	Long spore chains on aerial mycelia	LL- and *meso-*	Gal	MK–9(H$_{6,8}$)	PII	2c	66–77
Family *Micromonosporaceae*							
Micromonospora	Monomeric spore on substrate mycelia	*meso-* and/or 3-OH-DAP	Ara, Xyl	MK–10(H$_{4,6}$) MK–9(H$_{4,6}$)	PII	3b	71–73
Actinoplanes	Sporangia with motile spore	*meso-* and/or 3-OH-DAP	Ara, Xyl	MK–9(H$_{4,6}$)	PII	2d	72–73
Dactylosporangium	Finger or club shape sporangia with motile spore	*meso-*	Ara, Xyl	MK–9(H$_{4,6,8}$)	PII	3b	72–73
Plantactinospora	Single or cluster spores on substrate mycelia	*meso-*	Ara, Gal, Xyl	MK–10(H$_{4,6,8}$)	PII	2d	69.7
Actinocatenispora	Spore chain on aerial mycelia	*meso-*	Ara, Gal, Xyl	MK–9(H$_{4,6}$)	PII	3b	72
Family *Streptosporangiaceae*							
Microbispora	Longitudinal pair of spore on aerial hyphae	*meso-*	Madurose	MK–9(H$_{2,4}$)	IV	3c	71–73

TABLE 7.2 *(Continued)*

Genus	Morphology	Isomer of diaminopimilic acid	Diagnostic sugar	Major menaquinone	Major phospholipids[γ]	Fatty acid pattern[α]	G+C content (mol%)
Nonomuraea	Spore chain or pseudosporangia on aerial hyphae	*meso-*	Madurose	MK–9(H$_{2,4}$)	IV	3c	64–69
Microtetraspora	Spore chain with four spores (or more) on short aerial hyphae	*meso-* and a small amount of LL-	Variable (madurose, or no diagnostic sugar)	MK–9(H$_{2,4}$)	IV	3c	69–71
Family *Thermomonosporaceae*							
Actinomadura	Spore chain on aerial mycelia	*meso-*	Gal, Glu, Man, Rib, Mad	MK–9(H$_{4,6,8}$)	PI,PIV	3a	66–73
Family *Nocardiaceae*							
Nocardia	Mycelia fragmented into rod or cocci	*meso-*	Ara, Gal	MK–8(H$_4$)	PII	Simple fatty acid + Mycolic acid	63–72

*Data from Ørskov (1923); Couch (1950); Thiemann et al. (1976); Thawai et al. (2006); Qin et al. (2009); Franco (2012); Goodfellow and Maldonadov (2012); Kempfer (2012a,b); Trujillo and Goodfellow (2012a,b).

[α] Fatty acid type according to Kroppenstedt (1985)

[γ] Polar lipids type according to Lechevalier et al. (1977)

Recently, 3,4 dihydroxy-DAP was reported (Matsumoto et al., 2014). In addition, the acyl type of muramic acid, *acetyl type* or *glycolyl type*, are also used for the identification of actinomycetes.

According to Lechevalier and Lechevalier (1970), aerobic actinomycetes containing *meso*-diaminopimelic acid was recognized as the fourth type of whole-cell diagnostic sugar (Type A: galactose, arabinose as diagnostic sugars, without xylose; Type B: madurose as diagnostic sugar without arabinose or xylose; Type C: no diagnostic sugar; and Type D: xylose and arabinose as diagnostic sugars). These whole-cell sugars are determined by using both TLC (cellulose) (Staneck and Robert, 1974) and the HPLC method (Mikami and Ishida, 1983).

In 1997, Lechevalier et al. surveyed the phospholipids which located in the cell membrane (phospholipid bilayer) of 97 strains representing 20 genera of the actinobacteria. On the basis of presence and absence of certain nitrogenous phospholipids, the phospholipid composition was classified to five types [PI, no nitrogenous phospholipids; PII, with only one nitrogenous phospholipid, phosphatidyl ethanolamine; PIII, contained phosphatidylcholine as characteristics phospholipid; PIV, contained an unknown previously described phospholipids containing glucosamine; and PV, contained phosphatidyl glycerol in addition to unknown previously undescribed phospholipids containing glucosamine]. The extraction and analysis of phospholipids, using two-dimensional thin layer chromatography was previously described by Minnikin et al. (1984). According to this method (10×10 cm silica gel) was developed in chloroform:methanol:water (65:25:4, v/v) and chloroform:acetic acid:methanol:water (40:7.5:6:2, v/v), respectively. Several spraying reagents, phosphomolybdic acid, anisaldehyde, Dragendorff's reagent, ninhydrin, and molybdenum blue, were used to visualize the chromatogram of each TLC plates.

Isoprenoid quinones (menaquinones and ubiquinones) are constituents of the bacterial plasma membrane which play a role in electron transport and oxidative phosphorylation system and are widely distributed in aerobic bacteria. Collins et al. (1977) found that menaquinones were the only isoprenoid quinones found in actinobacteria. The variations in the length and degree of unsaturation on polyprenyl residue are useful for the criteria of identification of actinomycetes. The accurate method to identify menaquinone is liquid chromatography/mass spectrometry (LC/MS). Fatty acids in bacteria are the composition of the cell membrane. Gas-liquid chromatography is the advantage technique for cellular fatty acid identification (Sasser, 1990). Furthermore, the presence of mycolic acids, 2-hydroxy–3-alkyl fatty acids with a long alkyl chain, is the characteristics for identification of the actinomycetes especially family *Corynebacteriaceae*, *Mycobacteriaceae*, and *Nocardiaceae*.

For the identification of actinomycetes in the genus level, the observation of spore morphology, chemotaxonomic characteristics and 16S rRNA gene sequence analysis are the key characteristics. At present, most of the identification started with the 16S rRNA gene analysis. Ribosomal 16S rRNA is a part of the 30S small subunit of the bacterial ribosome. Because of its highly conserved regions in all bacteria, slow rate of evolution, the16S rRNA gene sequence is likely the barcode for identification of actinomycetes in the genus level. At present, Basic Local Alignment Search Tool (BLAST) search of 16S rRNA gene can be determined easily. BLAST analysis can be done online using the NCBI webpage at http://blast.ncbi.nlm.nih.gov/Blast.cgi or http://ezbiocloud.net/ (Kim et al., 2012). In peat swamp forest soils, *Streptomyces* are the most abundant population of actinomycetes. The other rare genera which have been often isolated from this environment were *Micromonospora*, *Microbispora*, *Microtetraspora*, *Actinomadura*, *Asanoa*, *Acrocarpospora*, *Nonomuraea*, *Actinocatenispora*, *Plantactinospora*, *Dactylosporangium*, *Actinoplanes*, and *Nocardia*. These genera are differentiated by their morphology and chemotaxonomic characteristics (Table 7.2). There are many novel actinomycete species distributed in peat swamp forest soils: *Streptomyces actinomycinicus* RCU-197[T] contained LL-diaminipimelic acid in cell-wall peptidoglycan. The major cellular fatty acids were iso-$C_{14:0}$, iso-$C_{15:0}$, anteiso-$C_{15:0}$ and iso-$C_{16:0}$. This strain contained type PII phospholipid which had phosphatidylethanolamine. The strain contained diphosphatidylglycerol, phosphatidylethanolamine, phosphatidylglycerol, phosphatidylinositol, phosphatidylinositol mannosides, an unknown aminolipid, and two unknown phospholipids. MK–9(H_6) and MK–9(H_8) were major components. The diagnostic sugars of this species were not observed, but only glucose and ribose were detected in the whole-cell analysis (Tanasupawat et al., 2016a).

Micromonospora species including *M. eburnea*, *M. siamensis*, *M. auratinigra* and *M. pattaloongensis* (Thawai et al., 2004b, 2005a,b, 2008; Songsumanus et al., 2011) produced the monomeric spore on substrate mycelia. These species exhibited type PII, and the phospholipid profiles were phosphatidyl ethanolamine, diphosphatidyl glycerol, phosphatidyl inositol, and phosphatidyl inositol mannosides. They have branched satu-rated fatty acid (iso-$C_{15:0}$, iso-$C_{16:0}$, iso-$_{C17:0}$, anteiso-$C_{15:0}$ and anteiso-$C_{17:0}$) and saturate fatty acid especially $C_{17:0}$ as a major component. The major menaquinones were MK–10(H_4), MK–10(H_6), MK–9(H_4) and MK–9(H_6) while xylose and arabinose were diagnostic sugars.

Acrocarpospora phusangensis contained *meso*-diaminopimelic acid, glutamic, and alanine in cell-wall peptidoglycan (Niemhom et al., 2013a). The predominant menaquinone was MK–9(H4). Whole-cell sugar contains glucose, mannose, rhamnose, ribose, and madurose (diagnostic sugar). Phospholipid profiles were diphosphatidylglycerol, phosphatidylglycerol, phosphatidylethanolamine, phosphatidylmethylethanolamine, hydroxy-phosphatidylethanolamine, phosphatidylinositol, phosphatidylinositol mannoside, ninhydrin-positive phosphoglycolipids, and two unidentified phospholipids. Major fatty acids were $C_{17:1}$ ω8c, iso-$C_{16:0}$, $C_{16:0}$, $C_{17:0}$, anteiso-$C_{15:0}$, iso-$C_{14:0}$, $C_{16:1}$ω7c, 10-methyl-$C_{17:0}$ and iso-$C_{15:0}$.

Asanoa siamensis (Niemhom et al., 2013b) contained *meso*-diami-nopimelic acid and 3-OH diaminopimelic acid, glutamic acid, glycine, and alanine in cell wall-peptidoglycan. Predominant menaquinones were MK–10(H_8) and MK–9(H_9). Whole-cell sugars contained glucose, mannose, rhamnose, ribose, and xylose. Major phospholipids were diphosphatidylglycerol, phosphatidylglycerol, phosphatidylinositol, and phosphatidylethanolamine. Major cellular fatty acids are iso-$C_{15:0}$, anteiso-$C_{15:0}$ and anteiso-$C_{17:0}$.

Actinomadura rayongensis (Phongsopitanun et al., 2015a) belonged to the family *Thermomonosporaceae*, was isolated from soil of Nong Jum Rung peat swamp forest, Rayong Province, Thailand. Its cell-wall peptidoglycan contained meso-diaminopimelic acid. Whole-cell hydroly-sate contains glucose, ribose, galactose, and madurose. Major polar lipids are diphosphatidylglycerol, phosphatidylinositol, phosphatidylinositol mannoside (type PI). Predominant menaquinone is MK–9(H_6), MK–9(H_8) and MK–9(H_4). Major cellular fatty acids are $C_{16:0}$ and iso-$C_{16:0}$. *Dactylo-sporangium sucinum* (Phongsopitanun et al., 2015b) was isolated from Nong-Jum-Rung peat swamp forest soil. Its cell-wall peptidoglycan contains 3-OH diaminopimelic acid and a small amount of meso-diaminopimelic acid. Major phospholipids are diphosphatidylglycerol, phosphatidylethanolamine, phosphatidylglycerol, phosphatidylinositol, and phosphatidylinositol mannosides. The major isoprenoid quinones were MK–9(H$_8$) and MK–9(H$_6$). The major cellular fatty acids were C$_{17:0}$, C$_{18:0}$, C$_{18:1}$ω9c, anteiso-C$_{15:0}$, iso-C$_{15:0}$, iso-C$_{16}$:0, iso-C$_{17:0}$ and anteiso-C$_{17:0}$.

Actinocatenispora thailandica (Thawai et al., 2006), novel genera which belong to the family *Micromonosporaceae*, was isolated from peat swamp forest in Pattaloong Province, Thailand. Cell-wall peptidoglycan contains glutamic acid, glycine, alanine, and *meso*-diaminopimelic acid. Galactose, xylose, glucose, arabinose, mannose, and ribose were observed in the whole-cell hydrolysate. Major phospholipids are phosphatidyl-ethanolamine, diphosphatidylglycerol, phosphatidylinositol, phospha-tidylinositol mannosides and phosphatidylglycerol (type PII). Major manaquinones were MK–9(H$_4$) and MK–9(H$_6$).

Actinaurispora siamensis (Thawai et al., 2010), a later reclassified as *Plantactinospora siamensis*, was isolated from soil collected form peat swamp forest in Chaing Mai Province, Thailand. It contained glutamic acid, glycine alanine and *meso*-diaminopimelic acid in the cell wall. Whole-cell hydrolysates contained glucose, xylose, mannose, and galactose. Major phospholipids were phosphatidylethanolamine, diphosphatidylglycerol, phosphatidylinositol, and phosphatidylinositol mannosides (type PII). Predominant menaquinones were MK–9(H$_6$), MK–10(H$_6$), MK–9(H$_8$) and MK–10(H$_8$). Major fatty acids were iso-C$_{15:0}$, iso-C$_{16:0}$, anteiso-C$_{17:0}$, anteiso-C$_{15:0}$ and iso-C$_{17:0}$.

Nocardia rayongensis (Tanasupawat et al., 2016b) isolated from Nong Jum Rung peat swamp forest in Rayong Province, Thailand contained *meso*-diaminopimelic acid and mycolic acid. Whole-cell sugars contained galactose and arabinose as diagnostic sugars. Major menaquinone was MK–8(H$_4$ω-cycl). Major phospholipids were diphosphatidylglycerol, phosphatidylethanolamine, phosphatidylinositol, and phosphatidylinositol mannosides. The major cellular fatty acids were C$_{16:0}$ and C$_{18:1}$ω9c.

7.4 DIVERSITY OF ACTINOMYCETES IN PEAT SWAMP FOREST SOILS

In Thailand, Thawai (2004a) studied the diversity of *Micromonospora* species of peat swamp forest in the southern part of Thailand including Trang, Pattaloong, Yala, and Narathiwat Provinces. In this study, totally 52 *Micromonospora* species were isolated from 17 soil samples. All isolates showed monomeric spores on the substrate hyphae having an approximate diameter of 0.5–0.6 μm. The spores were rough, nodular, and smooth on the surface and non-motile. Based on 16S rRNA gene analysis the author classified these strain in 11 groups. On the basis of a polyphasic approach including DNA-DNA hybridization, most of them could be classified in *M. chalcea* and *M. aurantiaca*. At that time many isolates was an unidentified species, which later were classified and proposed into the new species of the genus *Micromonospora* including *Micromonospora eburnea*, *Micromonospora aurantinigra*, *Micromonospora siamensis*, and *Micromonospora narathiwatensis*.

In 2010, the rare actinomycete species from 4 peat swamp forest soils in Phayao Province, Thailand were isolated by Songsumanus, and five isolates were obtained. Two isolates were closely related to *Nonomuraea helvata* JCM 3143[T] (98.6% similarity), others were identified as *Nonomuraea* based on 16S rRNA gene analysis. Three isolates, P1440, P1803, and P1605, produced extensively branched substrate hyphae that fragment into rod-shaped and non-motile spores. Aerial hyphae were powdery and yellowish white. The substrate mycelia were light orange yellow to brilliant yellow on YMA. Strains P1440 and P1803 exhibited the highest 16S rRNA gene sequence similarity to *Nocardia arthritidis* JCM 12120[T] (99.4%) and *N. araoensis* JCM 12118[T] (99.4%), respectively, while strain P1605 exhibited the highest 16S rRNA gene sequence similarity to *N. cyriacigeorgica* JCM 11763[T] (99.9%).

Niemhom (2012) reported the diversity of rare actinomycetes of temperate peat swamp forest soils collected from Phu Sang National Park (pH 6.5–8.0), Phayao Province and Doi Inthanon National Park (pH 3.5–6.5), Chiangmai Province, Thailand. Totally 109 actinomycetes were isolated (72 isolates from Phu Sang National park and 37 isolates from Doi Inthanon National park). On the basis of morphology, they were classified in 14 genera including *Acrocarpospora* (1 isolate), *Actinomadura* (1 isolate), *Actinoplanes* (14 isolates), *Amycolatopsis* (2 isolates), *Asanoa* (2 isolates), *Dactylosporangium* (18 isolates), *Herbidospora* (1 isolate), *Micromonospora* (34 isolates), *Nocardia* (10 isolates), *Planosporangium* (1 isolate), *Polymorphospora* (3 isolates), *Pseudonocardia* (2 isolates), *Sphaerisporangium* (4 isolates) and *Streptosporangium* (7 isolates) and 9 unidentified isolates.

In addition, 77 actinomycete isolates from 16 peat swamp forest soil samples collected in Rayong Province were isolated and they were identified as *Streptomyces* (35 isolates), *Micromonospora* (3 isolates),

Microbispora (9 isolates), *Microtetraspora* (6 isolates), *Actinomadura* (2 isolates), *Nocardia* (4 isolates), *Actinoallomurus* (1 isolate), *Dactylosporangium* (3 isolates) and unknown species (13 isolates) based on their morphology and16S rRNA gene sequence analyses (Tanasupawat and Suwanborirux, 2014).

Peat swamp forest showed high biodiversity of not only plants and animals but also microorganisms. In Thailand, since 2005, many of novel actinomycetes were isolated from peat swamp forest soils. Most novel actinomycetes discovered from peat swamp forest were found to be members of the family *Micromonosporaceae*, especially the genus *Micromonospora* including *M. humi*, *M. siamensis*, *M. aurantionigra*, *M. eburnean* and *M. narathiwatensis* reported from the peat swamp forest soil of Thailand. Other than *Micromonospora* species, *Asanoa siamensis*, *Dactylosporangium sucinum* and *Plantactinospora siamensis*, which belonged to this family, were isolated from Thailand. In 2006, Thawai, and colleagues described the novel genus of family *Micromonosporaceae* and *Actinocatenispora*. The type species of this genus are *Actinocatenispora thailandica*. The member of this genus found to be a Gram-positive, non-acid-fast, aerobic with branching substrate mycelia. Type species of this genus showed phylogenetically separated from the other genera in the family *Micromonosporaceae*. As it produced true aerial mycelia which formed and bear the spore chain (more than 10 spores), it can differentiate from other genera of family *Micromonosporaceae* using this morphology. Other novel actinomycetes isolated from Thai peat swamp forest are *Ac. phusangensis*, *A. rayongensis*, *N. rayongensis*, and *S. actinomycinicus*. The characteristics of these novel species are listed in Table 7.3. The scanning electron micrographs of some actinomycetes are showed in Figure 7.2.

7.5   SECONDARY METABOLITES AND BIOLOGICAL ACTIVITY

In Thailand, Thawai et al. (2000, 2004a) isolated actinomycetes from peat swamp forest soils in Trang and Pattaloong Province and these strains were identified as *Streptomyces*, *Actinomadura*, *Micromonospora*, *Nocardia*, and *Actinoplanes*. Some of them exhibited antibacterial activity against *Escherichia coli* ATCC 25922, *Kocuria rhizophila* ATCC9341, *Staphylococcus aureus* ATCC 25923, *Bacillus subtilis* ATCC 6633 and *Pseudomonas aeruginosa* ATCC 27853. *Micromonospora auratinigra*

TABLE 7.3 Characteristics and Distribution of Novel Actinomycetes Isolated from Peat Swamp Forest Soils

Species	Province/ Part of Thailand	Growth Temp. (°C)	NaCl (% w/v)	L-Arabinose	Cellobiose	D-Fructose	D-Galactose	D-Mannitol	Melibiose	Raffinose	L-Rhamnose	D-xylose	Glycerol	Milk peptonization	Gelatin liquefaction	Nitrate reduction	DNA G+C (mol %)	References
Actinocatenispora thailandica	Pattaloong/S	25–30	7	-	+	-	-	+	+	+	-	-	+	w	w	+	72	Thawai et al., 2006
Acrocarpospora phusangensis	Phayao/N	25–30	1	+	+	+	+	+	+	+	+	+	-	+	-	-	71	Niemhom et al., 2013a
Asanoa siamensis	Phayao/N	25–30	2	+	+	+	+	+	+	w	+	+	w	+	-	w	72.3	Niemhom et al., 2013b
Actinomadura rayongensis	Rayong/E	25–37	4	-	+	ND	ND	-	ND	-	ND	-	ND	+	+	w	73.7	Phongsopitanun et al., 2015a
Dactylosporangium sucinum	Rayong/E	25–37	3	+	+	ND	ND	+	ND	-	ND	+	ND	+	-	-	72.5	Phongsopitanun et al., 2015b
Micromonospora humi	Phayao/N	20–30	5	+	-	-	+	-	-	+	ND	ND	-	+	+	-	73.0	Songsumanus et al., 2011
Micromonospora siamensis	Pattaloong/S	25–30	5	+	+	w	+	-	+	+	-	+	-	+	+	-	73.0	Thawai et al., 2005b
Micromonospora auratinigra	Pattaloong/S	25–30	2	+	+	+	+	-	+	+	+	+	-	-	-	+	72.8	Thawai et al., 2004b
Micromonospora eburnea	Yala/S	25–30	4	w	+	-	w	w	-	+	+	+	+	+	+	+	71.5	Thawai et al., 2005a
Micromonospora narathiwatensis	Narathiwat/S	25–30	4	w	+	-	w	w	-	-	-	+	+	+	+	+	71.4	Thawai et al., 2007
Nocardia rayongensis	Rayong	20–37	6	-	w	ND	ND	w	-	-	ND	w	ND	-	-	+	71	Tanasupawat et al., 2016b
Plantactinospora siamensis	Chiang Mai/N	25–30	1.5	-	+	+	+	-	+	+	+	+	-	w	w	-	72.6	Thawai et al., 2010
Streptomyces actinomycinicus	Rayong/E	30–37	7	w	+-	+	ND	w	ND	w	ND	w	+	+	+	-	ND	Tanasupawat et al., 2016a

+, positive; w, weakly positive; -, negative; ND, no data; N, Northern; E, Eastern; S, Southern Thailand

TT1–11^T produced a novel 24-membered polyene lactam macrolide, micromonosporin A (Figure 7.3). The derivative of micromonosporin A using hydrogenation reaction finally produced a new stable derivative, 9,11,13-trihydroxy–14,19,24-trimethyl–1-azacyclotetracosan–2-one. This derivative displayed weak antimicrobial activity and also exhibited anti-malarial activity at IC$_{50}$of 3.1 μg/ml and antimycobacterial activity with the MIC value of 50 μg/ml (Thawai et al., 2004c).

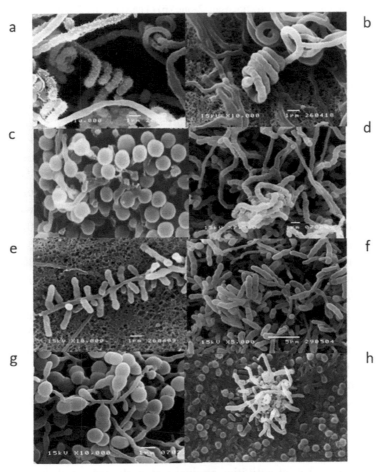

FIGURE 7.2 Scanning electron micrographs of *Streptomyces actinomycinicus* RCU–197^T (a), *Streptomyces* sp. RCU–043 (b), *Micromonospora* sp. KM4–29 (c), Thawai (2004a), *Nocardia rayongensis* RY45–3^T (d), *Actinomadura rayongensis* RCU–068^T (e), *Microtetraspora* sp. RCU–213 (f), *Microbispora* sp. CU–224 (g) and *Dactylosporangium sucinum* RY35–23^T (h).

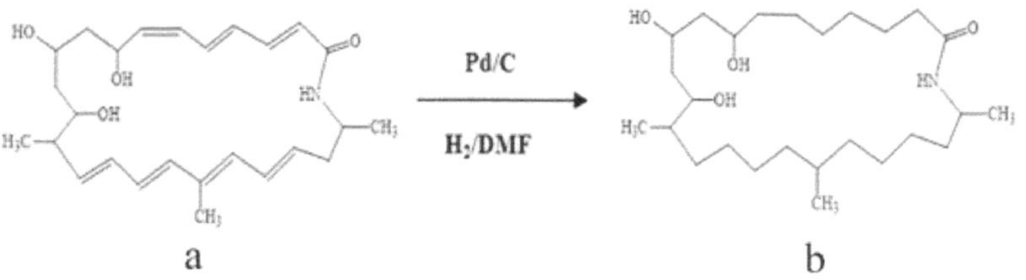

a   b

FIGURE 7.3 Structures of micromonosporin a (a) and it derivative 9,11,13-trihydroxy–14,19,24-trimethyl–1-azacyclotetracosan–2-one (b) (Thawai et al., 2004c).

Songsumanus et al. (2007) studied 67 actinomycete isolates from 17 soil samples collected from peat swamp forest in Phayao Province. Based on their morphological, cultural, physiological, and biochemical characteristics, they were identified as *Streptomyces* (62 isolates), *Micromonospora* (1 isolate) and *Nocardia* (4 isolates). On the primary screening for antimicrobial activities, 25 isolates including *Streptomyces* (24 isolates) and *Nocardia* (1 isolate) exhibited antimicrobial activities. Among them, seven isolates of *Streptomyces* showed the activities against *Staphylococcus aureus* ATCC 25923, *B. subtilis* ATCC 6633, *K. rhizophila* ATCC 9341, *E. coli* ATCC 25922, and *P. aerugenosa* ATCC 27853, while few isolates showed the activities against *Candida albicans* ATCC 10231. The 16S rDNA sequence analysis indicated the strain PY3–21 was similar to *Streptomyces pseudogriseolus* subsp. *glucofermentans* DSM 40026[T] and *S. griseorubens* DSM 40160[T] with 99.6% similarity. PY9–45 was similar to *S. endus* NRRL 2339[T], *S. sporocinereus* DSM 41460[T], and *S. hygroscopicus* DSM 40578[T] with 98.9% similarity. Strains RN3–4 and RN1–7 were closed to *S. parvulus* DSM 40048[T] and *S. rubrogriseus* DSM 41477[T] (99.7% and 98.8%, respectively). In addition, strains P17–3 and P18–5 showed 16S rDNA close to *Nocardia cyriacigeorgica* ATCC 14759[T] with 99.8% similarity.

Jeffery et al. (2011) isolated 40 actinomycetes from soil samples collected from MARDI Peat Land Research Station at Sessang, Sarawak. Based on this study, several isolates showed enzyme activities of cellulase, mannanase, xylanase, protease, and lipase. Moreover, some of them showed antimicrobial activities against test pathogens such as *Ralstonia solanacearum*, *Pantoea stewartii* and *B. subtilis*. The 16S rRNA gene analysis of eight isolates, which showed the best activities, revealed that seven were identified in the genus *Streptomyces* and one was identified in the genus *Kitasatospora*.

Some of *Acrocarpospora* (1 isolate), *Actinomadura* (1 isolate), *Actinoplanes* (2 isolates), *Asanoa* (1 isolate), *Dactylosporangium* (1 isolate), *Herbidospora* (1 isolate), *Micromonospora* (7 isolates), *Nocardia* (1 isolate), *Planosporangium* (1 isolate), *Polymorphospora* (3 isolates) and *Streptosporangium* (1 isolate) isolated from peat swamp forest soils collected from Phu Sang National park, Phayao Province and Doi Inthanon National Park, Chiangmai Province, Thailand, exhibited antimicrobial activity against *B. subtilis* ATCC 6633, *E. coli* ATCC 25922, *K. rhizophila* ATCC 9341, *S. aureus* ATCC 25923 and *S. aureus* DMST 20654 (MRSA) (Niemhom, 2012). The crude ethyl acetate extract of *Actinoplanes* sp. PS19–16, *Dactylosporangium* sp. PS19–2, *Micromonospora* sp. PS3–1, PS4–2 and PS6–1 showed anti-human small cell lung cancer activity (87.1–92.0% inhibition). *Micromonospora* sp. PS1–11, PS3–1 and PS6–1, and, *Polymorphospora* sp. PS6–6 and PS18–2 exhibited anti-*Mycobacterium tuberculosis* strain $H_{37}Ra$ with MIC 3.13–50.00 µg/ml. Only *Micromonospora* sp. PS3–1, PS4–2 and PS6–1 exhibited *Plasmodium falciparum* K1 with IC_{50} 2.00–3.23 µg/ml (Niemhom, 2012).

Niyasom et al. (2015) studied antibacterial activities of actinomycetes isolated from Phatthalung Botanical Garden, Phatthalung Province, Thailand, by using agar plug method. Three isolates exhibited antibacterial activity against all four tested bacteria, which were *Bacillus cereus*, *P. aeruginosa*, *E. coli*, and *S. aureus*. Furthermore, 10 isolates showed antifungal activity against *Fusarium* sp., *Curvularia* sp. and *Colletotrichum gloeosporioides*. The preliminary screening of novel *Streptomyces* species, *S. actinomycinicus* RCU–197[T], revealed that it showed antimicrobial activity against Gram-positive bacteria including *S. aureus* ATCC 25923 and *B. subtilis* ATCC 6633. The LC/MS screening of culture broth found that the strain produced Actinomycin D as its major secondary metabolite (Tanasupawat et al., 2016).

7.6 CONCLUSION

The isolation of actinomycetes from peat swamp forest soils was incorporated with several methods and using different selective media. In Thailand, the novel taxa, *Acrocarpospora phusangensis*, *Asanoa siamensis, Actinocatenispora thailandica, Plantactinospora siamensis, Actinomadura rayongensis, Dactylosporangium sucinum, Micromonospora chalcea, M. aurantiaca, M. eburnea, M. aurantinigra, M. siamensis,* and *M. narathiwatensis* are proposed. *Micromonospora* species, *Nonomuraea helvata, Nocardia rayongensis, N. arthritidis, N. cyriacigeorgica, Actinoplanes, Amycolatopsis, Asanoa, Herbidospora, Microbispora, Microtetraspora, Planosporangium, Polymorphospora, Pseudonocardia, Sphaerisporangium, Streptomyces actinomycinicus, Streptomyces* species and *Streptosporangium* strains were isolated. *M. aurantionigra* produced micromonosporin A while *S. actinomycinicus* produced actinomycin D. The peat swamp forest soils are rich of novel actinomycetes and the continuous investigation of them in this ecosystem should be further studied for their diversity and the discovery of novel compounds.

KEYWORDS

- *Acrocarpospora*
- *Actinocatenispora*
- *Actinomadura*
- **Actinomycetes**
- *Actinoplanes*
- *Amycolatopsis*
- *Asanoa*
- *Dactylosporangium*
- **diversity**
- *Herbidospora*
- **LC/MS**
- **metabolites**
- *Microbispora*
- *Micromonospora*
- *Microtetraspora*
- *Nocardia*
- *Nonomuraea*
- **peat swamp forest soil**

- *Planosporangium*
- *Plantactinospora*
- *Polymorphospora*
- *Pseudonocardia*
- *Sphaerisporangium*
- *Streptomyces*
- *Streptosporangium*

REFERENCES

Berdy, J., (2005). Bioactive microbial metabolites. *J. Antibiot.*, *58*, 1–26.

Burg, R. W., Miller, B. M., Baker, E. E., Birnbaum, J., Currie, S. A., Hartman, R., et al., (1979). Avermectins, new family of potent anthelmintic agents: Producing organism and fermentation. *Antimicrob. Agent Ch.*, *15*, 361–367.

Collins, M. D., Pirouz, T., Goodfellow, M., & Minnikin, D. E., (1977). Distribution of menauinones in actinomycetes and corynebacteria. *J. Gen. Microbiol.*, *100*, 221–230.

Couch, J. N., (1950). *Actinoplanes*, a new genus of the actinomycetales. *J. Elisha Mitchell Sci. Soc.*, *66*, 87–92.

Currie, C. R., Scott, J. A., Summerbell, R. C., & Malloch, D., (1999). Fungus-growing ants use antibiotic-producing bacteria to control garden parasites. *Nature*, *388*, 701–704.

El-Nakeeb, M. A., & Lechevalier, H. A., (1963). Selective isolation of aerobic actinomycetes. *Appl. Microbiol.*, *11*, 75–77.

Franco, C. M. M., & Genus, I. V., (2012). Microbispora. nonomura and ohara 1957, 307[AL] emend. Zhang, Wang, and Ruan 1998a, 418. In: Goodfellow, M., Kempfer, P., Busse, H. J., Trujillo, M. E., Suzuki, K., Ludwig, W., & Whitman, W. B., (eds.), *Bergey's Manual of Systematic Bacteriology Part B* (Vol. 5, pp. 1831–1838). Springer, New York.

Gavrish, E., Bollmann, A., Epstein, S., & Lewis, K., (2008). A trap for *in situ* cultivation of filamentous actinobacteria. *J. Microbiol. Methods*, *72*, 257–262.

Goodfellow, M., Maldonado, L. A., & Genus, I., (2012). Nocardia Trevisan 1889[AL]. In: Goodfellow, M., Kempfer, P., Busse, H. J., Trujillo, M. E., Suzuki, K., Ludwig, W., & Whitman, W. B., (eds.), *Bergey's Manual of Systematic Bacteriology Part* B (Vol. 5, pp. 377–419). Springer, New York.

Hayakawa, M., & Nonomura, H., (1987). Humic acid-vitamin agar, a new medium for the selective isolation of soil actinomycete. *Ferment. Technol.*, *65*, 501–509.

Hayakawa, M., Lino, H., Takeuchi, S., & Yamazaki, T., (1997). Application of a method incorporating treatment with Chloramine-T for the selective isolation of *Streptosporangiaceae* from soil. *J. Ferment. Bioeng.*, *84*, 599–602.

Hayakawa, M., Momose, Y., Kajiura, T., Yamazaki, T., Tamura, T., Hatano, K., & Nonomura, H., (1995). A selective isolation method for *Actinomadura viridis* in soil. *J. Ferment. Bioeng.*, *79*, 287–289.

Hayakawa, M., Momose, Y., Yamazaki, T., & Nonomura, H., (1996). A method for the selective isolation of *Microtetraspora glauca* and related four-spored actinomycetes from soil. *J. Appl. Bacteriol.*, *80*, 375–386.

Hayakawa, M., Otoguro, M., Takeuchi, T., Yamazaki, T., & Limura, Y., (2000). Application of a method incorporating differential centrifugation for selective isolation of motile actinomycetes in soil and plant litter. *Antonie Van Leeuwenhoek*, *78*, 171–185.

Hayakawa, M., Sadakata, E., Kajiura, T., Nonomura, H., (1991a). New methods for the highly selective isolation of *Micromonospora* and *Microbispora* from soil. *J. Ferment. Biotechnol.*, *72*, 320–326.

Hayakawa, M., Tamura, T., Lino, H., Nonomura, H., (1991b). New methods for the highly selective isolation of *Streptosporangium* and *Dactylosporangium* from soil. *J. Ferment. Bioeng.*, *72*, 327–333.

Hirsch, C. F., & Christensen, D. L., (1983). Novel method for selective isolation of actinomycetes. *Appl. Environ. Microbiol.*, *46*, 925–929.

Inahashi, Y., Matsumoto, A., Omura, S., & Takahashi, Y., (2011). *Streptosporangium oxazolinicum* sp. nov., a novel endophytic actinomycete producing new antitrypanosomal antibiotics, spoxazomicins. *J. Antibiot.*, *64*, 297–302.

Jeffery, J. S. H., Norzaimawati, A. N., & Rosnah, H., (2011). Prescreening of bioactivities from actinomycetes isolated from forest peat soil of Sarawak. *J. Trop. Agr. Food Sci.*, *39*, 245–253.

Kempfer, P., & Genus, I., (2012b). *Streptomyces* Waksman and Henrici 1943, 339[AL] emend. Witt and Stackebrandt 1990, 370 emend. Wellington, Stackebrandt, Sanders, Wolstrup, and Jorgensen 1992, 159. In: Goodfellow, M., Kempfer, P., Busse, H. J., Trujillo, M. E., Suzuki, K., Ludwig, W., & Whitman, W. B., (eds.), *Bergey's Manual of Systematic Bacteriology, Part B* (Vol. 5, pp. 1455–1767). Springer, New York.

Kempfer, P., & Genus, V. I., (2012a). Nonomuraea corrig. Zhang, Wang, and Ruan 1998b, 419VP. In: Goodfellow, M., Kempfer, P., Busse, H. J., Trujillo, M. E., Suzuki, K., Ludwig, W., & Whitman, W. B., (eds.), *Bergey's Manual of Systematic Bacteriology, Part B* (Vol. 5, pp. 1844–1861). Springer, New York.

Kim, O. S., Cho, Y. J., Lee, K., Yoon, S. H., Kim, M., Na, H., et al., (2012). Introducing EzTaxon-e: A prokaryotic 16S rRNA gene sequence database with phylotypes that represent uncultured species. *Int. J. Syst. Evol. Microbiol.*, *62*, 716–721.

Kroppenstedt, R. M., (1985). Fatty acid and menaquinone analysis of actinomycetes and related organisms. In: Goodfellow, M., & Minnikin, D. E., (eds.), *Chemical Methods in Bacterial Systematics* (pp. 173–199). Academic Press, London.

Küster, E., & Williams, S. T., (1964). Selection of media for isolation of *Streptomycetes*. *Nature*, *202*, 928–929.

Lechevalier, H. A., (1975). The 25 years of neomycin. *CRC Crit. Rev. Microbiol.*, *3*, 359–397.

Lechevalier, M. P., & Lechevalier, H., (1970). Chemical composition as a criterion in the classification of aerobic actinomycetes. *Int. J. Syst. Bacteriol.*, *20*, 435–443.

Lechevalier, M. P., Bievre, C. D., & Lechevalier, H., (1977). Chemotaxonomy of aerobic actinomycetes: Phospholipid composition. *Biochem. Syst. Ecol.*, *5*, 249–260.

Levine, D. P., (2006). Vancomycin: A history. *Clin. Infect. Dis.*, *42*, S5–12.

Matsumoto, A., Kawaguchi, Y., Nakashima, T., Iwatsuki, M., Omura, S., & Takahashi, Y., (2014). *Rhizocolahellebore* gen. nov., sp. nov., an actinomycete of the family

Micromonosporaceae containing 3,4-dihydroxydiaminopimelic acid in the cell-wall peptidoglycan. *Int. J. Syst. Evol. Microbiol.*, *64*, 2706–2711.

Mikami, H., & Ishida, Y., (1983). Post-column fluorometric detection of reducing sugar in high-performance liquid chromatography using arginine. *Bunseki. Kagaku.*, *32*, E207–E210.

Minnikin, D. E., O'Donnell, A. G., Goodfellow, M., Alderson, G., Athalye, M., Schaal, A., & Parlett, J. H., (1984). An integrated procedure for the extraction of bacterial isoprenoid quinones and polar lipids. *J. Microbiol. Methods*, *2*, 233–241.

Niemhom, N., (2012). Biological activities and taxonomy of rare actinomycetes isolated from temperate peat swamp forest in Thailand. *Master Thesis*, King Mongkut's Institute of Technology, Ladkrabang, Thailand (p. 246).

Niemhom, N., Suriyachadkun, C., Tamura, T., & Thawai, C., (2013a). *Acrocarpospora phusangensis* sp. nov., isolated from a temperate peat swamp forest soil. *Int. J. Syst. Evol. Microbiol.*, *63*, 2174–2179.

Niemhom, N., Suriyachadkun, C., Tamura, T., & Thawai, C., (2013b). *Asanoa siamensis* sp. nov., isolated from soil from a temperate peat swamp forest. *Int. J. Syst. Evol. Microbiol.*, *63*, 66–71.

Niyasom, C., Boonmak, S., & Meesei, N., (2015). Antimicrobial activity of acidophilic actinomycetes isolated from acidic soil. *KMITL Sci. Tech. J.*, *15*, 62–69.

Nonomura, H., & Ohara, Y., (1969a). Distribution of actinomycetes in soil. VI. A culture method effective for both preferential isolation and enumeration of *Microbispora* and *Streptosporangium* strains in soil (Part 1). *J. Ferment. Technol.*, *47*, 463–469.

Nonomura, H., & Ohara, Y., (1969b). Distribution of actinomycetes in soil. VII. A culture method effective for both preferential isolation and enumeration of *Microbispora* and *Streptosporangium* strains in soil (Part 2), classification of the isolates. *J. Ferment. Technol.*, *47*, 701–709.

Nonomura, H., & Ohara, Y., (1971a). Distribution of actinomycetes in soil. VIII. Green spore group of *Microtetraspora*, its preferential isolation, and taxonomic characteristics. *J. Ferment. Technol.*, *49*, 1–7.

Nonomura, H., & Ohara, Y., (1971b). Distribution of actinomycetes in soil. I. New species of the genera *Microbispora* and *Microtetraspora*, and their isolation method. *J. Ferment. Technol.*, *49*, 887–894.

Nonomura, H., & Ohara, Y., (1971c). Distribution of actinomycetes in soil. X. New genus and species of monosporic actinomycetes in soil. *J. Ferment. Technol.*, *49*, 895–903.

Nonomura, H., & Ohara, Y., (1971d). Distribution of actinomycetes in soil. XI. Some new species of the genus *Actinomadura* Lechevalier et al. *J. Ferment. Technol.*, *49*, 904–912.

Orskov, J., (1923). *Investigations in to the Morphology of Ray Fungi*. Copenhagen: Levin and Munksgaard, Denmark.

Pathom-Aree, W., Stach, J. E. M., Ward, A. C., Horikoshi, K., Bull, A. T., & Goodfellow, M., (2006). Diversity of actinomycetes isolated from challenger deep sediment (10,989 m) from the Mariana Trench. *Extremophiles*, *10*, 181–189.

Phongsopitanun, W., Kudo, T., Ohkuma, M., Suwanborirux, K., & Tanasupawat, S., (2015b). *Dactylosporangium sucinum* sp. nov., isolated from Thai peat swamp forest soil. *J. Antibiot.*, *68*, 379–384.

Phongsopitanun, W., Tanasupawat, S., Suwanborirux, K., Ohkuma, M., & Kudo, T., (2015a). *Actinomadura rayongensis* sp. nov., isolated from peat swamp forest. *Int. J. Syst. Evol. Microbiol.*, *65*, 890–895.

Posa, M. R. C., Wijedasa, L. S., & Corlett, R. T., (2011). Biodiversity and conservation of tropical peat swamp forests. *BioScience*, *61*, 49–57.

Qin, S., Chen, H. H., Zhao, G. Z., Li, J., Zhu, W. Y., Xu, L. H., *et al.*, (2012). Abundant and diverse endophytic actinobacteria associated with medicinal plant *Maytenus austroyunnanensis* in Xishuangbanna tropical rain forest revealed by culture-dependent and culture-independent methods. *Envion. Microbiol. Report*, *4*, 522–531.

Qin, S., Li, J., Zhang, Y. Q., Zhu, W. Y., Zhao, G. Z., Xu, L. H., & Li, W. J., (2009). *Plantactinospora mayteni* gen. nov., sp. nov., a member of the family *Micromonosporaceae*. *Int. J. Syst. Evol. Microbiol.*, *59*, 2527–2533.

Sasser, M., (1990). *Identification of Bacteria by Gas Chromatography of Cellular Fatty Acids*. Newark, DE: MIDI Inc (pp. 1-6).

Songsumanus, A., (2010). *Biodiversity and Secondary Metabolites of Rare Actinomycetes From Soils*. PhD Thesis, Chulalongkorn University, Bangkok, Thailand (p. 186).

Songsumanus, A., Tanasupawat, S., Tahwai, C., Suwanborirux, K., & Kudo, T., (2011). *Micromonospora humi* sp. nov., isolated from peat swamp forest soil. *Int. J. Syst. Evol. Microbiol.*, *61*, 1176–1181.

Songsumanus, A., Tanasupawat, S., Thawai, C., & Suwanborirux, K., (2007). Identification and antimicrobial activities of actinomycetes from peat swamp forest soil in Payao province, *The 24ᵗʰ Annual Research Conference in Pharmaceutical Sciences*, Bangkok, Thailand.

Staneck, J. L., & Robert, G. D., (1974). Simplified approach to identification of aerobic actinomycetes by thin-layer chromatography. *Appl. Microbiol.*, *28*, 226–231.

Steinberg, B. A., Jambor, W. P., & Suydan, L. O., (1955/1956). Amphotericins A and B: Two new antifungal antibiotics possessing high activity against deep-seated and superficial mycoses. *Antibiot. Annu.*, *3*, 574–578.

Suzuki, S., (2001). Establishment and use of gellan gum media for selective isolation and distribution survey of specific rare actinomycetes. *Actinomycetologica*, *15*, 55–60.

Tanasupawat, S., & Suwanborirux, K., (2014). *Taxonomy and Antimicrobial Activity of Actinomycetes From Peat Swamp Forest Soil in Rayong Province*. Chulalongkorn University, Bangkok, Thailand.

Tanasupawat, S., Phongsopitanun, W., Suwanborirux, K., Ohkuma, M., & Kudo, T., (2016b). *Nocardiarayongensis* sp. nov., isolated from Thai peat swamp forest soil. *Int. J. Syst. Evol. Microbiol.*, *66*, 1950–1955.

Tanasupawat, S., Phongsopitanun, W., Suwanborirux, K., Ohkuma, M., & Kudo, T., (2016a). *Streptomyces actinomycinicus* sp. nov., isolated from soil of a peat swamp forest. *Int. J. Syst. Evol. Microbiol.*, *66*, 290–295.

Thawai, C., (2004a). *Taxonomy of Micromonospora Strains From Thai Peat Swamp Forest Soils and Secondary Metabolites of a Selected Isolate*. PhD Thesis, Chulalongkorn University, Bangkok, Thailand.

Thawai, C., Kittakoop, P., Tanasupawat, S., Suwanborirux, K., Sriklung, K., Thebtaranonth, Y., & Micromonosporin, A., (2004c). A novel 24-membered polyene lactam macrolide from *Micromonospora* sp. isolated from peat swamp forest. *Chem. Biodiver.*, *1*, 640–645.

Thawai, C., Tanasupawat, S., & Kudo, T., (2008). *Micromonospora pattaloongensis* sp. nov., isolated from a Thai mangrove forest. *Int. J. Syst. Evol. Microbiol.*, *58*, 1516–1521.

Thawai, C., Tanasupawat, S., Itoh, T., & Kudo, T., (2006). *Actinocatenispora thailandica* gen. nov., sp. nov., a new member of the family *Micromonosporaceae*. *Int. J. Syst. Evol. Microbiol.*, *56*, 1789–1794.

Thawai, C., Tanasupawat, S., Itoh, T., Suwanborirux, K., & Kudo, T., (2004b). *Micromonosporaaurantionigra* sp. nov., isolated from a peat swamp forest in Thailand. *Actinomycetologica*, *18*, 8–14.

Thawai, C., Tanasupawat, S., Itoh, T., Suwanborirux, K., & Kudo, T., (2005b). *Micromonospora siamensis* sp. nov., isolated from Thai peat swamp forest. *J. Gen. Appl. Microbiol.*, *51*, 229–234.

Thawai, C., Tanasupawat, S., Itoh, T., Suwanborirux, K., Suzuki, K., & Kudo, T., (2005a). *Micromonospora eburnea* sp. nov., isolated from a Thai peat swamp forest. *Int. J. Syst. Evol. Microbiol.*, *55*, 417–422.

Thawai, C., Tanasupawat, S., Pongpech, P., & Suwanborirux, K., (2000). Antibiotic-producing actinomycetes from peat swamp forest soils in Trang. *Thai. J. Pharm. Sci.*, *24*, 40.

Thawai, C., Tanasupawat, S., Suwanborirux, K., & Kudo, T., (2010). *Actinaurisporasiamensis* gen. nov., sp. nov., a new member of the family *Micromonosporaceae*. *Int. J. Sys. Evol. Microbiol.*, *60*, 1660–1666.

Thiemann, J. E., Pagani, H., & Beretta, G., (1976). A new genus of the *Actinoplanaceae*: *Dactylosporangium*, gen. nov. *Arch. Mikrobiol.*, *58*, 42–52.

Trujillo, M. E., & Goodfellow, M., (2012a). Genus III. *Actinomadura* Lechevalier and Lechevalier 1979a, 400[AL] emend. Kropenstedt, Stackebrandt, and Goodfelllow 1990, 156. In: Goodfellow, M., Kempfer, P., Busse, H. J., Trujillo, M. E., Suzuki, K., Ludwig, W., & Whitman, W. B., (eds.), *Bergey's Manual of Systematic Bacteriology, Part B* (Vol. 5, pp. 1940–1959). Springer, New York.

Trujillo, M. E., Goodfellow, M., & Genus, V., (2012). *Microtetraspora* Thiemann, Pagani, and Beretta 1968b, 296AL emend. Zhang, Wang, and Ruan 1998a, 420VP. In: Goodfellow, M., Kempfer, P., Busse, H. J., Trujillo, M. E., Suzuki, K., Ludwig, W., & Whitman, W. B., (eds.), *Bergey's Manual of Systematic Bacteriology, Part B* (Vol. 5, pp. 1838–1844), Springer, New York.

UNDP, (2006). *Malaysia's Peat Swamp Forests - Conservation and Sustainable Use.* Kuala Lumpur, Malaysia.

Waksman, S. A., & Woodruff, H. B., (1940). Bacteriostatic and bacteriocidal substances produced by soil actinomycetes. *Proc. Soc. Exper. Biol.*, *45*, 609–661.

Wang, J., Soisson, S. M., Young, K., Shoop, W., Sodali, S., Galgoci, A., et al., (2006). Platensimycin is a selective FabF inhibitor with potent antibiotic properties. *Nature*, *441*, 358–361.

Williams, S. T., & Davies, F. L., (1965). Use of antibiotics for selective isolation and enumeration of actinomycetes in soil. *J. Gen. Microbiol.*, *38*, 251–261.

Wu, Z., Xie, L., Zhang, J., Nie, Y., Hu, J., Wang, S., & Zhang, R., (2005). A new tetrodotoxin-producing actinomycete, *Nocardiopsis dassonvillei*, isolated from the ovaries of puffer fish *Fugu rubripes*. *Toxicon.*, *45*(7), 851–859.

CHAPTER 8

BIODIVERSITY OF *TRICHODERMA* AND THEIR APPLICATIONS IN AGRICULTURE, INDUSTRY, AND MEDICINE

SHAFIQUZZAMAN SIDDIQUEE

Biotechnology Research Institute, University Malaysia Sabah, Jalan UMS, 88400 Kota Kinabalu, Sabah, Malaysia

8.1 INTRODUCTION

Globally almost 1.5 million of fungal species has distributed, but only 5% of species are discovered. This approximation is estimated from the ratio of fungal species to vascular plant species for diverse ecological defined groups of fungi in well-studied regions (Hawksworth, 1991; Hawksworth, 2001). The generic name of *Trichoderma* and *Hypocrea* are considered as different genera; several species have commonly linked as asexual (anamorph) and sexual (teleomorph) morphs of one and the same species. *Trichoderma* is free-living fungi and ubiquitously existed in all ecosystems, especially in soils. These filamentous fungi could be applied with the different way human activities, such as enzymes production; frequently used in the biofuel industry. Besides, these fungi are also widely applied in bioremediation of inorganic and organic wastes (heavy metals), biofungicides, plant growth promoters and biocontrol agents against plant pathogenic fungi. Furthermore, it can produce a large amount of secondary metabolites, some of which have clinical significances, and some species engineered for the heterologous proteins.

Weindling (1932, 1934) reported that the specific potential strains of *Trichoderma* have the ability to apply for human welfare; for examples, mycoparasite, antibiotic, and biocontrol agents against plant pathogenic diseases. Today this fungal is among the most valuable and widely rectify research microbes on the basis of the huge volume of published research works. Unfortunately, identification of important economic fungi only relies on their morphology characters that are mostly found to be highly variable, for a reason, the deposited in the culture collection has roughly distinguished about 50% wrongly labeled, even though wrongly deposited of many *Trichoderma* strains DNA sequencing in NCBI GenBank database. Consequently, the accurate identification of *Trichoderma* strains are economically and clinically more important, especially for two reasons: (i) most beneficial traits are occurred by species-specific, so it is important to identify the potential species/strain; and (ii) more important for properly handle and protect the personal protective equipment (PPE) safety procedures (as illiterate farmers/farm workers). Because some strains are produced mycotoxins, so it's are human pathogens, especially threatening immune-compromised. The potential *Trichoderma* species has applied multidisciplinary applications with different branches including in agriculture, industry, and medicine.

8.2 BIODIVERSITY OF *TRICHODERMA*

The common characters of generic *Trichoderma* are fast growing filamentous fungi; it found all types of soil such as temperate and tropical regions (Samuels, 1996; Kumar and Gupta, 1999). It also existed on the decaying wood plant, vegetables, compost, and other organic matters (Harman et al., 2004; Samuels, 2006; Schuster and Schmoll, 2010). Some researchers have discovered *Trichoderma* species in extreme environments like salt marshes, mangrove swamps, and estuarine sediment (Domsch et al., 1980). The author can honestly say that the genus *Trichoderma* never has limited habitat range, so it can be found different diversity area.

Trichoderma species are extensively colonized root surfaces and also associated as endophytes via penetrating between cells in root tissue. The first few layers of the root cortex interaction are restricted to cells due to the plant defense responses which frontier more internal growth (Yedidia et al., 1999; Harman et al., 2004). In order to provide positive effects on root development and nutrient uptake (Harman, 2011), plant growth, root colonization with *Trichoderma* isolates inhibited by plant pathogenic fungi. Disease suppression occurred by different mechanisms, including antibiotic production (Reino et al., 2008), parasitism (Druzhinina et al., 2011), competition for space and nutrient (Elad, 1996; Howell, 2003) and induction of localized and systemic disease resistance in the host plant (Harman et al., 2004; Hermosa et al., 2012).

Biodiversity study of *Trichoderma* community can be considered as the main prerequisite for exploitation of the genus in biological applications. Most studies of *Trichoderma* diversity in demarcated geographical regions have focused on isolates cultured from soils and plants, e.g., in Malays and Malaysian Borneo island (Sariah et al., 2005; Siddiquee et al., 2007, 2009, 2012a; Asis and Siddiquee, 2016; Cummings et al., 2016), Tenerife Island (Zachow et al., 2009), China (Zhang et al., 2005; Sun et al., 2012), Russia, and the Siberian Himalayas (Kullnig et al., 2000), India (Kamala et al., 2015), South-East Asia (Kubicek et al., 2003), Iran (Naeimi et al., 2011), Egypt (Gherbawy et al., 2004), Bangladesh (Mostafa Kamal and Shahjahan, 1995), Korea (Park et al., 2004; Park et al., 2006), Philippines (Nagamani and Mew, 1987), Canary Islands (Zachow et al., 2009), Europe (Wuczkowski et al., 2003; Jaklitsch, 2009, 2011; Jaklitsch and Voglmayr, 2015), Saudi Arabia (Abd-Elsalam et al., 2010), Canada (Widden and Abitbol, 1980), Sardinia (Migheli et al., 2009), Africa (Degenkolb et al., 2008; Samuels et al., 2012a,b; Kredics et al., 2014) and South America (Hoyos-Carvajal et al., 2009).

The distribution and colonization of *Trichoderma* strains are mostly influenced by the several ecological parameters such as pH, moisture, soil temperature, organic matter, atmosphere, nutrient content and plant types (Domsch et al., 1980; Klein and Eveleigh, 1998). Soil temperature is one of the most important factors that distributed at varying *Trichoderma* communities with different geographical regions. Some *Trichoderma* species such as *T. harzianum* is commonly found from warm tropical soil whereas others species such as *T. viride* and *T. polysporum* are frequently found in the cool temperate regions (Klein and Eveleigh, 1998). Tronsmo and Dennis (1978) mentioned that temperature does not only affect the *Trichoderma* growth of but also affect their metabolites activity, mainly production of enzymes and volatile antibiotics. *Trichoderma* species can grow at a temperature as low as 0°C and as high as 40°C (Domsch et al., 1980). At 35°C, altitudes above 1000 m have found two species namely *Hypocrea crystalligena* and *Hypocrea voglmayrii* (Jaklitsch et al., 2005, 2006).

Soil moisture is another important factor which influences the *Trichoderma* propagules in the soils (Lavelle and Spain, 2001). The soil moisture is closely related to the soil temperature. As temperature rises, the soil moisture will drop (Dix and Webster, 1995). The soil's moisture is mostly influenced by spore production, hyphal growth, and germination of *Trichoderma* communities (Clarkson et al., 2004). Globally distribution of *Trichoderma* species are affected by the acidic substrates, its optimal pH range from 3.5 to 5.6 (Domsch et al., 1980; Kredics et al., 2004; Gao et al., 2007) for spore germination and growth. The pH of the soil is greatly dependent on the soil carbon dioxide (CO_2) content. When carbon dioxides in the atmosphere combine with water, the reaction will form a weak acid called carbonic acid that will result in lowering the pH values of the soil (Killham, 1994). Steyaert et al. (2010) reported that the ambient pH effects on the conidiation of three *Trichoderma* species such as *T. pleuroticola, T. atroviride,* and *T. hamatum* are shown to a single blue-light burst or mycelia injury. *T. atroviride* conidiation clearly demonstrated that all cells are possibly skilled for photoconidiation. *T. hamatum* observed a clear injury response from pH 2.8–3.2. *T. pleuroticola* also showed three distinct pH-dependent colony morphologies from pH 2.8 to 5.2.

Antarctica is the coldest, driest, and windiest location in the whole world in 1983; the lowest temperature of Vostok recorded at −89.2°C. Most of Antarctica has roofed with ice; its thickness is more than 1.6 kilometers. Thus, do you believe any microorganism survive under the most extreme condition - between the limit of adaptability, near death, scarcely surviving, and hardly reproduced (Friedmann and Weed, 1987). Several researchers have distinguished with different *Trichoderma* species existence; namely, *T. koningii* by Hughes et al. (2007), *T. atroviride* by Holker et al. (2002), *T. citinoviride* by Jacklitsch (2011) and *T. asperellum* by Ren et al. (2009.

8.3 BIOTECHNOLOGICAL APPLICATIONS OF *TRICHODERMA* IN AGRICULTURE

8.3.1 *TRICHODERMA AS BIOCONTROL AGENTS*

Cook and Baker (1983) have defined biocontrol as "biological control is the reduction of the amount of inoculum or disease-producing activity of a pathogen accomplished by or through one or more organisms other than man." The wide definition has included the use of less virulent variants of the pathogen, more resistant cultivars of the host and microbial antagonists "that interfere with the survival or disease producing activities of the pathogen." Biological control has progressed rapidly towards agricultural pests and pathogens in the above 40 years (Jacobsen and Backman, 1993).

Trichoderma species have several outstanding mechanisms for survival and proliferation including a physical attack on other harmful fungi. The potential of *Trichoderma* strains firstly used as biocontrol agents against plant diseases previously reported by Weindling (1934). He found the mycoparasitic action of *Trichoderma* on *Rhizoctonia* and *Sclerotinia* and the positive effects achieved with plant disease control. The growth of fungal pathogens inhibited by the potential *Trichoderma* strains at least three mechanisms; first parasitism (killing and deriving nutrients from host), second competition (for space and nutrients), and third antibiotic production (Chet and Baker, 1981; Whipps and Lumsden, 1991) whether acting individually or all at the same time.

Trichoderma is beneficial fungi to mankind because it can apply numerous commercial applications. Currently, *Trichoderma* strains are commercially available for disease protection and high yield production of various crops. For examples, the products are commercially available as TrichoGreen™ (Malaysia), BioTrek 22G™, Promot™, T–22G™, SoilGuard™, RootShield™ (USA), Supresivit™ (Denmark), Binab™ (Sweden), Trichojet™, Trieco™, Trichoseal™ (New Zealand), Gliomix™ (Finland), Trichopel™ (India) and Trichodex™ (Israel). For your information, all of these products are not registered as biological control agents, however in commercial marketed uses as an alternative as plant strengtheners, plant growth promoters, or soil conditioners. These products have entered to the marketplace with less stringent toxicology for plant protectants (Paulitz and Belanger, 2001).

Trichoderma harzianum is a ubiquitous asexual fungus and applied as a biocontrol agent against several plant-pathogenic fungi (Adriana and Sergio, 2001; Krupke et al., 2003). *T. harzianum* (FA 1132) has reported more effective against *Ganoderma boninense* which caused basal stem rot of oil palm based on nursery trials (Abdullah et al., 2003), against *Cylindrocladium scoparium* which caused on damping-off of red pines (Yang et al., 1995), against *Botrytis cinerea* which caused on strawberry leaves (Batta, 1999), against *B. cinerea* which caused on grape berries

(Elad, 1994), against *Penicillium expansum* which caused on apple fruit (Batta, 2004) and against *Alternaria alternata* which caused on fig leaves (Batta, 2000).

Since the 1990s, several commercial *Trichoderma*-based formulations are available in market-placed; for example, TrichoGreen TM bio-product applied against *Ganoderma* on basal stem rot of oil palm (Abdullah et al., 2005), Trichodex TM applied against *Botrytis cinerea* on Grey mold of grape vines (O' Neill et al., 1996) and Supresivit® applied against *Pythium ultimum* on cucumber and *R. solani* on peas (Koach, 1999). *Trichoderma* based products existed up to one-third of all fungal biocontrol preparations and sold for the control of diseases on agricultural crops during cultivation and storage period (Chernin and Chet, 2002).

Some *Trichoderma* species have produced important volatile antifungal compounds (Bonnarme et al., 1997; Mumpuni et al., 1998; Siddiquee et al., 2012b) as well as non-volatile compounds inhibited towards *G. boninense* (Shamala and Idris, 2005; Siddiquee et al., 2009). A 6-pentyl-alpha pyrone (6-PAP) is a well established volatile antifungal compound; produced by *T. harzianum* (Scarselletti and Faull, 1994; Cooney et al., 1997; Siddiquee et al., 2012b). The characteristics of 6-PAP production have been reported by Rocha-Valadez et al. (2005). *T. harzianum* produced large amounts of volatile compounds (e.g., hydrogen cyanide, aldehydes, ethylene, alcohols, and ketones) and nonvolatile compounds (e.g., peptides) that are inhibited the mycelial growth of fungi. A novel compound of L-amino oxidase (Th-LAAO) is isolated from *T. harzianum* ETS 323 and applied antagonistic effect against *R. solani* (Yang et al., 2011). The most published research works rectified that *T. harzianum* species is more effective in reducing disease caused by soil-borne plant pathogens (Papavizas, 1985; Chet, 1987). Harziandione is derived from *T. harzianum* and *T. longibrachiatum* and applied as biocontrol agents against plant pathogens (*Colletotrichum lagenarium* and *F. oxysporum*) (Miao et al., 2012).

Mostly beneficial organisms have involved as a biological control to reduce the negative effects of plant pathogens and also promote positive effects to the plant. The potential *Trichoderma* strains have recognized as biocontrol agents, i.e., *T. longibrachiatum* (Andrade et al., 1996) for control of American leaf-spot disease in coffee, caused by *Mycena citricolor*, and *T. viride* and *T. harzianum* which are applied for control of myriad soil-borne and systemic phytopathogens. The main mechanisms are involved between biocontrol agents and plant pathogenic fungi, which are (i) volatile antibiotic compounds; (ii) mycoparasitism or hyperparasitism; (iii) competition for space and nutrients; and (iv) competition for niche colonization (Vinale et al., 2008; Lorito et al., 2010; Vinale et al., 2010). Outstanding of these beneficial effects, the specific potential strains of *T. asperellum, T. atroviride,* and *T. harzianum* are applied for the biocontrol of molds and as plant growth promoters in fruit growing, agricultural crops, and vegetable gardening (Harman et al., 2004; Harman, 2006; Verma et al., 2007).

8.3.2 APPLICATION OF TRICHODERMA IN INSECT PATHOGENICITY

Fungal volatile organic compounds are applied with different biological activities. Numerous fungal compounds can affect several behaviors of insect biology, such as development, fecundity, survival, and feeding activity (Vey et al., 2001). The author should discuss a new pest control strategy with low environmental impact as like as biological pest control, so it is a promising field for natural compounds proficiently interfering with the insect's processes of host plant selection. However, the fungal compounds can influence antifeedant activity toward insect pests still sparse (Quesada-Moraga et al., 2006; Ganassi et al., 2007), and so far a very few fungal compounds are isolated and characterized for the uses of antifeedant activity. The feeding preference of the aphid *Schizaphis graminum* has influenced by the potential *Trichoderma* strains previously reported by Ganassi et al. (2007). It is one of the most important pests of cereal crops, and a potential antifeedant activity toward aphids.

Two new fungal compounds (namely, citrantifidiene, and citrantifidiol) are cultured from *T. citrinoviride* ITEM 4484 and showed antifeedant activity (Evidente et al., 2008). Citrantifidiene is quite different from citrantifidiol. The assay of antifeedant activity found by citrantifidiol and citrantifidiene, it is emerging perspective of these compounds for practical application of biocontrol against aphid pest. The main function of volatile organic compounds attracts to other food sources insect also. For example, 1-octen–3-ol compound isolated by *T. harzianum* T32 (Siddiquee et al., 2012b) when came from human skin serves as a host odor cue that attracts bloodsucking insects, such as the mosquito *Anopheles gambiae* (Kline et al., 2007).

8.3.3 APPLICATION OF TRICHODERMA IN NEMATICIDAL ACTIVITIES

The several important compounds have produced by *Trichoderma* fungal that harmful to nematodes or use as a biocontrol agent. Three volatile organic compounds are identified by *Trichoderma* species (strain YMF 1.00416), namely 1β-vinylcyclopentane–1α,3α-diol, and two known compounds, 6-pentyl–2H-pyran–2-one and 4-(2-hydroxyethyl)phenol. The nematicidal compound of 6-pentyl–2H-pyran–2-one is killed above 85% of *Bursaphelenchus xylophilus, Caenorhabditis elegans*, and *Panagrellus redivivus*, in 48 h at 200 mg L^{-1} in a 2 mL vial showed by nematicidal activity assays (Yang et al., 2012). A nematicidal compound of trichodermin is isolated from ethyl acetate extraction of *Trichoderma* YMF1.02647 study on bioassay-guided fractionation (Yang et al., 2010) which can kill more than 95% of both *C. elegans* and *P. redivivus* in 72 h at 0.4 g L^{-1}. Acetic acid of the nematicidal compound is identified by the culture of *T. longibrachiatum* (Djian et al., 1991). Nematicidal activity of gliotoxin compound is found by *T. virens* strain (Anitha and Murugesan, 2005). A peptide cyclosporin A obtained from *T. polysporum* that applied nematicidal activity against *M. incognita* (Li et al., 2007). Viridian compound isolated by *Trichoderma* spp. and showed weak activity against *Anguillula aceti* (Watanabe et al., 2004; Anitha and Murugesan, 2005).

Numerous *Trichoderma* species such as *T. harzianum* (Siddiqui and Shaukat, 2004), *T. longibrachiatum* (Djian et al., 1991), *T. hamatum* (Girlanda et al., 2001), *T. viride* (Zhang and Zhang, 2009), *T. virens* (Meyer et al., 2001), *T. koningii* (Sankaranarayanan et al., 1997) and *T. compactus* (Yang et al., 2010) have applied nematicidal activity. Strong nematicidal activity showed by *T. compactus*, causing above 80% mortality towards the nematodes. The most potential strain of *T. harzianum* YMF1.02647 is killed < 95%, and also *C. elegans* and *P. redivivus* are showed more positive component of the nematicidal within 48 h (Yang et al., 2010).

Trichodermin is an important compound which is isolated from several species of *Trichoderma* including *T. reesei, T. viride, T. longibrachiatum, T. harzianum*, and other fungi such as *Memnoniella echinata* and *Stachybotrys cylindrospora* (Watts et al., 1988; Nielsen et al., 1998; Reino et al., 2008). Trichodermin compound is identified from *T. harzianum* strain and has applied against three nematodes, namely *C. elegans, P. redivevus,* and *B. xylophilus* which can kill more than 95% both *C. elegans* and *P. redivivus* in 72 h at 400 mg L^{-1}, however, only 54.2% mortality found against *B. xylophilus* in 72 h.

Several approaches are currently used to control these nematodes, including chemical nematicides and biological nematicidal agents (Gu et al., 2007; Yang et al., 2012). Chemical nematicides have significant environmental problems due to their existing toxic residues. Several *Trichoderma* species have produced volatile organic compounds that could be used as nematode biocontrol agents (Yagi et al., 1993; Ristaino and Thomas, 1998). Several nematicidal compounds are trichodermin (Yang et al., 2010), gliotoxin (Anitha and Murugesan, 2005), acetic acid (Djian et al., 1991), and the peptide cyclosporin A (Li et al., 2007), all of these compounds are identified by *Trichoderma* species. A renowned nematicidal compound of 6-pentyl–2H-pyran–2-one is identified by *Trichoderma* sp. YMF 1.00416 (Yang et al., 2012). In the same compound is isolated previously from different fungal cultures including *T. harzianum* (Mourad et al., 2009), *T. viride* (Collins and Halim, 1972), *T. koningii* (Pratt et al., 1972), and *Myrothecium* sp. (Li et al., 2005). This compound inhibited wider range of phytopathogenic fungal growth, including *Botrytis cinerea, R. solani, Armillaria mellea* (Claydon et al., 1987), *P. expansium,* and *F. graminearum* (Cooney and Lauren, 1999) and also affects biomass accumulation of *Trichoderma* species (Serrano-Carreon et al., 2002).

8.3.4 APPLICATION OF VOLATILE COMPOUNDS FROM TRICHODERMA IN PLANT GROWTH AND DEVELOPMENT

Trichoderma species have several beneficial effects on plant growth development to both abiotic and biotic stresses. Some researchers have discovered the specific potential *Trichoderma* strains which promoted growth responses in tomato, cucumber pepper, and radish (Baker et al., 1984; Chang et al., 1986). In similar studies conducted by Harman (2000) and mentioned that the specific potential *Trichoderma* strain is enabled to enhance the root development and high yield productions, the increases of secondary roots, and foliar seedling area and fresh weight. The interaction outcome between different tomato lines and biocontrol strains of *T. harzianum* and *T. atroviride* are showing the effects of the plant genetic background, for example, one tomato cultivar, in which, *Trichoderma* treatment did not employ any plant growth promotion effect and even seen to be detrimental (Tucci et al. 2011). The growth-enhancing activity of *T. atroviride* on tomato seedlings is linked with the reducing ethylene production from a decrease in its precursor 1-aminocyclopropane–1-carboxylic acid (ACC) through the microbe degradation of indole–3-acetic acid (IAA) in the rhizosphere, and/or through the ACC deaminase (ACCD) activity existence in the microorganisms (Gravel et al., 2007). ACCD sequencing has obtained from *Trichoderma* genomes (Kubicek et al., 2011), and RNAi silencing of the ACCD gene obtained from *T. asperellum* which showed a weak ability of the mutants to promote the canola seedlings root elongation (Viterbo et al., 2010). Moreover, Auxins produced by some *Trichoderma* species that enhanced the plant growth and root development (Contreras-Cornejo et al., 2009). An auxin-like effect has observed in etiolated pea stems treated with 6-pentyl-α-pyrone and harzianolide; it is the major volatile organic compounds obtained from different *Trichoderma* strains (Vinale et al., 2008). Maize rhizosphere colonization by *T. virens* also induced higher photosynthetic rates and systemic increases in the uptake of CO_2 in leaves (Vargas et al., 2009). The enhancement of plant growth occurred with different treatments of harzianolide (1 mg L^{-1}), anthraquinones (10 mg L^{-1}) and T39 butenolide (1 mg L^{-1}), previously reported by Vinale et al. (2008). Harzianic acid is enhanced the plant growth development evaluated by measuring the length of canola (*Brassica napus*) seedlings whose seeds are coated with different amounts of *T. harzianum* metabolites. Stem length increased by 42, 44, and 52% at concentrations of 100, 10, and 1 ng per seed of harzianic acid compound, as compared to the control (Vinale et al., 2009).

Treatment on tomato seeds with *T. harzianum* accelerates seed germination, salinity, osmotic, chilling, increases seedling vigor and ameliorates water, and heat stresses by inducing physiological protection in plants against oxidative damage (Mastouri et al., 2010). These responses are comparable with the effects induced in plants of *Paravalsa indica*, which showed strong growth-promoting activity during its symbiosis with the wider spectrum of plants and induces resistance to fungal diseases and tolerance to salt stress (Vadassery et al., 2009).

8.4   BIOTECHNOLOGICAL APPLICATIONS OF *TRICHODERMA* IN INDUSTRY

An interesting chance discovery that changed the face of the enzyme industry happened during the Second World War; a filamentous fungus thriving in its tropical home of the Solomon Islands made a crucial mistake. It happily degraded tents and uniforms of the military, which happened to be using its habitat. Initially, research on the organisms focused on preventing the damage conducted by the hydrolytic enzymes of fungi. However, numerous applications, especially for cellulases, advanced research on the degradation of plant material and fungus, together with other species of the genus; offer a proper toolbox from plant protection of enzymes and chemicals, including biofuels. Many fungi are evaluated for the proper degradation of plant cell wall material along with *T. reesei* (Siu and Reese, 1953). At first *T. reesei* (*T. viride*) was not even among the isolates studies for efficient cellulose production, later its efficient enzyme system resulted in *T. reesei* becoming a model system for plant cell wall degradation.

Some *Trichoderma* species produced cellulolytic and hemicellulolytic enzymes; it can be used in soil bioremediation and the biodegradation of chlorophenolic compounds (Sivasithamparam and Ghisalberti, 1998). *T. harzianum* is commercially used for the production of cellulases and for heterologous protein expressions (Mach and Zeilinger, 2003; Schmoll and Kubicek, 2003). Marquesa et al. (2002) have extracted endoglucanase, and extracellular endoxylanase enzymes from *A. terreus* CCMI498 and cellulose have grown *T. viride* CCMI84, and both enzymes improved the paper strength properties as well as the potential use of these enzymes for drinking mixed office wastepaper (MOW).

As we are now entered in the 21st century, so we are going to face one of the biggest challenges of society due to the growing demand of energy for transportation, heating, and industrial processes, and providing the raw material for the industry in a sustainable way. The European Commission plans to substitute progressively 20% of conventional fossil fuels with alternative fuels in the transport sector by 2020. Hemicelluloses are the second most common polysaccharides in nature and represent about 20–35% of lignocellulosic biomass. Hemicelluloses are structurally heterogeneous polymers of hexoses (mannose, glucose, galactose), pentoses (arabinose, xylose), and sugar acids. Xylans mostly found hardwood hemicelluloses, whereas glucomannans are prosperous found softwood hemicelluloses. Xylans obtained from different sources such as cereals, grasses, hardwood, and softwood with a significantly different composition. For examples, birch wood (Roth) xylan contains 89.3% xylose, 1.4% glucose, 1% arabinose, and 8.3% anhydrobiotic acid; while rice bran contains 46% xylose, 6.1% galactose, 44.9% arabinose, 1.9% glucose, and 1.1% anhydrobiotic acid (Shibuya and Iwasaki, 1985; Kormelink and Voragen, 1993). *T. reesei* produced endoxylanase during its cultivations on cellulose. It has low inventory of hemicellulases (16 hemicellulase genes), variable regulation of enzymes like xylanases (GH11 and GH30), β-xylosidases (GH52, GH43), arabinofuranosidases (GH54 and GH62), β-glucuronidase (GH79), acetylesterase, acetyl xylan esterase; iTRAQ quantified when this fungus is cultivated with cellulose, sawdust, and corn stover.

The common volatile compound of 3-methyl–1-butanol is the target for biofuel alternatives uses. Volatile compounds correspondent to the current fossil fuel formulations components are previously reported from some filamentous fungi such as *T. harzianum* FA1132, *T. longibrachiatium* T128, and *A. niger* SS10 isolated from oil palm plantation soils in Malaysia (Siddiquee et al., 2012b, 2015). The profiles of the volatile compound have included medium-chain-length branched hydrocarbons and other organic compounds derived from the above fungi. The fossil fuel alternative is termed of 'myco-diesel' because of their similarity compounds found in diesel fuel. The hydrocarbons comprised on the primary branched-chain aromatics and alkanes range from 4 to 12 carbons in length. Diesel fuel is a mixture of hydrocarbons from 9 to 23 carbons in length, with an average length of 16 carbons. The types of hydrocarbons in diesel have a strong impact on the properties of the fuel. For example, unsaturation, and branching leads to better octane numbers in gasoline and lower octane numbers in diesel. According to these criteria, above 278 and 295 volatile compounds are identified respectively from *T. harzianum* strain FA1132 and *A. niger* strain SS10 (Siddiquee et al., 2012b, 2015). The volatile compounds discovered are hydrocarbons, benzene, alcohols, alkanes, ketones, furanes, pyrones (lactones), sesquiterpenes, and monoterpenes. Volatiles compounds are mostly identified in the C–C range of n-hydrocarbons with a predominance of C presented by *T. harzianum* (Siddiquee et al., 2012b). Accordingly, alkane hydrocarbons are found in the C7 to C25 range with a predominance of C12 to C16. The content of C and 7- and 8-methyl-C is dominated and found from 60 up to 97% of the total hydrocarbons. Many of these hydrocarbons have included as octane; 1-octene; heptane, 2-methyl; hexadecane; undecane, 4-methyl; nonane, 3-methyl; and benzene, 1,3-dimethyl. *T. reesei* is strong cellulose promoters and the versatile tools developed for genetic engineering (Kubicek et al., 2009; Schuster et al., 2012), *T. reesei* is used as a host for heterologous protein production. A proof-of-principle study revealed the possible use of *T. reesei* as a cell factory for the production of valuable chemical compounds.

Investigation of secondary metabolism with *Trichoderma* species is the main role of bioactive compounds in biocontrol, although these compounds have industrial value as well. Metabolites of *Trichoderma* have applied as aroma compounds in the food industry. While *T. reesei* is not commonly known to secrete appreciable amounts of secondary metabolites, other species of the *Trichoderma* genus are important producers of metabolites with antibiotic activity including polyketides, pyrones, terpenes, and polypeptides (Sivasithamparam and Ghisalberti, 1998). With the advent of challenges in factious disease, it can be expected that this research field will increase in importance in the near future while *Trichoderma* species become a valuable source of new antibiotics. Besides the industrial enzyme products and metabolites are involved some fungi in nanotechnology also. Application of fungal biotechnology opened up intriguing possibilities with *Trichoderma* species being important organisms for its development of nanoparticles.

8.5 BIOTECHNOLOGICAL APPLICATIONS OF *TRICHODERMA* IN HUMAN HEALTH AND THE ENVIRONMENT

Apart from their use in agriculture and industry, *Trichoderma* species have several advantages in the environment and health perspective. These fungi are playing important roles in biodegradation and recycling of complex polymers such as the lignocellulosic wastes and chitins (Schmoll and Schuster, 2010). It is capable of remediating heavy metals (Siddiquee et al., 2013), toxins (such as cyanides) and xenobiotics, thus reducing the residue loads in the environment (Harman, 2006). Tobacco plants expressing *T. virens* GST took up and converted anthracene to non-toxic naphthalenes (Dixit et al., 2011). This is one of an example showing that *Trichoderma* genes can be used for bioremediation of polluted area. Anthraquinones have established compounds obtained from *Trichoderma* species. Normally anthraquinones function have been found as diuretics, phytoestrogens, laxatives, immune stimulators, antiviral agents, antifungal agents, and anticancer agents (Perchellet et al., 2000; Matsuda et al., 2001; Semple et al., 2001; Liu et al., 2009). Polycyclic aromatic hydrocarbons (PAHs) are a bigger group of pollutants and found as common constituents of petroleum, shale oil and coal tar; mostly formed as by-products of incomplete combustion. Studies conducted by Sarawathy and Hallberg (2002), four strains of *Penicillium* and *T. harzianum* have found pyrene as the only source of carbon and energy, and have several potentials for bioremediation of PAH-contaminated waste.

The direct comportment of *Trichoderma* in human health is complicated. These fungi have numerous potential sources of medicine; some are producers of mycotoxins, and some are human pathogens. Some secondary metabolites of *Trichoderma* are applied potential anticancer and antimicrobial drugs (Shi et al., 2010) by the viridian-analog wortmannin (Smith et al., 2009; Viswanathan et al., 2012), trichosetin, and peptaibiotics. Several medicinal compounds are isolated from marine fungi or the endophytic *Trichoderma* marine fungi (Bae et al., 2009; Ding et al., 2012) and continue novel secondary metabolites are necessary for identifying. Gliotoxin is the first antibiotic compound identified from *T. virens*. This compound is not only found medicinal properties but also can apply as biocontrol agent at the same time produced by the human pathogen *A. fumigatus*; is branded as mycotoxin (Scharf et al., 2012). Trichothecins groups of mycotoxins are produced by *T. brevicompactum* (Tijerino et al., 2011). Other species such as *T. longibrachiatum* and *Hyprocrea orientalis* are well-known to be pathogens in immune-suppressed humans. Some cases are reported that *Trichoderma* is causing infection in humans with normal immunity. To date, the majority of *Trichoderma* species are not reported from human infections. A prerequisite of causing opportunistic infection in humans is the ability to grow at 37°C. It can be recommended that all *Trichoderma* strains intended for agricultural and industrial applications should be tested for growth at 37°C as a safety precaution, in order to minimize potential health risks. Furthermore, the most frequent human pathogen within the genus *Trichoderma* is *T. longibrachiatum*, while the application of *T. longibrachiatum* species in biotechnology and agriculture should be handled with special care or rather ignored (Druzhinina et al., 2008).

8.6 CONCLUSION

Species-level identification of *Trichoderma* strains is crucial because of similar morphology characters; however, need to find the proper way to identify the classification chaotically. The genus *Trichoderma* grow rapidly in artificial culture and produce large numbers of green or white conidia from conidiogenous cells located at the ends of conidiophores. Morphology characters are to be variable to a certain degree in their color, conidiophore, and shape of conidia, phialade, and pustules. These types of characteristics allow to identify until the genus level, but the species level identification of *Trichoderma* very difficult to deduce and considerable confusion. When the identification of *Trichoderma* isolates obtained from clinical specimens, so it needs to be confirmed by combine morphology characters and reliable molecular methods such as gene sequencing of the internal transcribed spacer 1 and 2 regions, translational elongation factor 1-α (tef1) gene, and calmodulin gene. As furthermore, the most frequent human pathogen of the genus *Trichoderma* is very closely related and belongs to section *"Longibrachiatum"* (e.g., *T. longibrachiatum, T. citrinoviride*). Thus the author recommends that the occurrence of pathogenic *Trichoderma* strains may be restricted to the species of section *"Longibrachiatum."* Application of *Trichoderma* section *"Longibrachiatum"* species in biotechnology, agriculture, industry, and medicine should be handled with special care or rather ignored.

KEYWORDS

- biocontrol agents
- biodegradation
- biodiversity
- fungal diversity
- genus *Trichoderma*
- human health and the environment

- insect pathogenicity
- nematicidal activities
- plant growth enhancements
- plant pathogens
- rhizosphere
- *Trichoderma* applications
- *Trichoderma* in industry
- volatile compounds

REFERENCES

Abd-Elsalam, K. A., Almohimeed, I., Moslem, M. A., & Bahkali, A. H., (2010). M13-microsatellite PCR and rDNA sequence markers for identification of *Trichoderma* (*Hypocreaceae*) species in Saudi Arabian soil. *Genet Mol. Res.*, *9*, 2016–2024.

Abdullah, F., Ilias, G. N. M., Nelson, M., Izzati, N. A. M., & Umi, K. Y., (2003). Disease assessment and the efficacy of *Trichoderma* as a biocontrol agent of basal stem rot of oil palms. *Sci. Putra. Res. Bull.*, *11*(2), 31–33.

Abdullah, F., Nagappan, J., & Sebran, N. H., (2005). Biomass production of *Trichoderma harzianum* (Rifai) in the palm oil mill effluents (POME). *Int. J. Biol. Biotechnol.*, *2*(3), 571–575.

Adriana, O., & Sergio, O., (2001). *In vitro* evaluation of *Trichoderma* and *Gliocladium* antagonism against the symbiotic fungus of the leaf-cutting and *Atta cephalotes*. *Mycopathologia*, *150*, 53–60.

Andrade, R., Ayer, W. A., & Trifonov, L. S., (1996). The metabolites of *Trichoderma longibrachiatum*. Part, I. I., The structures of trichodermolide and sorbiquinol. *Can. J. Chem.*, *74*, 371–379.

Anitha, R., & Murugesan, K., (2005). Production of gliotoxin on natural substrates by *Trichoderma virens*. *J. Basic. Microbiol.*, *45*, 12–19.

Asis, A., & Siddiquee, S., (2016). Identification of *Trichoderma* species from wet paddy field soil samples. *Trans. Sci. Technol.*, *3*(1), 1–7.

Bae, H., Sicher, R. C., Kim, M. S., Kim, S. H., Strem, M. D., & Melnick, R. L., (2009). The beneficial endophyte *Trichoderma hamatum* isolate dis 219b promotes growth and delays the onset of the drought response in *Theobroma cacao*. *J. Exp. Bot.*, *60*(11), 3279–3295.

Baker, R., Elad, Y., & Chet, I., (1984). The controlled experiment in the scientific method with special emphasis on biological control. *Phytopathol.*, *74*, 1019–1021.

Batta, Y. A., (1999). Biological effect of two strains of microorganisms antagonistic to *Botrytis cinerea*: Causal organism of gray mold on strawberry. *An-Najah Uni. J. Res. Nat. Sci.*, *13*, 67–83.

Batta, Y. A., (2000). Alternaria leaf spot disease on fig trees: Varietals susceptibility and effect of some fungicides and *Trichoderma*. *Islamic. Uni. J. Gaza.*, *8*, 83–97.

Batta, Y. A., (2004). Effect of treatment with *Trichoderma harzianum* Rifai formulated in invert emulsion on postharvest decay of apple blue mold. *Int. J. Food Microbiol.*, *96*, 281–288.

Bonnarme, P., Djian, A., Feron, G., Ginies, C., Durand, A., & Le Quere, J. L., (1997). Production of 6-pentyl-pyrone (6-PAP) by *Trichoderma* sp. from vegetable oils. *J. Biotechnol.*, *56*, 143–150.

Chang, Y. C., Baker, R., Kleifeld, O., & Chet, I., (1986). Increased growth of plants in the presence of the biological control agent *Trichoderma harzianum*. *Plant Dis.*, *70*, 145–148.

Chernin, L., & Chet, I., (2002). Microbial enzymes in the biocontrol of plant pathogens and pests. In: Burns, R. G., & Dick, R. P., (eds.), *Enzymes in the Environment: Activity, Ecology, and Applications* (pp. 171–226). Marcel Dekker, New York.

Chet, I., & Baker, R., (1981). Isolation and biological potential of *Trichoderma hamatum* from soil naturally suppress of *Rhizoctonia solani*. *J. Phytopathol.*, *71*, 268–290.

Chet, I., (1987). *Trichoderma* - application, mode of action and potential as a biocontrol agent of soil-borne plant pathogenic fungi. In: Chet, I., (ed.), *Innovative Approaches to Plant Disease Control* (pp. 137–160). John Wiley and Sons, New York.

Clarkson, J. P., Mead, A., Payne, T., & Whipps, J. M., (2004). Effect of environmental factors and *Sclerotium cepivorum* isolate on sclerotial degradation and biological control of white rot by *Trichoderma*. *Plant Pathol.*, *53*, 353–362.

Claydon, N., Allan, M., Hanson, J. R., & Avent, A. G., (1987). Antifungal alkyl pyrones of *Trichoderma harzianum*. *Trans Br. Mycol. Soc.*, *88*, 503–513.

Collins, R. P., & Halim, A. F., (1972). Characterization of the major aroma constituent of the fungus *Trichoderma viride*. *J. Agric. Food Chem.*, *20*, 437–438.

Contreras-Cornejo, H. A., Macias-Rodriguez, L., Cortes-Penagos, C., & Lopez-Bucio, J., (2009). *Trichoderma virens*, a plant beneficial fungus, enhances biomass production and promotes lateral root growth through an auxin-dependent mechanism in *Arabidopsis*. *Plant Physiol.*, *149*, 1579–1592.

Cook, R. J., & Baker, K. F., (1983). *The Nature and Practice of Biological Control of Plant Pathogens* (p. 539). American Phytopathological Society, Minnesota, St. Paul.

Cooney, J. M., & Lauren, D. R., (1999). Biotransformation of the *Trichoderma* metabolite 6-npentyl- 2H-pyran-2-one by selected fungal isolates. *J. Nat. Prod.*, *62*, 681–683.

Cooney, J. M., Lauren, D. R., Poole, P. R., & Whitaker, G., (1997). Microbial transformation of the *Trichoderma* metabolite 6-n pentyl–2H pyran–2-one. *J. Nat. Prod.*, *60*, 1242–1244.

Cummings, N. J., Ambrose, A., Braithwaite, M., Bissett, J., Roslan, H. A., Abdullah, J., et al., (2016). Diversity of root-endophytic *Trichoderma* from Malaysian Borneo. *Mycol. Prog.*, *15*, 50.

Degenkolb, T., Dieckmann, R., Nielsen, K. F., Gräfenhan, T., Theis, C., & Zafari, D., (2008). The *Trichoderma brevicompactum* clade: A separate lineage with new species, new peptaibiotics, and mycotoxins. *Mycol. Prog.*, *7*, 177–219.

Ding, G., Chen, I., Chen, A., Tian, X., Zhang, H., Chen, H., Liu, X. Z., Zhang, Y., & Zhou, Z. M., (2012). Trichalasins C and D from the plant endophytic fungus *Trichoderma gansii*. *Fitoterapia*, *83*, 541–544.

Dix, N. J., & Webster, J., (1995). *Fungal Ecology* (pp. 1–594). Chapman and Hall, London.

Dixit, P., Mukherjee, P. K., Ramachandran, V., & Eapen, S., (2011). Glutathione transferase from *Trichoderma virens* enhances cadmium tolerance without enhancing its accumulation in transgenic *Nicotiana tabacum*. *PLoS One, 6*(1), e16360.

Djian, C., Pijarouvski, L., Ponchet, M., & Arpin, N., (1991). Acetic acid, a selective nematicidal metabolite from culture filtrates of *Paecilomyces lilacinus* (Thom) Samsan and *Trichoderma longibrachiatum* Rifai. *Nematologica, 37*, 101–102.

Domsch, K. H., Gams, W., & Anderson, T., (1980). *Compendium of Soil Fungi* (pp. 1–859). Academic Press, London.

Druzhinina, I. S., Komon-Zelazowsk, M., Kredics, L., Hatvani, L., Antal, Z., Belayneh, T., & Kubicek, C. P., (2008). Alternative reproductive strategies of *Hypocrea orientalis* and genetically close but clonal *Trichoderma longibrachiatum*, both capable of causing invasive mycoses of humans. *Microbiol., 154*, 3447–3459.

Druzhinina, I. S., Seidl-Seiboth, V., Herrera-Estrella, A., Horwitz, B. A., Kenerley, C. M., Monte, E., et al., (2011). *Trichoderma*: The genomics of opportunistic success. *Nat. Rev. Microbiol., 9*, 749–759.

Elad, Y., (1994). Biological control of grape gray mold by *Trichoderma harzianum*. *Crop Protect, 13*, 35–38.

Elad, Y., (1996). Mechanisms involved in the biological control of *Botrytis cinerea* incited diseases. *Eur. J. Plant Pathol., 102*, 719–732.

Evidente, A., Ricciardiello, G., Andolfi, A., Sabatini, M. A., Ganassi, S., Altomare, C., Favilla, M., & Melck, D., (2008). Citrantifidiene and citrantifidiol: Bioactive metabolites produced by *Trichoderma citrinoviride* with potential antifeedant activity toward aphids. *J. Agric. Food Chem., 56*, 3569–3573.

Friedmann, E. I., & Weed, R., (1987). Microbial trace-fossil formation, biogenous, and abiotic weathering in the Antarctic cold desert. *Science, 236*, 703–705.

Ganassi, S., De Cristofaro, A., Grazioso, P., Altomare, C., Logrieco, A., & Sabatini, M. A., (2007). Detection of fungal metabolites of various *Trichoderma* species by the aphid *Schizaphis graminum*. *Entomol. Exp. Appl., 122*, 77–86.

Gao, L., Sun, M. H., Liu, X. Z., & Che, Y. S., (2007). Effects of carbon concentration and carbon to nitrogen ratio on the growth and sporulation of several biocontrol fungi. *Mycol. Res., 111*, 87–92.

Gherbawy, Y., Druzhinina, I., Shaban, G. M., Wuczkowsky, M., Yaser, M., & El-Naghy, M. A., (2004). *Trichoderma* populations from alkaline agricultural soil in the Nile valley, Egypt, consist of only two species. *Mycol. Prog., 3*, 211–218.

Girlanda, M., Perotto, S., Moenne-Loccoz, Y., Bergero, R., Lazzari, A., Defago, G., Bonfante, P., & Luppi, A. M., (2001). Impact of biocontrol *Pseudomonas fluorescens* CHA0 and a genetically modified derivative on the diversity of culturable fungi in the cucumber rhizosphere. *Appl. Environ. Microb., 67*, 1851–1864.

Gravel, V., Antoun, V., & Tweddell, R. J., (2007). Growth stimulation and fruit yield improvement of greenhouse tomato plants by inoculation with *Pseudomonas putida* or *Trichoderma atroviride*: Possible role of indoleacetic acid (IAA). *Soil Biol. Biochem., 39*, 1968–1977.

Gu, Y. Q., Mo, M. H., Zhou, J. P., Zou, C. S., & Zhang, K. Q., (2007). Evaluation and identification of potential organic nematicidal volatiles from soil bacteria. *Soil Biol. Biochem., 39*, 2567–2575.

Harman, G. E., (2000). Myths and dogmas of biocontrol - changes in perceptions derived from research on *Trichoderma harzianum* T–22. *Plant Dis.*, *84*, 377–393.

Harman, G. E., (2006). Overview of mechanisms and uses of *Trichoderma* spp. *Phytopathol.*, *96*, 190–194.

Harman, G. E., (2011). Multifunctional fungal plant symbionts: New tools to enhance plant growth and productivity. *New Phytol.*, *189*, 647–649.

Harman, G. E., Howell, C. R., Viterbo, A., Chet, I., & Lorito, M., (2004). *Trichoderma* species - opportunistic, virulent plant symbionts. *Nat. Rev. Microbiol.*, *2*, 43–56.

Hawksworth, D. L., (1991). The fungal dimension of biodiversity: Magnitude, significance, and conservation. *Mycol. Res.*, *95*, 641–655.

Hawksworth, D. L., (2001). The magnitude of fungal diversity: The 1.5 million species estimate revisited. *Mycol. Res.*, *105*, 1422–1432.

Hermosa, R., Viterbo, A., Chet, I., & Monte, E., (2012). Plant-beneficial effects of *Trichoderma* and of its genes. *Microbiol.*, *158*, 17–25.

Holker, U., Schmiers, H., Grobe, S., Winkelhofer, M., & Polsakiewicz, M., (2002). Solubilization of low-rank coal by *Trichoderma atroviride*: Evidence for the involvement of hydrolytic and oxidative enzymes by using [sup14] C-labelled lignite. *J. Ind. Microbiol. Biotechnol.*, *4*, 207.

Howell, C. R., (2003). Mechanisms employed by *Trichoderma* species in the biological control of plant diseases: The history and evolution of current concepts. *Plant Dis.*, *87*, 4–10.

Hoyos-Carvajal, L., Orduz, S., & Bissett, J., (2009). Genetic and metabolic biodiversity of *Trichoderma* from Colombia and adjacent neotropic regions. *Fungal Genet Biol.*, *46*, 615–631.

Hughes, K. A., Bridge, P., & Clark, M. S., (2007). Tolerance of Antarctic soil fungi to hydrocarbons. *Sci. Total Environ.*, *372*, 539–548.

Jacobsen, B. J., & Backman, A., (1993). Biological and cultural plant disease controls: Alternatives and supplements to chemicals in IPM systems. *Plant Dis.*, *77*, 311–315.

Jaklitsch, W. M., & Voglmayr, H., (2015). Biodiversity of *Trichoderma* (*Hypocreaceae*) in Southern Europe and Macaronesia. *Stud. Mycol.*, *80*, 1–87.

Jaklitsch, W. M., (2009). European species of *Hypocrea*, Part, I., The green-spored species. *Stud. Mycol.*, *63*(1), 1–91.

Jaklitsch, W. M., (2011). European species of *Hypocrea*, Part II: Species with hyaline ascospores. *Fungal Divers*, *48*, 1–250.

Jaklitsch, W. M., Komon, M., Kubicek, C. P., & Druzhinina, I. S., (2005). *Hypocrea voglmayrii* sp. nov. from the Austrian alps represents a new phylogenetic clade in *Hypocrea/Trichoderma*. *Mycologia*, *97*, 1365–1378.

Jaklitsch, W. M., Komon, M., Kubicek, C. P., & Druzhinina, I. S., (2006). *Hypocrea crystalligena* sp. nov., a common European species with a white-spored *Trichoderma* anamorph. *Mycologia*, *98*, 499–513.

Kamala, T. H., Devi, S. I., Sharma, K. C., & Kennedy, K., (2015). Phylogeny and taxonomical investigation of *Trichoderma* spp. from Indian region of Indo-Burma biodiversity hot spot region with special reference to Manipur. *Biomed Res. Int.*, 1–21.

Killham, K., (1994). *Soil Ecology*. Cambridge University Press, Cambridge.

Klein, D., & Eveleigh, D. E., (1998). Ecology of *Trichoderma*. In: Kubicek, C. P., & Harman, G. E., (eds.), *Trichoderma and Gliocladium, Basic Biology, Taxonomy, and Genetics* (pp. 57–74). Taylor, and Francis, London, UK.

Kline, D., Allan, S. A., Bernier, U. R., & Welch, C. H., (2007). Evaluation of the enantiomers of 1-octen-3-ol and 1-octyn-3-ol as attractants for mosquitoes associated with a freshwater swamp in Florida, USA. *Med. Vet. Entomol., 21*, 323–331.

Koach, E., (1999). Evaluation of chemical products for microbial control of soil-borne plant diseases. *J. Crop Protect., 18*, 119–125.

Kormelink, F. J. M., & Voragen, A. G. J., (1993). Degradation of different [(glucurono) arabino]xylans by a combination of purified xylan degrading enzymes. *Appl. Microbiol. Biotechnol., 38*, 688–695.

Kredics, L., Hatvani, L., Naeimi, S., Körmöczi, P., Manczinger, L., & Vágvölgyi, C., (2014). Biodiversity of the genus *Hypocrea/Trichoderma* in different habitats. In: Gupta, V. K., Schmoll, M., Herrera-Estrella, A., Upadhyay, R. S., Druzhinina, I., & Tuohy, M. G., (eds.), *Biotechnology and Biology of Trichoderma* (pp. 3–24). Elsevier, London.

Kredics, L., Manczinger, L., Antal, Z., Penzes, Z., Szekeres, A., & Kevei, F., (2004). *In vitro* water activity and pH dependence of mycelial growth and extracellular enzyme activities of *Trichoderma* strains with biocontrol potential. *J. Appl. Microbiol., 96*, 491–498.

Krupke, A. O., Castle, A. J., & Rinker, D. L., (2003). The North American mushroom competitor, *Trichoderma aggressivum* f. *agressivum*, produces antifungal compounds in mushroom compost that inhibit mycelial growth of the commercial mushroom *Agaricus bisporus*. *Mycol. Res., 107*, 1467–1475.

Kubicek, C. P., Bissett, J., Druzhinina, I. S., Kullnig-Gradinger, C., & Szakacs, G., (2003). Genetic and metabolic diversity of *Trichoderma*: A case study on South-East Asian isolates. *Fungal Genet Biol., 38*, 310–319.

Kubicek, C. P., Herrera-Estrella, A., Seidl-Seiboth, V., Martinez, D. A., Druzhinina, I. S., & Thon, M., (2011). Comparative genome sequence analysis underscores mycoparasitism as the ancestral life style of *Trichoderma*. *Genome Biol., 12*(4), R40.

Kubicek, C. P., Mikus, M., Schuster, A., Schmoll, M., & Seiboth, B., (2009). Metabolic engineering strategies for the improvement of cellulase production by *Hypocrea jecorina*. *Biotechnology for Biofuels, 2*, 19.

Kullnig, C. M., Szakacs, G., & Kubicek, C. P., (2000). Molecular identification of *Trichoderma* species from Russia, Siberia, and the Himalaya. *Mycol. Res., 104*(9), 1117–1125.

Kumar, A., & Gupta, J. P., (1999). Variation in enzyme activity of tebuconazole tolerant biotypes of *Trichoderma viride*. *Ind. Phytopath., 53*, 263–266.

Lavelle, P., & Spain, A. V., (2001). *Soil Ecology*. Kluwer Academic Publishers, Dordrecht (p. 654).

Li, G. H., Zhang, K. Q., Xu, J. P., Dong, J. Y., & Liu, Y. J., (2007). Nematicidal substances from fungi. *Recent Pat. Biotechnol., 1*, 212–233.

Li, X., Kim, S. K., Jung, J. H., Kang, J. S., Choi, H. D., Son, B. W., & Korean, B., (2005). Biological synthesis of polyketides from 6-n-pentyl-α-pyrone by *Streptomyces* sp. *Chem. Soc., 26*, 1889–1890.

Liu, S. Y., Lo, C. T., Shibu, M. A., Leu, Y. L., Jen, B. Y., & Peng, K. C., (2009). Study on the anthraquinones separated from the cultivation of *Trichoderma harzianum* strain Th-R16 and their biological activity. *J. Agric. Food Chem., 57*, 7288–7292.

Lorito, M., Woo, S. L., Harman, G. E., & Monte, E., (2010). Translational research on *Trichoderma*: From omics to the field. *Annu. Rev. Phytopathol., 48*, 395–417.

Mach, L., & Zeilinger, S., (2003). Regulation of gene expression in industrial fungi, *Trichoderma. Appl. Microbiol. Biotechnol., 60*, 515–522.

Marquesa, S., Palab, H., Alvesa, L., Ameral-Collacoa, M. T., Gamab, F. M., & Girio, F. M., (2002). Characterization and application of glycanase secreted by *Aspergillus terrus* CCMI 498 and *Trichoderma viride* CCMI84 for enzymatic deinking of mixed office wastepaper. *J. Biotechnol., 100*, 209–219.

Mastouri, F., Bjorkman, T., & Harman, G. E., (2010). Seed treatment with *Trichoderma harzianum* alleviates biotic, abiotic, and physiological stresses in germinating seeds and seedlings. *Phytopathol., 100*, 1213–1221.

Matsuda, H., Shimoda, H., Morikawa, T., & Yoshikawa, M., (2001). Phytoestrogens from the roots of *Polygonum cuspidatum* (Polygonaceae): Structure-requirement of hydroxyanthraquinones for estrogenic activity. *Bioorg. Med. Chem. Lett., 11*, 1839–1842.

Meyer, S. L. F., Roberts, D. P., Chitwood, D. J., Carta, L. K., Lumsden, R. D., & Mao, W., (2001). Application of *Burkholderia cepacia* and *Trichoderma virens*, alone, and in combinations, against *Meloidogyne incognita* on bell pepper. *Nematropica., 31*, 75–86.

Miao, F. P., Liang, X. R., Yin, X. L., Wang, G., & Ji, N. Y., (2012). Absolute configurations of unique harziane diterpenes from *Trichoderma* species. *Org. Lett., 14*(15), 3815–3817.

Migheli, Q., Balmas, V., Komoń-Zelazowska, M., Scherm, B., Fiori, S., & Kopchinskiy, A., (2009). Soils of a Mediterranean hot spot of biodiversity and endemism (Sardinia, Tyrrhenian Islands) are inhabited by pan-European, invasive species of *Hypocrea/ Trichoderma. Environ. Microbiol., 11*, 35–46.

Mostafa, K. M. D., & Shahjahan, A. K. M., (1995). *Trichoderma* in rice field soils and their effect on *Rhizoctonia solani. Bangladesh J. Bot., 24*, 75–79.

Mourad, D., Cristina, P. R. M., Belen, R., Rosa, H., Enrique, M., Josefina, A., & Isidro, G. C., (2009). Hemisynthesis and absolute configuration of novel 6-pentyl–2H-pyran–2-one derivatives from *Trichoderma* spp. *Tetrahedron, 65*, 4834–4840.

Mumpuni, A., Sharma, H. S. S., & Brown, A. E., (1998). Effect of metabolites produced by *Trichoderma harzianum* biotypes and *Agaricus bisporus* on their respective growth radii on culture. *Appl. Environ. Microbiol., 64*, 5053–5056.

Naeimi, S., Khodaparast, S. A., Javan-Nikkhah, M., Vágvölgyi, C. S., & Kredics, L., (2011). Species patterns and phylogenetic relationships of *Trichoderma* strains in rice fields of southern Caspian Sea, Iran. *Cereal Res. Commun., 39*, 560–568.

Nagamani, A., & Mew, T. W., (1987). *Trichoderma* in Philippine rice field soils. *Int. Rice Res. Newslett., 12*, 25.

Nielsen, K. F., Hansen, M. O., Larsen, T. O., & Thrane, U., (1998). Production of trichothecene mycotoxins on water damaged gypsum boards in Danish buildings. *Int. Biodeter. Biodegr., 42*, 1–7.

O'Neill, T. M., Elad, Y., Shtienberg, D., & Cohen, A., (1996). Control of grapevine grey mold with *Trichoderma harzianum* T39. *Biocontrol. Sci. Technol., 6*, 139–146.

Papavizas, G. C., (1985). *Trichoderma* and *Gliocladium*: Biology, ecology, and potential for biocontrol. *Ann. Rev. Phytopath., 23*, 23–54.

Park, M. S., Bae, K. S., & Yu, S. H., (2004). Molecular and morphological analysis of *Trichoderma* isolates associated with green mold epidemic of oyster mushroom in Korea. *J. Huazhong. Agric. Univ., 23*, 157–164.

Park, M. S., Bae, K. S., & Yu, S. H., (2006). Two new species of *Trichoderma* associated with green mold of oyster mushroom cultivation in Korea. *Mycobiol.*, *34*, 111–113.

Paulitz, T. C., & Belanger, R. R., (2001). Biological control in greenhouse systems. *Ann. Rev. Phytopath.*, *39*, 103–133.

Perchellet, E. M., Magill, M. J., Huang, X., Dalke, D. M., Hua, D. H., & Perchellet, J. P., (2000). 1,4-Anthraquinone: An anticancer drug that blocks nucleoside transport, inhibits macromolecule synthesis, induces DNA fragmentation, and decreases the growth and viability of L1210 leukemic cells in the same nanomolar range as daunorubicin *in vitro*. *Anticancer Drugs*, *11*, 339–352.

Pratt, B. H., Sedgley, J. H., Heather, W. A., & Shepherd, C. J., (1972). Oospore production in *Phytophthora cinnamomi* in the presence of *Trichoderma koningii*. *Aust. J. Biol. Sci.*, *25*, 861–863.

Quesada-Moraga, E., Carrasco-Diaz, J. A., & Santiago-Alvarez, C., (2006). Insecticidal and antifeedant activities of protein secreted by entomopathogenic fungi against *Spodoptera littoralis* (Lep, Noctuidae). *J. Appl. Entomol.*, *130*, 442–452.

Reino, J. L., Guerrero, R. F., Hernandez-Galan, R., & Collado, I. G., (2008). Secondary metabolites from species of the biocontrol agent *Trichoderma*. *Phytochem. Rev.*, *7*, 89–123.

Ren, J., Xue, C., Tian, L., Xu, M., Chen, J., Deng, Z., Proksch, P., Lin, W., & Asperelines, A. F., (2009). Peptaibols from the marine-derived fungus *Trichoderma asperellum*. *J. Nat. Prod.*, *72*, 1036–1044.

Ristaino, J., & Thomas, W., (1998). Agriculture, methyl bromide and the ozone hole: can we fill the gap? *Plant Dis.*, *81*, 964–997.

Rocha-Valadez, J. A., Hassan, M., Corkidi, G., Flores, C., Galindo, E., & Serrano-Carreon, L., (2005). 6-Pentyl-alpha-pyrone production by *Trichoderma harzianum*: The influence of energy dissipation rate and its implications on fungal physiology. *J. Biotechnol. Bioeng.*, *91*, 54–61.

Samuels, G. J., (1996). *Trichoderma* - a review of biology and systematic of the Genus. *Mycol. Res.*, *100*(8), 923–935.

Samuels, G. J., (2006). *Trichoderma*: Systematics, the sexual state, and ecology. *Phytopathol.*, *96*, 195–206.

Samuels, G. J., Ismaiel, A., De Souza, J., & Chaverri, P., (2012b). *Trichoderma stromaticum* and its overseas relatives. *Mycol. Prog.*, *11*, 215–254.

Samuels, G. J., Ismaiel, A., Mulaw, T. B., Szakacs, G., Druzhinina, I. S., Kubicek, C. P., & Jaklitsch, W. M., (2012a). The *Longibrachiatum* clade of *Trichoderma*: A revision with new species. *Fungal Divers*, *55*, 77–108.

Sankaranarayanan, C., Hussaini, S., Sreeramakumar, S., & Prasad, R. D., (1997). Nematicidal effect of fungal filtrates against root-knot nematodes. *J. Biol. Control*, *11*, 37–41.

Sarawathy, A., & Hallberg, R., (2002). *Principles of Genetics* (pp. 517–585). John Wiley and Sons, Inc., New York.

Sariah, M., Choo, C. W., Zakaria, H., & Norihan, M. S., (2005). Quantification and characterization of *Trichoderma* spp. from different ecosystems. *Mycopathologia*, *159*, 113–117.

Scarselletti, R., & Faull, J. L., (1994). *In vitro* activity of 6-pentyl-α-pyrone, metabolite of *Trichoderma harzianum*, in the inhibition of *Rhizoctonia solani* and *Fusarium oxysporum* f. sp. *lycopersici. J. Mycol. Res.*, *98*, 1207–1209.

Scharf, D. H., Heinekamp, T., Remme, N., Hortschansky, P., Brakhage, A. A., & Hertweck, C., (2012). Biosynthesis and function of gliotoxin in *Aspergillus fumigatus. Appl. Microbiol. Biotechnol.*, *93*, 467–472.

Schmoll, M., & Kubicek, C. P., (2003). Regulation of *Trichoderma* cellulase formation: Lessons in molecular biology from an industrial fungus. *Acta Microbiologica et Immunologica Hungarica*, *50*, 125–145.

Schmoll, M., & Schuster, A., (2010). Biology and biotechnology of *Trichoderma. Appl. Microbiol. Biotechnol.*, *87*(3), 787–799.

Schuster, A., Bruno, K. S., Collett, J. R., Baker, S. E., Selboth, B., Kubicek, C. P., & Schmoll, M., (2012). A versatile toolkit for high throughput functional genomics with *Trichoderma reesei. Biotechnol. Biofuels*, *5*(1), 1.

Semple, S. J., Pyke, S. M., Reynolds, G. D., & Flower, R. L., (2001). *In vitro* antiviral activity of the anthraquinone chrysophanic acid against poliovirus. *Antiviral. Res.*, *49*, 169–178.

Serrano-Carreon, L., Balderas-Ruiz, K., Galindo, E., & Rito-Palomares, M., (2002). Production and biotransformation of 6-pentyl-α-pyrone by *Trichoderma harzianum* in two phase culture systems. *Appl. Microbiol. Biotechnol.*, *58*, 170–174.

Shamala, S., & Idris, A. S., (2005). Antifungal properties in diffusible and non-diffusible compounds of *Trichoderma* isolates on the growth of *Ganoderma boninense*. In: *27ᵗʰ Symposium of the Malaysian Society for Microbiology* (pp. 24–27).

Shi, M., Wang, H. N., Xie, S. T., & Zang, Y. Z., (2010). Antimicrobial peptaibols, novel suppressors of tumor cells, targeted calcium-mediated apoptosis and autophagy in human hepatocellular carcinoma cells. *Molecular Cancer*, *9*(26), 26.

Shibuya, N., & Iwasaki, T., (1985). Structural features of rice bran hemicellulose. *Phytochemistry*, *24*, 285–289.

Siddiquee, S., Abdullah, F., Tan, S. G., & Aziz, E. R., (2007). Phylogenetic relationships of *Trichoderma harzianum* based on the sequence analysis of the Internal Transcribed Spacer Region–1 of the rDNA. *J. Appl. Sci. Res.*, *3*(9), 896–903.

Siddiquee, S., Aishah, S. N., Azad, S. A., Shafawati, S. N., & Naher, L., (2013). Tolerance and biosorption capacity of Zn^{2+}, Pb^{2+}, Ni^{3+} and Cu^{2+} by filamentous fungi (*Trichoderma harzianum*, *T. aureoviride* and, *T.*, *virens*). *Adv. Biosci. Biotechnol.*, *4*, 570–583.

Siddiquee, S., Al Azad, S., Fatimah, A. B., Laila, N., & Kumar, V. S., (2015). Separation and identification of hydrocarbons and other volatile compounds from cultures of *Aspergillus niger* by GC-MS using two different capillary columns and solvents. *J. Saudi Chem. Soc.*, *19*(3), 243–256.

Siddiquee, S., Cheong, B. E., Taslima, K., Hossain, K., & Hasan, M. M., (2012b). Separation and identification of volatile compounds from liquid cultures of *Trichoderma harzianum* by GC-MS using three different capillary columns. *J. Chromatogr. Sci.*, *50*, 358–367.

Siddiquee, S., Tan, S. G., Yusuf, U. K., Fatihah, N. H. N., & Hasan, M. M., (2012a). Characterization of Malaysian *Trichoderma* isolates using random amplified microsatellites (RAMS). *Mol. Biol. Rep.*, *39*, 725–722.

Siddiquee, S., Umi, K. Y., Hossain, K., & Jahan, S., (2009). *In vitro* studies on the potential *Trichoderma harzianum* for antagonistic properties against *Ganoderma boninense. J. Food Agric. Environ., 7*, 970–976.

Siddiqui, I. A., & Shaukat, S. S., (2004). *Trichoderma harzianum* enhances the production of nematicidal compounds *in vitro* and improves biocontrol of *Meloidogyne javanica* by *Pseudomonas fluorescens* in tomato. *Lett. Appl. Microbiol., 38*, 169–175.

Siu, R. G. H., & Reese, E. T., (1953). Decomposition of cellulose by microorganism. *The Botanical Rev., 19*, 377–416.

Sivasithamparam, K., & Ghisalberti, E. L., (1998). Secondary metabolism in *Trichoderma* and *Gliocladium*. In: Kubicek, C. P., & Harman, G. E., (eds.), *Trichoderma and Gliocladium: Basic Biology, Taxonomy, and Genetics* (Vol. 1, pp. 139–191). Taylor and Francis Ltd., London.

Smith, R. A., Yuan, H., Weissleder, R., Cantley, L. V., & Josephson, L., (2009). A wortmannin-cetuximab as a double drug. *Bioconj. Chem., 20*, 2185–2189.

Steyaert, J. M., Weld, R. J., Mendoza-Mendoza, A., & Stewart, A., (2010). Reproduction without sex: Conidiation in the filamentous fungus *Trichoderma. Microbiol., 156*, 2887–2900.

Sun, R., Liu, Z., Fu, K., Fan, L., & Chen, J., (2012). *Trichoderma* biodiversity in China. *J. Appl. Genet, 53*, 343–354.

Tijerino, A., Hermosa, R., Cardoza, R. E., Moraga, J., Malmierca, M. G., Aleu, J., Collado, I. G., Monte, E., & Gutierrez, S., (2011). Overexpression of the *Trichoderma brevicompactum* tri5 gene: Effect on the expression of the trichodermin biosynthetic genes and on tomato seedlings. *Toxin (Basel), 3*, 1220–1232.

Tronsmo, A., & Dennis, C., (1978). Effect of temperature on antagonistic properties of *Trichoderma* species. *Trans Br. Mycol. Soc., 71*, 469–474.

Tucci, M., Ruocco, M., De Masi, L., De Palma, M., & Lorito, M., (2011). The beneficial effect of *Trichoderma* spp. On tomato is modulated by the plant genotype. *Mol. Plant Pathol., 12*, 341–354.

Vadassery, J., Tripathi, S., Prasad, R., Varma, A., & Oelmuller, A., (2009). Monodehydroascorbate reductase 2 and dehydroascorbate reductase 5 are crucial for a mutualistic interaction between *Piriformospora indica* and *Arabidopsis. J. Plant Physiol., 166*, 1263–1274.

Vargas, W. A., Mandawe, J. C., & Kenerley, C. M., (2009). Plant-derived sucrose is a key element in the symbiotic association between *Trichoderma virens* and maize plants. *Plant Physiol., 151*, 792–808.

Verma, M., Brar, S. K., Tyagi, R. D., Surampalli, R. Y., & Valéro, J. R., (2007). Antagonistic fungi, *Trichoderma* spp.: Panoply of biological control. *Biochem. Engineer. J., 37*, 1–20.

Vey, A., Hoag, I. R. E., & Butt, T. M., (2001). Toxic metabolites of fungal biocontrol agents. In: Butt, T. M., Jackson, C., & Magan, N., (eds.), *Fungi as Biocontrol Agents: Progress, Problems, and Potential* (pp. 1–311). CAB International Publishing, Bristol, USA.

Vinale, F., Flematti, G., Sivasithamparam, K., Lorito, M., Marra, R., Skelton, B. W., & Ghisalberti, E. L., (2009). Harzianic acid, an antifungal and plant growth promoting metabolite from *Trichoderma harzianum. J. Nat. Prod., 72*, 2032–2035.

Vinale, F., Ghisalberti, E. L., Flematti, G., Marra, R., Lorito, M., & Sivasithamparam, K., (2010). Secondary metabolites produced by a root-inhabiting sterile fungus antagonistic towards pathogenic fungi. *Lett. Appl. Microbiol.*, *50*, 380–385.

Vinale, F., Sivasithamparam, K., Ghisalberti, E. L., Marra, R., Barbetti, M. J., Li, H., Woo, S. L., & Lorito, M., (2008). A novel role for *Trichoderma* secondary metabolites in the interactions with plants. *Physiol. Mol. Plant Pathol.*, *72*, 80–86.

Viswanathan, K., Ononye, S., Cooper, H., Kyle, H. M., Anderson, A., & Wright, D., (2012). Viridin analogs derived from steroidal building blocks. *Bioorg. Med. Chem. Lett.*, *22*, 6919–6922.

Viterbo, A., Landau, U., Kim, S., Chernin, L., & Chet, I., (2010). Characterization of ACC deaminase from the biocontrol and plant growth-promoting agent *Trichoderma asperellum* T203. *FEMS Microbiol. Lett.*, *305*(1), 42–48.

Watanabe, A., Kamei, K., Sekine, T., Waku, M., Nishimura, K., Miyaji, M., Tatsumi, K., & Kuriyama, T., (2004). Effect of aeration on gliotoxin production by *Aspergillus fumigatus* in its culture filtrates. *Mycopathol.*, *157*, 19–27.

Watts, R., Dahiya, J., Chaudary, K., & Tauro, P., (1988). Isolation and characterization of a new antifungal metabolite of *Trichoderma reesei*. *Plant and Soil*, *107*(1), 81–84.

Weindling, R., (1932). *Trichoderma lignorum* as a parasite of other soil fungi. *Phytopathol.*, *22*, 837–845.

Weindling, R., (1934). Studies on a lethal principle effective in the parasitic action of *Trichoderma lignorum* on *Rhizoctonia solani* and other soil fungi. *Phytopathol.*, *24*, 1153–1179.

Whipps, J. M., & Lumsden, R. D., (1991). Biological control of *Pythium* species. *Biocontrol. Sci. Techn.*, *1*, 75–90.

Widden, P., & Abitbol, J. J., (1980). Seasonality of *Trichoderma* species in aspruce-forest soil. *Mycologia*, *72*, 775–784.

Wuczkowski, M., Druzhinina, I. S., Gherbawy, Y., Klug, B., Prillinger, H., & Kubicek, C. P., (2003). Species pattern and genetic diversity of *Trichoderma* in a mid-European, primeval floodplain-forest. *Microbiol. Res.*, *158*, 125–133.

Yagi, K., Williams, N. Y., Wang, N. Y., & Cicerone, J., (1993). Agricultural soil fumigation as a source of atmospheric methyl bromide. *Proc. Natl. Acad. Sci. USA*, *90*, 8420–8423.

Yang, C. A., Cheng, C. H., Lo, C. T., Liu, S. Y., Lee, J. W., & Peng, K. C., (2011). A novel L-amino acid oxidase from *Trichoderma harzianum* ETS 323 associated with antagonism of *Rhizoctonia solani*. *J. Agric. Food Chem.*, *59*, 4519–4526.

Yang, D., Bernier, L., & Dessureault, M., (1995). *Phaeotheca dimorphospora* increases *Trichoderma harzianum* density in soil and suppresses red pine damping-off caused by *Cylindrocladium scoparium*. *Can. J. Bot.*, *73*, 693–700.

Yang, Z. S., Li, G. H., Zhao, P. J., Zheng, X., Luo, S. L., Li, L., Niu, X. M., & Zhang, K. Q., (2010). Nematicidal activity of *Trichoderma* spp. And isolation of an active compound. *World J. Microbiol. Bioetchnol.*, *26*, 2297–2302.

Yang, Z. S., Yu, Z., Lei, L., Xia, Z. Y., Shao, L., Zhang, K., & Li, G. H., (2012). Nematicidal effect of volatiles produced by *Trichoderma* sp. *J. Asia-Pacific. Entomol.*, *15*, 647–650.

Yedidia, I., Benhamou, N., & Chet, I., (1999). Induction of defense responses in cucumber plants (*Cucumis sativus*, L.,) by the biocontrol agent *Trichoderma harzianum*. *Appl. Environ. Microbiol.*, *65*, 1061–1070.

Zachow, C., Berg, C., Müller, H., Meincke, R., Komon-Zelazowska, M., & Druzhinina, I. S., (2009). Fungal diversity in the rhizosphere of endemic plant species of Tenerife (Canary Islands): Relationship to vegetation zones and environmental factors. *ISME J.*, *3*, 79–92.

Zhang, C. L., Druzhinina, I. S., Kubicek, C. P., & Xu, T., (2005). *Trichoderma* biodiversity in China: Evidence for a North to South distribution of species in East Asia. *FEMS Microbiol. Lett.*, *251*, 251–257.

Zhang, S. X., & Zhang, X., (2009). Effects of two composted plant pesticide residues, incorporated with *Trichoderma viride*, on root-knot nematode in Balloon flower. *Agric. Sci. China*, *8*, 447–454.

CHAPTER 9

BIODIVERSITY AND CONSERVATION OF HIGHER PLANTS IN BANGLADESH: PRESENT STATUS AND FUTURE PROSPECTS

ABUL KHAYER MOHAMMAD GOLAM SARWAR

Laboratory of Plant Systematics, Department of Crop Botany, Bangladesh Agricultural University, Mymensingh 2202, Bangladesh

9.1 INTRODUCTION

'Biological diversity' and shortly biodiversity, was first used by Raymond F. Dasmann, a wildlife scientist and conservationist in 1968 in his book titled *A Different Kind of Country* advocating conservation (https://en.wikipedia.org/wiki/Biodiversity). 'The term biological diversity' means the variety amongst living organisms from all sources comprising of inter-alia, terrestrial, marine, and different aquatic ecosystems, and the ecological complexes of which they are a part; it includes variety within species, between species, and of ecosystems (https://www.cbd.int/convention/articles/default.shtml?a=cbd-02). Recently, the diversity of indigenous knowledge is also included as a component of biodiversity. Biological diversity is of basic importance to the working of all natural and human-engineered ecosystems, and by extension to the surroundings services that nature provides freed from fee to human society. Living organisms, many interrelating species, play vital roles in the cycles of key elements (carbon, nitrogen, and others) and water in the environment. Plant genetic resources additionally provide fundamental material for choice and improvement through breeding to make sure food and nutritional safety needs of world's hastily growing populace in a meet.

Bangladesh, which lies in the northeastern part of South Asia between 20°34′ and 26°38′ N latitude and between 88°01′ and 92°41′ E longitude, is one of the biggest deltaic plains in the world. It has a total area of 147,570 sq km, and about 80% of the total area formed by the confluence of the Ganges (Padma), the Jamuna, the Meghna and the Brahmaputra rivers. The residual 20% of the land location is made from the undulating, forested hill tracts. Geologically Bangladesh occupies a bigger a part of the Bengal Basin, and the country is blanketed by way of tertiary folded sedimentary rocks (12%) within the north, northeastern and eastern portions; elevated Pleistocene residuum (8%) in the northwestern, mid-northern and eastern portions; and Holocene deposits (80%) together with unconsolidated sand, silt, and clay. The sub-tropical monsoon weather prevails all through the country with temperately warm temperatures, high relative humidity, and heavy rainfall at some point of the rainy season. Three seasons are generally identified: a hot, humid summertime from March to June; a fab, wet monsoon from June to October; and a fab, dry winter from October to March. In general, the most summertime temperature ranges between 32°C and 38°C. April is the warmest month in most parts of the country and January is the coldest when the average temperature for most of the country is 10°C. Most parts of the country receive annual rainfall, at least, 200cm except in Rajshahi, where the annual rainfall is about 160cm. About 80% of the total rainfall of Bangladesh falls during the monsoon season. Natural disasters, which include floods, tropical cyclones, tornadoes, and tidal bores – unfavorable waves or floods due to flood tides speeding up estuaries - ravage the country, specifically the coastal belt nearly every year. Wonderful physiographic traits, differences in hydrological and meteorological situations, and the difference in the soil houses in Bangladesh make contributions in growing various varieties of ecosystems enriched with an enormous range of flora and fauna (Nishat et al., 2002).

The eastern part of Bangladesh biogeographically falls in the Indo-Burma part that's one of the 25 known biodiversity hotspot regions of the world and thought to have 7,000 endemic plant species (Myers et al., 2000; http://www.eoearth.org/view/article/150621/). The country as an entire is a transition between the Indo-Gangetic plains and the eastern Himalayas and, in flip, a part of the Indo-Chinese language sub-vicinity of the oriental realm (MoEF, 2004). Thus, this country acts as an important merging and sharing habitat, land bridge and biological corridors of flora and fauna between these sub-regions. This strategic position makes Bangladesh as one of the greatest ecologically important and biologically varied landscapes in phrases of migratory species, stepping stones, staging floor and flyways for wildlife activities of this region (DoE, 2015). On the basis of both biotic and abiotic components of ecosystems, the world is divided into five Global Ecological Domains (GED) and 20 Global Ecological Zones (GEZs). Bangladesh belongs to two of these zones, *viz.* Tropical Rain Forest GEZ (33%) and Tropical Moist Deciduous Forest GEZ (67%) of the Tropical Domain of the GED. The central, north-eastern, and south-eastern forests belong to Tropical Rain Forest GEZ (http://www.fao.org/forestry/17847/en/bgd/). Bangladesh is divided into 25 bio-ecological zones on the basis of biological and physical parameters (Nishat et al., 2002). Additionally, Bangladesh has been divided into 30 Agro-Ecological Zones, on the basis of physiography, soil properties, soil salinity, depth, and duration of flooding (FAO-UNDP, 1988), which also indicates country's richness in ecosystem diversity. In a nutshell, Bangladesh, although occupies quite small geographical vicinity, is wealthy in both floral and faunal diversities obtrusive in a numerous variety of ecosystems beginning from the northern and eastern hills to the southern seas; maximum deciduous forests to the mangroves and different agro-ecosystems unfold over the wetlands, floodplains in addition to the hills.

In this chapter, present state of knowledge on plant biodiversity in Bangladesh territory and the strategies followed for their conservation are discussed. The future prospect of biodiversity and conservation has also been emphasized here.

9.2 STATE OF FORESTS AND FORESTRY IN BANGLADESH

The extent of forest area as reported by the Department of Forest (DoF) is 2.53 million ha (Table 9.1). Out of this 2.53 million ha of forests, the DoF manages 1.53 million ha (Ahmed et al., 2007; Table 9.2). It has been reported by the *National Forest Inventory in Bangladesh* that about 48% region of the country has vegetative cover in some form or the other (http://www.fao.org/forestry/17847/en/bgd/). The area of different tree cover categories used in this inventory is shown in Table 9.3.

Natural forests are deteriorating, and their depletion is a continuous process. The well-being and general structure of the forests are 'below average' (Choudhury and Hossain, 2011). Nevertheless, plantations (afforestation) have a distinct situation. Almost all plantations are mounted underneath a few sorts of improvement programmes underneath different schemes. Reasonable funding is presented from projects coffers for establishing the plantations. Funds range are also made available from projects for the upkeep of the plantations for about three years (DoE, 2015).

The protection and subsequent care of those plantations are normally vested with the DoF from the fourth year of establishment. In the fourth year, the seedlings have reached pole length, and they start attracting nearby human beings. Pilferage and illicit felling start at this level. It is a 'Herculean task' for the DoF to guarantee protection with the small amount of DoF staff at the ratio of 1:14,000 people. The resulting effect is a constant reduction of the stock from the plantations, as they reached pole length (Choudhury and Hossain, 2011). As this lingers, the plantations have a "scattered tree" look at 15 to 20 years. The plantations that looked favorable at the early phase, usually fail at 15 to 20 years of age. This situation, nevertheless, is not binding for the plantations which have been established beneath social forestry programmes involving the contributors. In the case of participatory plantations, the contributor's labor as a strong force to fight theft and pilferage. The growth rate (MAI13) in these participatory plantations ranges between 0.5 to 7.5 $m^3ha^{-1}year^{-1}$. In most cases, the Mean Annual Increment (MAI) is about 4 $m^3ha^{-1}year^{-1}$, which is acceptable. In brief, the general state of forests is not good in Bangladesh (Choudhury and Hossain, 2011). The central factors undermining the sustainability of forests and forest administration in Bangladesh were enumerated and discussed by Choudhury and Hossain (2011).

TABLE 9.1   Extent of Forest Area as Reported by Department of Forest (2007) (http://www.fao.org/3/a-am628e.pdf)

Forest category	Area (million ha according to DoF)	Area (million ha according to Forest Resources Management Project)
Hill forest	1.40	0.40
Sal forest	0.12	0.12
Village forest	0.27	N.A.
Mangrove forest	0.74	0.48
Total	2.53	

TABLE 9.2 Forest Area under Different Management Categories (DoE, 2015)

Category of Forest	Area (million ha)	% Total Land
Managed Forests (Bangladesh Forest Department)	1.53	10.54
Unclassed State Forests (Ministry of Land)	0.73	5.07
Village Forests (Private)	0.27	1.88
Total	2.53	17.49

TABLE 9.3 Tree Cover in Bangladesh (http://www.fao.org/forestry/17847/en/bgd/)

Tree cover categories	Area (in million ha)
No tree-cover (mostly arable land)	7.60
Less than 5% tree-cover	2.89
5 to 10% tree-cover	1.43
10 to 30% tree-cover	1.27
30 to 70% tree-cover	1.23
Over 70% tree-cover	0.33
Total	14.75

9.3 HISTORY OF FLORISTIC WORKS IN BANGLADESH

Flora is the plant life occurring in a particular region or time, generally the naturally occurring or indigenous (native) plant life. Complete inventory and production of flora with plant resources are very important for each and every country for planning and implementing national conservation strategy to save threatened taxa. The study of plants began as early as 4,000 years BC in some ancient cultures of Egypt, Babylon, China, and India. The ancient scriptures on plants in Indian sub-continent were mostly related to medicine, horticulture, and agriculture (Pasha, 2014). The *Charaka Samhita* (about 700 BC) is the earliest surviving example, written by the physician Charaka, who described 1,500 plants and identified 350 as medicinal.

A giant step in the exploration and production of flora of the territory of the British India including the area now in Bangladesh had begun with *Hortus Bengalensis* (Roxburgh, 1814) and *Flora Indica* (Roxburgh, 1820–1832), and thereafter the *Flora of British India* (Hooker, 1872–1897), *Forest Flora of British Burma* (Kurz, 1877), *Bengal Plants* (Prain,

1903a), *The Flora of Sundarbans* (Prain, 1903b), *Flora of Assam* (Kanjilal et al., 1934–1940), *Grasses of Burma, Ceylon, India, and Pakistan* (Bor, 1960) and so on. Moreover, five local flora of Bangladesh namely, *List of Plants of Chittagong Collectorate and Hill Tracts* (Heinig, 1925), *Flora of the Chakaria Sundarbans* (Cowan, 1928), *On the flora of Chittagong* (Raizada, 1941), *Common plants in and around Dacca* (presently Dhaka) (Datta and Mitra, 1953) and *Flora of Cox's Bazar* (Sinclair, 1955) have been published. These publications have been used as the baseline of the floristic study of Bangladesh. The plant exploration history of Bangladesh, including Indian sub-continent, could be divided into seven phases – Portuguese period, Dutch period, Linnaean period, Post-Linnaean period, British period in Far Eastern India, Pakistan Era and Bangladesh Era (Pasha, 2014). The exploration history of this area was elaborately recorded by Burkill (1965), Pasha and Uddin (2007) and Pasha (2014). The trends and opportunities of plant taxonomic and biodiversity research in Bangladesh have also been discussed in detail by Rahman (2014a).

9.4 PRESENT STATUS OF KNOWLEDGE ON PLANT DIVERSITY OF BANGLADESH

Bangladesh is one of the richest biodiversity areas in the Indian subcontinent estimating approximately *c.* 5,700 vascular angiosperms, *c.* 1,700 pteridophytes and 4 gymnosperms of which more than 80% species are Indian elements (Khan, 1991). The flora shows a considerable admixture of Cacher and Khasia elements. The *Encyclopedia of Flora and Fauna of Bangladesh* included 3611, 195 and 7 species of angiosperms, pteridophytes, and gymnosperms, respectively (Siddiqui et al., 2007; Ahmed et al., 2008a,b; Ahmed et al., 2009a-d). It has been revealed that out of 3,611 species of angiosperms, 2,623 species belong to 158 families of dicotyledons, and 988 species belong to 41 families of monocotyledons; although the number of species diversity is still a point of debate and should be increased as many new records have been reported in recent publications (Barbhuiya and Gogoi, 2010). In the national context, between June 2009 and June 2013, 64 angiosperm species were recorded from Bangladesh, and 8 were described as new to science (Irfanullah, 2013). Moreover, Pasha and Uddin (2013) have listed 5016, 311 and 21 species of angiosperms, pteridophytes, and gymnosperms, respectively in their publication

Dictionary of Plant Names of Bangladesh. A total of 777 genera of 160 families of flowering plants was represented by single species in the flora of the country (Sarwar, 2013); 280 genera represented by single species are used for medicinal purposes (Sarwar, 2015). Among them, 22 genera are monotypic in strict sense, i.e., represented worldwide only by the 'Type species'; 19 taxa are threatened as per the International Union for Conservation of Nature (IUCN) Red List categories (Sarwar, 2015). Endemism is not widely expressed amongst the flora, and *c.* 25 species are known as endemic to Bangladesh (Pasha, 2012). Recently, Bangladesh National Herbarium has also reported 50 angiosperm species as new records for the country (Ara and Khan, 2015). Among the recorded species in the flora of Bangladesh, at least, 130 genera (8%) and 400 species (about 8%) have been identified as exotic plants (Hossain and Pasha, 2004). Some largest families of Bangladesh flora are Poaceae, Fabaceae, Orchidaceae, Rubiaceae, Asteraceae, and Cyperaceae which include about 50.25% and 55.51% of the total number of genera and species, respectively (Table 9.4). But in the global context, the richest floral diversity in Bangladesh flora is Lauraceae (43.74%), Cyperaceae (34.29%) and Convolvulaceae (32.73%) (Pasha and Uddin, 2013).

In the case of other plant groups, the *Encyclopedia of Flora and Fauna of Bangladesh* reported 3,002 algal taxa (including cyanobacteria) under 424 genera and 127 families. Being microscopic, a tremendous possibility exists to discover new algal species in Bangladesh. For example, a four-season sampling of a couple of water bodies in the tea gardens of Moulvibazar district revealed 421 algal taxa of which 130 were found for the first time in Bangladesh and 3 were new to the biologists (Irfanullah, 2013). Continuous new records of both macro- and microscopic flora justify the need for a complete inventory of Bangladesh flora in order to have a good understanding of the biodiversity trend.

Inventory and documentation of the *Flora of Bangladesh* were initiated in May 1970 when a project was approved by the Government entitled "Botanical Survey of Bangladesh (the then East Pakistan)" under the Agricultural Research Council of Bangladesh (the then Pakistan). The project was sustained until 1973 while it was taken up as a contact project under the newly named Bangladesh Agricultural Research Council. The first volume (Fasc. 1) of the *Flora of Bangladesh* has been published in 1972 including five angiosperm families, *viz.* Caricaceae, Casuarinaceae, Hydrophyllaceae, Martyniaceae, and Phytolaccaceae (Khan, 1972). So far only 74 families have been completely inventoried and published in 62 fascicles (Khan, 1972–1987; Khan and Rahman, 1989–2002; Rahman, 2003; Rahman and Khanam, 2003; Khanam and Ara, 2007–2008; Ara and Khan, 2009–2015) from Bangladesh National Herbarium. However, these 74 families composed of 596 species under 278 genera, many more species need to be inventoried.

There is no exclusive account of the forest flora of Bangladesh. List of important tree species is found in management plans of respective forest divisions. Checklists on the woody flora of Sylhet forests and Sal forests were published by Bangladesh Forest Research Institute (BFRI) (Alam, 1988, 1999). Illustrated accounts of 148 tree species from homesteads of Bangladesh are found in Khan and Alam (1996). Das and Alam (2001) made an account of 342 tree species comprising of indigenous and introduced trees occurring in the forests, villages, and homesteads of Bangladesh. Recently, Basak and Alam (2015) made a checklist of 1,048 tree species (including Gymnospermae, Dicotyledons, and Monocotyledons) under 432 genera in 99 families from Bangladesh.

Like in other regions around the globe, the biodiversity of Bangladesh is also entering a critical period. The present status of plant biodiversity is a more complicated issue than the debate on the total number of species present in this territory. Many of the taxa recorded from Bangladesh territory have not been collected again since their original collections since Hooker (1872–1897) or have not been collected after its second collection by Prain in 1903a; therefore, their physical existence in nature at present is very much questionable. The first volume of *Red Data Book of Vascular Plants of Bangladesh* has been published in 2001 which included 106 species, although only four angiosperm species are categorized as threatened (1 Critically Endangered, 1 Endangered and 2 Vulnerable) in Bangladesh (Khan et al., 2001).

The information presented in the Encyclopedia can be considered as the most recent update for Bangladesh. After consulting eight volumes of *Encyclopedia of Flora and Fauna of Bangladesh* on vascular plants, Irfanullah (2011) has reported that about 13% species of Bangladesh flora is designated as threatened (Table 9.5). A few families are significantly threatened; for example, about 53% species of Orchidaceae are threatened (94 species out of 179), whereas, in Lamiaceae, it is more than 30% (26 species out of 86). Needless to say, these threatened statuses are purely in the national context.

The 2nd volume of *Red Data Book of Vascular Plants of Bangladesh* has recently been published including 120 species with varying levels of threats (Ara et al., 2013). Out of 13 angiosperm families containing 520 species assessed by Rahman (DoE, 2015), 235 species were categorized into different threatened categories. Moreover, Rahman (2014b) reported a total of 214 taxa presumed extinct from Bangladesh flora. The challenge remains to have a clear, acceptable, and comprehensive evaluation of the flora of Bangladesh following the latest guidelines of the IUCN Red List. In the meanwhile, the 3rd volume of *Red Data Book of Vascular Plants of Bangladesh* is under preparation, which includes 260, 37 and 2 species of angiosperms, pteridophytes, and gymnosperms, respectively. Although there is no detailed information on the loss of plant species diversity, it appears from the literature that about 225 vascular plant species of those that were reported about 50 to 100 years ago are not found nowadays in the territory of Bangladesh (DoE, 2016).

Not only the forest genetic resources, but the number of local cultivars of various crops is also steadily declining due to the promotion and economic consideration of only a few selected cultivars, especially of the modern, both the high yielding and hybrids, cultivars of rice, wheat, pulses, oilseeds, and other crops. Local cultivars made significant contributions which still make an essential contribution to food security in many countries as well as providing the genetic diversity for modern plant breeding. They are well adapted to marginal production environments, fit in with local farming systems and meet local tastes and nutritional preferences. For example, nearly 12,500 traditional cultivars of rice were cultivated in different seasons of Bangladesh (BRRI, 1982). In Genetic Resource and Seed Division of Bangladesh Rice Research Institute (BRRI), more than 8,000 landraces of rice were collected and conserved as long medium, and short-term storage (Table 9.6). And we do not know the fate of rest of the cultivars/landraces/germplasms; these might be extinct from nature. This is causing irreparable loss to the genetic diversity of our crop plants and their wild relatives.

TABLE 9.4   Twenty Species-Rich Families in Flora of Bangladesh (Pasha and Uddin, 2013)

Sl. No.	Family name	Genera	Species
1	Poaceae	132	342
2	Fabaceae	80	301
3	Orchidaceae	77	242
4	Rubiaceae	68	217
5	Asteraceae	104	197
6	Cyperaceae	24	166
7	Acanthaceae	40	161
8	Euphorbiaceae	50	159
9	Lamiaceae	38	104
10	Araceae	30	90
11	Scrophulariaceae	30	87
12	Convolvulaceae	20	87
13	Asclepiadaceae	35	82
14	Verbenaceae	19	80
15	Zinziberaceae	18	80
16	Caesalpiniaceae	18	76
17	Lauraceae	15	76
18	Moraceae	7	75
19	Malvaceae	16	66
20	Mimosaceae	16	60
Total		837	2748

TABLE 9.5   Number of Threatened Species, in Parenthesis Percentage of Total Species, in Major Vascular Plant Groups (Irfanullah, 2011)

Vascular plant groups	Number of species	Critically Endangered (CR)	Endangered (EN)	Vulnerable (VU)	Total number of threatened species
Pteridophytes	195	0	0	36	36 (18.46)
Gymnosperms	7	0	1	0	1 (14.29)
Angiosperms	3,611	30	126	293	449 (12.43)
Dicotyledons	2,623	8	80	179	267 (10.18)
Monocotyledons	988	22	46	114	182 (18.42)
Total	3,813	30	127	329	486 (12.75)

TABLE 9.6   Rice Genetic Resources in the BRRI Gene Bank (DoE, 2015)

Variety/Line	Registered in accession
Indigenous *indica*	
Local landraces	5202
Pure line selection	1030
Exotic *indica* landraces (IRRI, China, USA, Turkey)	790
Exotic/breeding lines	968
Wild Rice of Bangladesh (*Oryza rufipogon, O. officinalis, O. nivara,* and *O. sativa* f. *spontanea*)	42
Wild rice from IRRI	12
Total	8044

9.5   CAUSES OF DEPLETION OF PLANT BIODIVERSITY IN BANGLADESH

The main reason behind the depletion of plant/forest biodiversity can be attributed to the anthropogenic activities on the world's ecosystem. Some of the major reasons that causes biodiversity reduction in Bangladesh are high population density, extreme poverty and unemployment, low levels of education, habitat destruction, degradation, and fragmentation of land, changes of land use pattern, over-exploitation, and illegal collection, changes in hydrological regime, environmental pollution and degradation, natural calamities – floods, cyclones, increase in soil salinity and so on, related to sea level rise and global climate changes, invasive alien species and others. The causes of forest biodiversity depletion in Bangladesh have been elaborately addressed by Rahman (http://cdn.intechopen.com/pdfs-wm/47712.pdf).

There are three major causes for which this serious condition of crop plant genetic resources has been elevated in Bangladesh. These include: (i) the absence of sufficient conservation efforts to keep and protect native crop genetic resources, (ii) absence of gene bank and software for the conservation and propagation of the several native crop cultivars, and (iii) unstable rivalry between the local cultivars consequent to the introduction, adoption, and promotion of the exotic and locally established modern, both inbred and hybrid, cultivars (http://cdn.intechopen.com/pdfs-wm/47712.pdf). The causes and consequences of biodiversity, both flora, and fauna, losses have described and discussed in detail in the revised National Biodiversity Strategy and Action Plan (NBSAP) (DoE, 2016).

9.6   CONSERVATION STRATEGIES ON PLANT GENETIC RESOURCES OF BANGLADESH

The loss of plant biodiversity threatened agriculture, forestry, environment, and in the long run humanity itself (FAO, 1984). Therefore, the Convention on Biological Diversity (CBD) emphasized the need for protection of the same through the Article 6 placed the obligation squarely on the countries to develop their own national strategies, plans, and programmes for conservation and sustainable use of biological diversity (https://www.cbd.int/convention/articles/default.shtml?a=cbd-06). The Rio Convention explicitly has a provision stating that 'timely, reliable and accurate information on forests and forest ecosystems is essential for public understanding and informed decision-making' should be made available by the participating countries. The NBSAP is a national document that answers to the responsibilities of the CBD. It includes strategies and action plans for the conservation, sustainable use and unbiased sharing of biodiversity benefits. With the establishment of the NBSAP in 2004, Bangladesh has made a significant step forward to fulfilling the global commitment of the country to the CBD. The NBSAP of Bangladesh has identified 16 strategies, including 128 action programmes (MoEF, 2004). The action plans are further categorized as short-term (0–3 yrs), medium-term (4–7 yrs) and long-term (8–10 yrs) programmes. Since its formulation, the Government of Bangladesh has taken a number of steps to implement the NBSAP.

A detailed assessment of the mainstreaming and implementation of the NBSAP and its cross-sectoral integration had been presented in the Fourth National Report of Bangladesh (DoE, 2010). Progress like preparation of *Encyclopedia of Flora and Fauna of Bangladesh* has been made over the years. New rules and acts have been developed; in some cases, existing legislative mechanisms are reviewed and updated in implementing the NBSAP objectives (DoE, 2015). For example, the *Wildlife Protection Act 1974* has been updated to *Bangladesh Wildlife (Conservation and Security) Act, 2012*. The traditional form of government-owned wetland leasing out systems has already been updated in 2005. Community-based management approaches instead of traditional leasing system in some government-owned wetlands have been introduced in some cases. The *Forest Policy of 1994* had explicit aim of bringing 25% lands of the country under forest cover by 2015, although the present situation is very far from the target point. Similar development initiatives have taken place in Bangladesh in recent times that could be treated as the success in achieving NBSAP principles. In addition, species conservation issues are taken as forefront activities by many concerned agencies and affiliated organizations of the government. The Government, in dealing with climate change induced threats to biodiversity, also put sufficient emphasis to uphold the principles of CBD while developing *National Adaptation Plan of Action and Bangladesh Climate Change Strategy and Action Plan 2009* (DoE, 2015).

As a part of conservation programmes, ecosystem (*in situ*) conservation has a direct benefit at all levels of biodiversity – from genetic to the ecosystem and shows the government's commitment to conserving its natural habitats. The important biodiversity areas of Bangladesh are conserved in different forms, like protected areas (PAs), ecologically critical areas (ECAs), botanical gardens, eco-parks, safari parks and fish sanctuaries (DoE, 2016).

9.6.1 IN SITU CONSERVATION

In situ conservation is carried out in the following areas: nature reserves, national parks, wildlife sanctuaries, and world heritage sites. PAs are "areas especially dedicated to the protection and maintenance of biological diversity and associated cultural resources, which are managed through legal or other effective means." These have long been core elements of national as well as regional conservation strategies. With the objective of preserving biodiversity (flora in addition to fauna) and the natural environment inside diverse forest sorts, three types of protected area under different IUCN protected area management groups are demarcated in the Bangladesh Wildlife Preservation Act, 1974.

Wildlife Sanctuary, an area, maintained as an undisturbed breeding ground for wild fauna and in which the habitat is included for the continued well-being of the resident or migratory fauna. Countrywide park, a comparatively large place of natural splendor to which the members of the general public have right of entry for leisure, education, and research, and wherein wildlife are secured. Sports reserve commonly incorporates a relatively remote area supposed for protection of the natural world in fashionable and to boom the populace of specific species. A listing of the PAs (for *in situ* conservation) declared so far under diverse forest types of Bangladesh is given below (Table 9.7). The total area of PAs is 456,086 hectares (without the Marine PA) which are c. 18% of the total forest areas of the country.

Bangladesh Government has set up and affirmed numerous eco-parks and two safari parks to preserve biodiversity and genetic resources for research and other purposes. Table 9.8 lists the name and area of eco-parks and safari parks in Bangladesh. Although these sites are mostly used for recreational purposes, they also harbor a diverse community of flora and fauna. Both *in situ* and *ex situ* conservation approaches have been approved here to preserve and keep biodiversity in sound condition.

Department of Environment (DoE) has also declared 13 sites as Ecologically Critical Areas (ECAs) under section 5 of the *Bangladesh Environment Conservation Act, 1995* (Table 9.9). The total area coverage of ECAs is 384,529 hectares or 2.60% of the total country. A project for Coastal and Wetland Biodiversity Management at Cox's Bazar and Hakaluki Haor which is being executed by the DoE; the project has been undertaking various programmes towards conservation of the biological diversities of 4 ECAs namely, Teknaf Peninsula, Sonadia Island, and St. Martin's Island of Cox's Bazar, and Hakaluki Haor, Sunamganj. The aim is to ensure conservation, management, and sustainable use of the biological and other resources of the ECAs through establishing institutional arrangement.

The Arannayk Foundation, a joint initiative of the Governments of the People's Republic of Bangladesh and the United States of America based on the provisions of the US Tropical Forest Act of 1998, which seeks to contribute to the *in situ* conservation of biodiversity assets of tropical forests. The natural ecosystem and biodiversity resources of the forests in Bangladesh are maintained in a sustainable manner by responsible stakeholders through collective efforts - providing access and benefits to local communities, national economy and the human kind at large in an equitable manner (http://www.arannayk.org/index.php?option=com_cont ent&view=featured&Itemid=101).

TABLE 9.7 List of Protected Areas of Bangladesh (http://www.bforest.gov.bd/; Accessed on 14 November 2017)

Sl. No.	Protected Areas	Location	Area (ha)	Established
A) National Parks (IUCN category II and V)				
1	Himchari National Park	Cox's Bazar	1,729.00	15–02–1980
2	Bhawal National Park	Gazipur	5,022.00	11–05–1982
3	Madhupur National Park	Tangail/ Mymensingh	8,436.00	24–02–1982
4	Lawachara National Park	Moulvibazar	1,250.00	07–07–1996
5	Kaptai National Park	Chittagong Hill Tracts	5,464.00	09–09–1999
6	Nijhum Dweep National Park	Noakhali	16,352.23	08–04–2001
7	Ramsagar National Park	Dinajpur	27.75	30–04–2001
8	Satchari National Park	Habigonj	242.91	15–10–2005
9	Khadimnagar National Park	Sylhet	678.80	13–04–2006
10	Medhakachhapia National Park	Cox's Bazar	395.92	08–08–2008
11	Baroiyadhala National Park	Chittagong	2,933.61	06–04–2010
12	Kuakata National Park	Patuakhali	1,613.00	24–10–2010
13	Nababgonj National Park	Dinajpur	517.61	24–10–2010
14	Singra National Park	Dinajpur	305.69	24–10–2010
15	Kadigarh National Park	Mymensingh	344.13	24–10–2010
16	Altadighi National Park	Naogaon	264.12	24–12–2011
17	Birgonj National Park	Dinajpur	168.56	24–12–2011
B) Wildlife Sanctuaries (IUCN category IV)				
18	Char Kukri-Mukri Wildlife Sanctuary	Bhola	40.00	19–12–1981
19	Pablakhali Wildlife Sanctuary	Chittagong Hill Tracts	42,087.00	20–09–1983
20	Chunati Wildlife Sanctuary	Chittagong	7,763.97	18–03–1986
21	Rema-Kalenga Wildlife Sanctuary	Hobigonj	1,795.54	07–07–1996
22	Sundarban (East) Wildlife Sanctuary	Bagerhat	122,920.90	29–06–2017
23	Sundarban (West) Wildlife Sanctuary	Satkhira	119,718.88	29–06–2017

TABLE 9.7   *(Continued)*

Sl. No.	Protected Areas	Location	Area (ha)	Established
24	Sundarban (South) Wildlife Sanctuary	Khulna	75,310.30	29–06–2017
25	Fashiakhali Wildlife Sanctuary	Cox's Bazar	1,302.43	11–04–2007
26	Dudpukuria-Dhopachari Wildlife Sanctuary	Chittagong	4,716.57	06–04–2010
27	Hajarikhil Wildlife Sanctuary	Chittagong	1,177.53	06–04–2010
28	Sangu Wildlife Sanctuary	Bandarban	2,331.98	06–04–2010
29	Teknaf Wildlife Sanctuary	Cox's Bazar	11,615.00	24–03–2010
30	Tengragiri Wildlife Sanctuary	Barguna	4,048.58	24–10–2010
31	Sonarchar Wildlife Sanctuary	Patuakhali	2,026.48	24–12–2011
32	Dudhmukhi Wildlife Sanctuary	Bagerhat	170.00	29–01–2012
33	Chadpai Wildlife Sanctuary	Bagerhat	560.00	29–01–2012
34	Dhangmari Wildlife Sanctuary	Bagerhat	340.00	29–01–2012
35	Nazirganj Wildlife (Dolphin) Sanctuary	Pabna	146.00	01–12–2013
36	Shilanda-Nagdemra Wildlife (Dolphin) Sanctuary	Pabna	24.17	01–12–2013
37	Nagarbari-Mohanganj Dolphin Sanctuary	Pabna	408.11	01–12–2013
D) Game Reserve				
38	Teknaf GR	Cox's Bazar	11,615.00	1983
E) Marine Protected Area				
39	Swatch of No-Ground	Bay of Bengal	173,800.00	27–10–2014
F) Special Biodiversity Conservation Area				
40	Special Biodiversity Conservation Area (Ratargul)	Sylhet	204.25	31.05.2015
41	Altadighi Special Biodiversity Conservation Area	Naogaon	17.34	09.06.2016
Total			629,885.36	

TABLE 9.8 Botanical Gardens and Eco-Parks (Both *In Situ* and *Ex Situ*) in Bangladesh Managed by Department of Forest (http://www.bforest.gov.bd/; Accessed on 14 November 2017)

Protected Areas	Ecosystem	Conservation Focus	Location	Area (ha)	Date
Baldha Garden	Man-made	Education, Plant species	Dhaka	1.37	1909
National Botanical Garden, Mirpur	Man-made	Plant species, Education	Dhaka	84.21	1961
Sitakunda Botanical Garden and Eco-Park	Mixed Evergreen	Plant species	Chittagong	808.00	1998
Bangabandhu Sheikh Mujib Safari Park	Mixed Evergreen	Wildlife species	Cox's Bazar	600.00	1999
Madhutila Eco-Park	Deciduous Forest	Natural habitat, vegetation, wildlife	Sherpur	100.00	1999
Madhabkunda Eco-Park	Mixed Evergreen	Natural habitat, vegetation, wildlife	Moulvibazar	265.68	2001
Banshkhali Eco-Park	Mixed Evergreen	Natural habitat, vegetation, wildlife	Chittagong	1,200.00	2003
Kuakata Eco-Park		Natural habitat, vegetation, wildlife	Patuakhali	5,661.00	2005
Tilagar Eco-Park	Mixed Evergreen	Natural habitat, vegetation, wildlife	Sylhet	45.34	2006
Borshijora Eco-Park	Mixed Evergreen	Natural habitat, vegetation, wildlife	Moulvibazar	326.07	2006
Bangabandhu Sheikh Mujib Safari Park	Deciduous Forest	Education, wildlife species, recreation	Gazipur	1,493.93	2013
Rajeshpur Eco-Park			Comilla	185.90	
Total				9,620.08	

TABLE 9.9 Ecologically Critical Areas (ECAs) of Bangladesh (DoE, 2015)

Name of the ECA	Type of Ecosystem	Location	Areas (hectare)	Year of Declaration
Cox's Bazar-Teknaf Peninsula	Coastal-Marine	Cox's Bazar	20,373	1999

TABLE 9.9   *(Continued)*

Name of the ECA	Type of Ecosystem	Location	Areas (hectare)	Year of Declaration
Sundarbans (10 km landward periphery)	Coastal-Marine	Bagerhat, Khulna, and Satkhira	292,926	1999
St. Martin's Island	Marine Island with coral reefs	Teknaf, Cox's Bazar	1,214	1999
Hakaluki Haor	Inland Freshwater Wetland	Sylhet and Moulvibazar	40,466	1999
Sonadia Island	Marine Island	Moheshkhali, Cox's Bazar	10,298	1999
Tanguar Haor*	Inland Freshwater Wetland	Moulvi Bazar	9,727	1999
Marjat Baor	Oxbow Lake	Kaliganj, Jhenaidah, and Chaugacha, Jessore	325	1999
Gulshan-Baridhara Lake	Urban Wetland	Dhaka city	101	2001
Buriganga	River	Around Dhaka	1336	2009
Turag	River	Around Dhaka	1184	2009
Sitalakhya	River	Narayanganj, Dhaka, Gazipur	3771	2009
Balu including Tongi canal	River	Around Dhaka	1315	2009
Jaflong-Dawki	River	Jaflong, Sylhet	1493	2015
Total			384,529	

*also designated a RAMSAR site.

9.6.2   EX SITU CONSERVATION

In contrast to *in situ* conservation, *ex situ* conservation includes any practices that preserve biodiversity (or genetic resources) outside the natural habitat the parental populace, for example, botanical garden, arboretum, field gene bank/preservation plot, and gene/seed bank. The area of National Botanical Garden at Mirpur is 84.2 ha consisting of 306 tree species (total

33,413 plants), 441 shrub species (20,746 plants), 201 herb species (13,092 plants) and 62 vine/climber species (1,190 plants). The total number of families of trees, herbs, and shrubs are 134 (Anonymous, 2014). Boldha garden covers an area of 1.15 ha with a total of 18,000 trees, herbs, and shrubs from 820 species and 92 families. Along with these two botanical gardens maintained by DoF, different public universities of Bangladesh are maintaining botanical gardens as their teaching and research facilities. These gardens also act as a conservatory of rare and endangered plant species from home and abroad.

The botanical garden of Bangladesh Agricultural University (BAU) is the largest of this category with an area of 9.8 ha. Along with teaching, research, and extension activities, one of the major objectives of this botanical garden is to collect, conserve, and the multiplication of rare and endangered plant species of Bangladesh. The botanical garden has so far developed 4 medicinal plant gardens, 3 endemic and exotic fruit gardens, vegetable garden, water, and rock gardens, palm, and cycads gardens, bamboo, and rattan gardens, fern, orchid, and cactus houses, and miscellaneous plant garden. A Mangrove forest (garden), a Chittagong Hill-tracts forest zone and a Tea garden have also been developed here. From Bangladesh, this botanical garden is the only member of the Botanic Gardens Conservation International (BGCI) and the only registered participant in the worldwide implementation of the International Agenda in support of plant conservation, environmental awareness and sustainable development to implement the International Agenda for Botanic Gardens in Conservation and the Global Strategy for Plant Conservation (GSPC).

Since its establishment in 1963, the Botanical Garden has been conducting collection and conservation activities, and presently *c.* 6,000 trees, herbs, and shrubs from 1,346 species, more than 20% of Bangladesh flora, under 327 genera and 215 families are harbored at this garden. Among the collections, 159 species, mostly medicinal, are rare in the wild, and 58 species are near threatened and/or threatened of different categories according to IUCN Red List (Table 9.10). Recently, the World Bank–University Grants Commission of Bangladesh funded a project entitled "*Strengthening postgraduate research capability on the collection, characterization, and conservation of plant genetic resources*" has started at this botanical garden in collaboration with the Department of Crop Botany, Bangladesh Agricultural University. The main objective of the project is to enhance capacity for collection, propagation, and conservation of threatened plant species of Bangladesh. Researchers are working on relocation, collection, and conservation of endangered plant species of Bangladesh, thus, the capability of plant genetic resources conservation of this garden will be increased many folds. So far, 35 species of threatened plant species are relocated in the wild and collected for conservation and multiplication at this garden through this project initiative.

Beside botanical garden, some other types of *ex situ* conservation activities are mainly limited to BFRI. The BFRI has established five preservation plots at different hill forest parts and 27 at the Sundarbans (mangrove) forest. The BFRI has also set up two clonal banks, one at Hyako, Chittagong (4 ha) and another at Ukhia, Cox's Bazar (4 ha). Seven tree species (*Tectona grandis, Gmelina arborea, Bombax ceiba, Diptero-carpus turbinatus, Syzygium grande, Swietenia mahagoni* and *Paraseri-anthes falcataria*) are preserved in these two locations. One bambusetum (1.5 ha) has been set up at the BFRI campus. This arboretum comprises of 27 bamboo species with 6 exotic species. One arboretum of medicinal plants (1 ha) has also been set up at the BFRI campus with a collection of 40 species and one cane arboretum (0.5 ha) with seven species. Three arboreta of tree species have been established at the BFRI-HQ with 56 species, Keochia Forest Research Station with 56 species and Charaljani Silviculture Research Station with 52 species.

For the agricultural crop genetic resources, Bangladesh Agricultural Research Institute (BARI), BRRI, and Bangladesh Jute Research Institute have been maintaining one Gene/Seed Bank at each. Moreover, Bangla-desh Sugarcrop Research Institute, BARI, Bangladesh Tea Research Institute (BTRI), Bangladesh Institute of Nuclear Agriculture, and Fruit Tree Improvement Project Germplasm Centre (FTIP-GPC) of Bangladesh Agricultural University, the 2nd largest repository of this type in the world, are maintaining Conservation Plots/Breeding Gardens to conserve the collections of their interest, which they have collected over the period of time. Thus, some institutes developed themselves as an important element of biodiversity conservation, and also as a unique center for education, research, and information relating to plant diversity. For example, BTRI harbored 199 angiosperm species under 155 genera and 69 families along with a total of 320 tea germplasms, clone, and seed stocks (Sarwar et al., 2008). The FTIP-GPC of Bangladesh Agricultural University possesses 688 varieties of fruit, both major and minor, trees under 216 species; and a total of 12 PhD and 120 MSc students have completed their degree, and 11 PhD and 70 MSc researches are going on (Rahim et al., 2012). Social forestry, agro-forestry, and community forestry programmes are also playing important roles in *ex situ* conservation of both forest and crop biodiversity in Bangladesh (DoE, 2015).

Nevertheless, new approaches to agrobiodiversity conservation are getting into the scenes. For example, community biodiversity management (CBM), an *in situ* conservation on-farm approach, is going well and getting popularity day by day (de Boef et al., 2013; DoE, 2016). The community seed bank, managed by Nayakriski Seed Network, is contributing to the on-farm management of 2,300 rice cultivars, a unique success in an era of high genetic erosion of local crop cultivars/land-races (Shrestha et al., 2013).

TABLE 9.10 List of Threatened Species Conserved in Bangladesh Agricultural University Botanical Garden (Khan et al., 2001; Siddiqui et al., 2007; Ahmed et al., 2008–2009; Ara et al., 2013)

Botanical Name	Family	Uses	Status*
Abroma augusta (L.) L. f.	Sterculiace	Medicine	NT
Acorus calamus L.	Araceae	Medicine	VU
Aglaonema clarkei Hook. f.	Araceae		VU
Asparagus racemosus Willd.	Liliaceae	Medicine	NT
Agrostophyllum khasianum Griff.	Orchidaceae		VU
Amomum costaum (Roxb.) Benth. *ex* Baker	Zingiberaceae	Medicine	EN
Antidesma lanceolatum Tul.	Euphorbiaceae	Fruit	VU
Bulbphyllum roxburghii (Lindl.) Rchb. f.	Orchidaceae		EN
Butea listeri (Prain) Blatt.	Fabaceae		VU
Calamus erectus Roxb.	Palmae		VU
Calamus guruba Buch.-Ham.	Palmae		VU
Calamus latifolius Roxb.	Palmae		VU
Calamus longisetus Griff.	Palmae		VU
Careya arborea Roxb.	Lecythidaceae	Medicine	VU
Caulokaemferia secunda (Wall.) K. Larsen	Zingiberaceae		VU
Cephalanthus naucleoides DC.	Rubiaceae		VU
Crepidium biauritum (Lindley) Szlach.	Orchidaceae		EN
Curcuma ferruginea Roxb.	Zingiberaceae	Medicine	EN
Cycas pectinata Buch.-Ham.	Cycadaceae	Medicine	CR

TABLE 9.10   *(Continued)*

Botanical Name	Family	Uses	Status*
Cyclobalanopsis oxydon (Mig.) Oerst	Fagaceae		EN
Dendrobium formusum Roxb. *ex* Lindl.	Orchidaceae		VU
Dendrobium longicornu Lindl.	Orchidaceae		CR
Elaeocarpus rugosus Lindl. *ex* G. Don	Elaeocarpaceae		VU
Elaeocarpus serratus L.	Elaeocarpaceae	Medicine	NT
Elaeocarpus sphaericus Gaertn. K. Schum	Elaeocarpaceae	Medicine	EN
Hedychium aureum C.B. Clarke & Mann. *ex* Baker	Zingiberaceae		VU
Hedychium glaucum Rosc.	Zingiberaceae		VU
Hedychium gracile Roxb.	Zingiberaceae		VU
Hedychium griffithianum Wall.	Zingiberaceae		VU
Hedychium speciosum Wall. *ex* Roxb.	Zingiberaceae		EN
Ipomoea mauritiana Jacq.	Convolvulaceae	Medicine	VU
Kigelia pinnata (Jacq.) DC.	Bignoniaceae	Medicine	EN
Lagenandra gomezii (Schott) Bogner & Jacobson	Araceae		VU
Licuala peltata Roxb.	Palmae		VU
Lithocarpus acuminata (Roxb.) Rehder	Fagaceae		EN
Litsea clarkei Prain	Lauraceae		CR
Mangifera sylvatica Roxb.	Anacardiaceae		VU
Paedaria foetida L.	Rubiaceae	Medicine	EN
Picrasma javanica Blume	Simarubaceae	Medicine	VU
Plumbago indica L.	Plumbaginaceae	Medicine	EN
Podocarpus nelifolius (Thunb.) R.Br. *ex* Mirb.	Coniferae		EN
Quercus acumunata Roxb.	Fagaceae		EN
Randia uligenosa (Retz.) DC.	Rubiaceae	Medicine	VU
Rauvolfia serpentiana (L.) Benth. *ex* Kurz.	Apocynaceae	Medicine	CR
Sapindus mukorrossi Gaertn.	Sapindaceae	Medicine	NT
Scaphium scaphigerum (Wall. *ex* G. Don) Guibourt & Planch.	Sterculiaceae		VU
Sonerila maculata Roxb.	Melatomataceae		CR
Spatholobus listeria Prain	Fabaceae	Medicine	VU

TABLE 9.10 *(Continued)*

Botanical Name	Family	Uses	Status*
Steudnera colocasioides Hook. f.	Araceae		VU
Steudnera virosa (Roxb.) Prain	Araceae	Medicine	VU
Strychnos nux-vomica L.	Loganiaceae	Medicine	VU
Symplocos macrophylla Wall. *ex* DC.	Symplocaceae		VU
Syzygium diospyrifolium (Wall. *ex* Duthie) S.N. Mitra	Myrtaceae	Fruit	VU
Tetradium glabrifolium (Champ. *ex* Benth.) T.G. Hartley	Rutaceae		EN
Tinospora crispa (L.) Hook. f. & Thoms.	Menispermaceae	Medicine	NT
Tinospora sinensis (Lour.) Merr.	Menispermaceae	Medicine	NT
Trichosanthes sp.	Cucurbitaceae	Medicine	VU
Xerospermum noronhianum (Blume) Blume	Sapindaceae		VU

NT: Near threatened; VU: Vulnerable; EN: Endangered; CR: Critically endangered.

9.7 CONCLUSION AND FUTURE PROSPECTS

Plants are one of the most significant elements of biodiversity which aid life systems on earth. They are the sources of improved yield and quality factors, and in all aspects, represent the very foundation of human existence. Moreover, herbal/traditional medicine, with no or a little side effect, is gaining wide popularity day by day. Over 50% of prescription drugs are derived from chemicals of those first identified in plants (Ghani, 1998); and more than 70% of rural people of Bangladesh depend on medicinal plants directly or indirectly for their primary health care and treatment of different diseases. In the flora of Bangladesh, more than 650 plant species possess medicinal properties (Uddin, 2013). As the completion of inventory and documentation of the *Flora of Bangladesh* is far beyond, so there are huge prospects to work with this as shortly stated below:

1. Rigorous exploration, inventory, and documentation of the flora.
2. Identification, documentation, and centralization of wild crop relatives, medicinal, and other economically important plants and their relatives.

3. Domestication and cultivation of under-utilized food plants like wild yams, aroids, vegetables [bathua (*Chenopodium album*), kalmi (*Ipomoea reptans, I. aquatica*), napha shak (*Malva verticillata*), dhemshi (*Fagopyrum esculentum*), helencha (*Enhydra fluctuans*), dhekii shak (*Diplazium frondosum*), dhundul (*Luffa cylindrica*), punarnaba (*Boerhavia diffusa*), thankuni (*Centella asiatica*), telakucha (*Coccinia grandis, C. indica*), gandhabhadali (*Paederia foetida*), shusni (*Marsilea quadrifolia*) and so on].

4. Wild crop relatives and under-utilized crops could be utilized in the crop improvement programmes to face the challenging demand of agricultural production with a rapidly growing population and with the alarming climate change.

5. Sustainable and judicious use of both timber and non-timber forest products.

6. In addition to Protected Areas, define, and identify special areas with biodiversity interest like small wetlands, fallow meadows with herbs, grass, and sedges.

7. Regular monitoring the change of floristic elements in different ecosystems due to climate changes vulnerabilities.

8. Networking (putting people from different interest groups together, allow database accession) and data exchange between botanical gardens and other biodiversity conservation related activities.

9. Information and material exchange – plant holdings at accession level and involving geographic origins and genetic history, availability of material for exchange, and so on, information about *ex situ* conservation activities, e.g., techniques been applied and systematics.

10. International collaboration to expedite the speed of exploration and documentation of the *Flora of Bangladesh* before it's extinction from the wild.

11. Eco-tourism – Bangladesh enjoys a wide range of forest type from evergreen and mixed evergreen tropical hill forests to the Sundarbans and freshwater swamp forests in the *Haor* basins. It is also rich in ethnic biodiversity with about 45 different ethnic communities widely spread over forested areas from hills to the plain land *Sal* forests of central, northern, and north-western part of the country. The pristine beauties of the forested lands in

Bangladesh and cultural diversities of the people could provide exciting destinations to travelers.

ACKNOWLEDGEMENTS

Author thanks Prof. M.M. Rahman, Prof. Dr. A.K.M. Azad-ud-Doula Prodhan, Dr. M.S. Haque, Dr. M.K. Alam (BFRI), Dr. S.B. Uddin (CU) and anonymous reviewers for their valuable comments, constructive criticism, improvement suggestions and/or sharing some unpublished data or publications.

KEYWORDS

- angiosperms
- Bangladesh
- biodiversity
- biological diversity
- ecosystem
- *ex situ* conservation
- gymnosperms
- herbarium
- *in situ* conservation
- IUCN
- National park
- pteridophytes
- red data book
- wildlife

REFERENCES

Ahmed, I., Sharma, R., & Decosse, P., (2007). Biodiversity protection through community initiatives. In: *National Tree Planting Brushier* (pp. 30–39). Department of Forest, Ministry of Environment and Forest, Government of Bangladesh, Dhaka.

Ahmed, Z. U., Begum, Z. N. T., Hassan, M. A., Khondker, M., Kabir, S. M. H., Ahmad, M., et al., (2008a). *Encyclopedia of Flora and Fauna of Bangladesh* (Vol. 6). *Angiosperms: Dicotyledons (Acanthaceae-Asteraceae).* Asiatic Society of Bangladesh, Dhaka (pp. 1–408).

Ahmed, Z. U., Begum, Z. N. T., Hassan, M. A., Khondker, M., Kabir, S. M. H., Ahmad, M., et al., (2008b). *Encyclopedia of Flora and Fauna of Bangladesh* (Vol. 12). *Angiosperms: Monocotyledons (Orchidaceae-Zingiberaceae).* Asiatic Society of Bangladesh, Dhaka (pp. 1–552).

Ahmed, Z. U., Hassan, M. A., Begum, Z. N. T., Khondker, M., Kabir, S. M. H., Ahmad, M., Ahmed, A. T. A., (2009a). *Encyclopedia of Flora and Fauna of Bangladesh* (Vol. 7). *Angiosperms: Dicotyledons (Balsaminaceae-Euphorbiaceae).* Asiatic Society of Bangladesh, Dhaka (pp. 1–546).

Ahmed, Z. U., Hassan, M. A., Begum, Z. N. T., Khondker, M., Kabir, S. M. H., Ahmad, M., Ahmed, A. T. A., (2009b). *Encyclopedia of Flora and Fauna of Bangladesh* (Vol. 8). *Angiosperms: Dicotyledons (Fabaceae-Lythraceae).* Asiatic Society of Bangladesh, Dhaka (pp. 1–474).

Ahmed, Z. U., Hassan, M. A., Begum, Z. N. T., Khondker, M., Kabir, S. M. H., Ahmad, M., Ahmed, A. T. A., (2009c). *Encyclopedia of Flora and Fauna of Bangladesh* (Vol. 9). *Angiosperms: Dicotyledons (Magnoliaceae-Punicaceae).* Asiatic Society of Bangladesh, Dhaka (pp. 1–488).

Ahmed, Z. U., Hassan, M. A., Begum, Z. N. T., Khondker, M., Kabir, S. M. H., Ahmad, M., Ahmed, A. T. A., (2009d). *Encyclopedia of Flora and Fauna of Bangladesh* (Vol. 10). *Angiosperms: Dicotyledons (Ranunculaceae-Zygophyllaceae).* Asiatic Society of Bangladesh, Dhaka (pp. 1–580).

Alam, M. K., (1988). *Annotated Checklist of the Woody Flora of Sylhet Forests, Bulletin 5, Plant Taxonomy Series.* Bangladesh Forest Research Institute, Chittagong.

Alam, M. K., (1999). *Annotated Checklist of the Woody Flora of Sal Forests of Bangladesh, Bulletin 8, Plant Taxonomy Series.* Bangladesh Forest Research Institute, Chittagong.

Anonymous, (2014). *Jatyo Udvid Uddan (National Botanical Garden), Mirpur, Dhaka.* Department of Forest, Ministry of Environment and Forest, Government of Bangladesh, Dhaka.

Ara, H., & Khan, B., (2009–2015). *Flora of Bangladesh, Fascicles 59–62.* Bangladesh National Herbarium, Dhaka.

Ara, H., & Khan, B., (2015). *Bulletin of the Bangladesh National Herbarium* (Vol. 4). Bangladesh National Herbarium, Dhaka (pp. 1–117).

Ara, H., Khan, B., & Uddin, S. N., (2013). *Red Data Book of Vascular Plants of Bangladesh* (Vol. 2). Bangladesh National Herbarium, Dhaka (pp. 1–280).

Barbhuiya, H. A., & Gogoi, R., (2010). Plant collections from Bangladesh in the Herbarium at Shillong (Assam), India. *Bangladesh J. Plant Taxon., 17,* 141–165.

Basak, S. R., & Alam, M. K., (2015). *Annotated Checklist of the Tree Flora of Bangladesh.* Bangladesh Forest Research Institute, Chittagong.

Bor, N. L., (1960). *The Grasses of Burma, Ceylon, India, and Pakistan (Excluding Bambusaceae).* Pergamon Press Ltd., London.

BRRI, (1982). *Deshi Dhaner Jat (in Bangla).* Bangladesh Rice Research Institute, Gazipur.

Burkill, I. H., (1965). *Chapters on the History of Botany in India.* Manager of Publications, Government of India, Delhi.

Choudhury, J. K., & Hossain, M. A. A., (2011). *Bangladesh Forestry Outlook Study*. Asia-Pacific forestry sector outlook study, I. I., Working paper no. APFSOS II/WP/2011/33, Food, and Agriculture Organization of the United Nations, Regional Office for Asia and the Pacific, Bangkok.

Cowan, J. M., (1928). The flora of the Chakaria Sundarbans. *Res. Bot. Surv. India, 11*, 197–225.

Das, D. K., & Alam, M. K., (2001). *Trees of Bangladesh*. Bangladesh Forest Research Institute, Chittagong.

Dasmann, R. F., (1968). *A Different Kind of Country*. MacMillan Company, New York.

Datta, R. M., & Mitra, J. N., (1953). Common plants in and around Dacca. *Bull. Bot. Soc. Bengal, 7*, 1–110.

De Boef, W. S., Subedi, A., Peroni, N., Thijssen, M., & O'Keeffe, E., (2013). *Community Biodiversity Management*. Routledge, London (pp. 1–418).

DoE, (2010). *Fourth National Report to the Convention on Biological Diversity*. Department of Environment, Ministry of Environment and Forest, Government of Bangladesh, Dhaka.

DoE, (2015). *Fifth National Report to the Convention on Biological Diversity*. Department of Environment, Ministry of Environment and Forest, Government of Bangladesh, Dhaka.

DoE, (2016). *National Biodiversity Strategy and Action Plan of Bangladesh 2016–2021*. Department of Environment, Ministry of Environment and Forest, Government of Bangladesh, Dhaka.

FAO, (1984). *In Situ Conservation of Wild Plant Genetic Resources: A Status Review and Action Plan* (pp. 1–83). Food and Agriculture Organization of the United Nations and International Union for Conservation of Nature, Rome.

FAO-UNDP, (1988). *Land Resources Appraisal of Bangladesh for Agricultural Development, Report 2: Agroecological Regions of Bangladesh*. Food and Agriculture Organization of United Nations (FAO), United Nations Development Programme (UNDP), Rome.

Ghani, A., (1998). *Medicinal Plants of Bangladesh: Chemical Constituents and Uses*. Asiatic Society of Bangladesh, Dhaka.

Heinig, R. I., (1925). *List of Plants of Chittagong Collectorate and Hill Tracts*. The Bengal Government Branch Press, Darjeeling, India.

Hooker, J. D., (1872–1897). The Flora of British India (Vol. I–VII). Reeve and Co., Kent, England.

Hossain, M. K., & Pasha, M. K., (2004). An account of the exotic flora of Bangladesh. *J. For. Env., 2*, 99–115.

Irfanullah, H. M., (2011). Conservation of threatened plants of Bangladesh: Miles to go before we start? *Bangladesh J. Plant Taxon., 18*, 81–91.

Irfanullah, H. M., (2013). Plant taxonomic research in Bangladesh (1972–2012): A critical review. *Bangladesh J. Plant Taxon., 20*, 267–279.

Kanjilal, U. N., Kanjilal, P. C., De, R. N., Das, A. K., & Bor, N. L., (1934–1940). Flora of Assam (Vol. 1–5), Government of Assam, India.

Khan, M. S., (1972–1987). Flora of Bangladesh, Fascicles 1–39. Bangladesh National Herbarium, Dhaka.

Khan, M. S., & Alam, M. K., (1996). *Homestead Flora of Bangladesh*. Forestry Division, Bangladesh Agricultural Research Council, Farm Gate, Dhaka (pp. 1–275).

Khan, M. S., & Rahman, M. M., (1989–2002). Flora of Bangladesh, Fascicles 40–53, Bangladesh National Herbarium, Dhaka.

Khan, M. S., (1991). Angiosperms. In: Nurul, I. A. K. M., (ed.), *Two Centuries of Plant Studies in Bangladesh and Adjacent Regions* (pp. 175–194). Asiatic Society of Bangladesh, Dhaka.

Khan, M. S., Rahman, M. M., & Ali, M. A., (2001). *Red Data Book of Vascular Plants of Bangladesh*. Bangladesh National Herbarium, Dhaka.

Khanam, M., & Ara, H., (2007–2008). *Flora of Bangladesh, Fascicles 56–58*. Bangladesh National Herbarium, Dhaka.

Kurz, S., (1877). *Forest Flora of British Burma* (Vol. 1 & 2). Office of the Superintendent of Government Printing, Calcutta.

MoEF, (2004). *National Biodiversity Strategy and Action Plan for Bangladesh*. Ministry of Environment and Forest, Government of Bangladesh, Dhaka.

Myers, N., Mittermeier, R. A., Mittermeier, C. G., Da Fonseca, G. A. B., & Kent, J., (2000). Biodiversity hotspots for conservation priorities. *Nature, 403*, 853–858.

Nishat, A., Huq, S. M. I., Barua, S. P., Khan, A. H. M., & Moniruzzaman, A. S., (2002). *Bio-Ecological Zones of Bangladesh*. IUCN Bangladesh Country Office, Dhaka.

Pasha, M. K., & Uddin, M. G., (2007). An account of the exploration and taxonomic studies of pteridophytes in Bangladesh. *J. Taxon. Biodiv. Res., 1*, 41–48.

Pasha, M. K., & Uddin, S. B., (2013). *Dictionary of Plant Names of Bangladesh*. Janokalyan Prokashani, Chittagong.

Pasha, M. K., (2012). An evaluation of endemism and endemics in Bangladesh flora. In: Roskaft, E., & Chivers, D. J., (eds.), *Proceedings of the International Conference on Biodiversity – Present State, Problems, and Prospects of its Conservation* (pp. 57–75). Norwegian Centre for International Cooperation in Education, Bergen, Norway.

Pasha, M. K., (2014). Flowering plant exploration history: In quest of Bangladesh flora. *J. Taxon. Biodiv. Res., 6*, 21–40.

Prain, D., (1903a). *Bengal Plants* (Vol. 1 & 2). Botanical Survey of India, Calcutta.

Prain, D., (1903b). The Flora of Sundarbans. *Rec. Bot. Surv. India, 2*, 231–270.

Rahim, M. A., Fakir, M. S. A., Hossain, M. M., Alam, M. S., Anwar, M. M., Alam, A. K. M. A., et al., (2012). BAU germplasm center – the largest fruit repository in Bangladesh-one-stop service for quality planting materials of fruits, conservation, development, production, diversity, research, and extension. In: Ghosh, S. N., (ed.), *Proceedings of the International Symposium on Minor Fruits and Medicinal Plants for Health and Ecological Security* (pp. 3–6). West Bengal, India.

Rahman, M. A., (2014a). Plant taxonomy and biodiversity researches in Bangladesh: Trends and opportunities. *Int. J. Environ., 3*, 324–344.

Rahman, M. A., (2014b). Plants of Bangladesh – Presumed extinct from its flora. *Abstracts 6th International Botanical Conference* (p. 63). Dhaka.

Rahman, M. M., & Khanam, M., (2003). *Flora of Bangladesh, Fascicle 55*. Bangladesh National Herbarium, Dhaka.

Rahman, M. M., (2003). *Flora of Bangladesh, Fascicle 54*. Bangladesh National Herbarium, Dhaka.

Raizada, M. B., (1941). On the flora of Chittagong. *Indian Fores., 67*, 245–254.

Roxburgh, W., (1814). Hortus Bengalensis (pp. 1-105). Royal Botanical Garden, Calcutta.

Roxburgh, W., (1820–1832). In: Carey, W., (ed.), Flora Indica (Vol. I–III, pp. 1–875). Mission Press, Calcutta.

Sarwar, A. K. M. G., (2013). Genera represented by single species in the flora of Bangladesh and their conservation needs – A review. *J. Today's Biol. Sci. Res. Rev.*, 2, 68–83.

Sarwar, A. K. M. G., (2015). Medicinal plant genetic resources of Bangladesh – Genera represented by single species and their conservation needs. *J. Med. Plants Stud.*, 3, 65–74.

Sarwar, A. K. M. G., Malaker, J. C., & Dutta, M. J., (2008). Floristic composition in the campus of Bangladesh Tea Research Institute – I. Angiosperms. *J. Agrof. Env.*, 2, 147–152.

Shrestha, P., Gezu, G., Swain, S., Lassaigne, B., Subedi, A., & De Boef, W. S., (2013). The community seed bank. In: De Boef, W. S., Subedi, A., Peroni, N., Thijssen, M., & O'Keeffe, E., (eds.), *Community Biodiversity Management* (p. 112). Routledge, London.

Siddiqui, K. U., Islam, M. A., Ahmed, Z. U., Begum, Z. N. T., Hassan, M. A., Khondker, M., et al., (2007). *Encyclopedia of Flora and Fauna of Bangladesh* (Vol. 11). *Angiosperms: Monocotyledons (Agavaceae - Najadaceae)*. Asiatic Society of Bangladesh, Dhaka (pp. 1–399).

Sinclair, J., (1955). Flora of Cox's Bazar. *Bull. Bot. Soc. Bengal*, 9, 84–116.

Uddin, S. N., (2013). *Traditional Uses of Ethnomedicinal Plants of the Chittagong Hill Tracts*. Bangladesh National Herbarium, Dhaka.

CHAPTER 10

IN SITU AND EX SITU CONSERVATION OF BIODIVERSITY IN INDIA: STRATEGIES AND APPROACHES FOR ENVIRONMENTAL SUSTAINABILITY

JEYABALAN SANGEETHA[1], DEVARAJAN THANGADURAI[2], PURUSHOTHAM PRATHIMA[2], MAXIM STEFFI SIMMI[1], PANNERI SREESHMA[1], MUNISWAMY DAVID[3], and RAVICHANDRA HOSPET[2]

[1]Department of Environmental Science, Central University of Kerala, Periye, Kasaragod 671316, Kerala, India

[2]Department of Botany, Karnatak University, Dharwad 580003, Karnataka, India

[3]Department of Zoology, Karnatak University, Dharwad 580003, Karnataka, India

10.1 INTRODUCTION

The term biodiversity or biological diversity refers to all life forms together on earth such as unicellular fungi, protozoa, bacteria, and multicellular organisms such as flora and fauna. In Asia and Pacific region there are three bio-geographic realms include major ecosystems such as mountain, forest, grassland, desert, wetland, and seas which are significant to the whole (Mutia, 2009). The Convention on Biological Diversity (CBD) defined biodiversity as "the variability among living organisms from all source including, inter alia, terrestrial, marine, and other aquatic ecosystems and ecological complexes of which the part of this includes diversity within species, between species and of ecosystems." Globally, biodiversity is unevenly distributed, and some places have greater diversity than others, for example, tropics have higher species diversity than poles. Factors such as soil type, climate, altitude, availability of water and species interaction has great influence on floral and faunal diversity. Certain areas which are not only biologically rich but also have endemic species and are under constant threat by human activities recognized as hotspots of biodiversity (Fisher and Christopher, 2007). At present, there are 35 biodiversity hotspots in the world. They include around 50% and 42% of endemic plant species and terrestrial vertebrates respectively. Human beings directly or indirectly depend on biodiversity for their food, fodder, fuel, leather, timber, rubber, medicines, liquor, fertilizer, and for several other raw materials. Generally, there are three levels of biodiversity which are closely interlinked because biological organisms are inseparable. These include genetic diversity (variety of genes and chromosomes among different species), species diversity (variation among the number and species richness of a region) and ecosystem diversity (a different ecosystem of a region). Biodiversity plays a vital role in existence, well being and also make the world a more attractive place to live (Chiras, 2012).

Ecosystem performs a vital role in human survival by providing many goods and services. Especially rural and some indigenous community is directly dependent on the ecosystem for their livelihood. Ecosystems which are biologically rich will have high productivity and maintenance through nutrient cycling which provides physical protection from various environmental conditions and detoxification (Kumar and Asija, 2007). Biodiversity and ecosystem functioning is a controversial topic because they are complicated in many ways including their types and diversity measurements. Biodiversity has great ecological value, when biodiversity is affected by any threat that will disturb the structure and functioning of the ecosystem, on the other hand, ecological balance is being maintained by biodiversity (Loreau et al., 2001). World population has been tremendously increasing since the 20th century, by 2045 it is expected to reach 9 million, and collectively the demand of resources will also increase. Reports show that approximately 10 million species will be eliminated by 2050. As a part of reducing biodiversity loss at global, regional, and national level CBD carry out target by 2010, which also includes poverty alleviation and providing better goods and services in all form of life on

Earth. Human history shows that the ecosystem has been highly inter-rupted by several anthropogenic activities. In recent years, species extinction is 1000 times greater than the past 50 years. The main reasons of these impacts are habitat loss, the introduction of invasive alien species, overexploitation, nutrient loading, and climate change. Climate change has a significant role in changing diversity in an ecosystem. Due to global warming, ecosystem shift into higher altitudes from the equatorial region which can alter the food chain and food web in the local and global level. These are not only the cause for the destruction of biodiversity but natural calamities such as volcanic eruption, earthquake, and landslides can also induce biodiversity loss (Dobson et al., 1997).

During Rio Earth Summit (1992) many nations signed biodiversity conservation treaty. At present, more than 187 nations are included in this treaty, and every nation develops a national conservation strategy and action plans (NCSAPs) for protecting and preserving the biodiversity. Many nations are following community-based Integrated Conservation and Development Program (ICDP) and Community - Based Natural Resource Management for protection of biodiversity at the local level, in the same way, help the rural community to use benefits sustainably. CBD followed by mainly three equally important objectives: sustainable development, conservation of biodiversity and equity in the sharing of benefits. Conservation of biodiversity can be carried out in many ways; generally, two strategies are followed for the conservation of biodiversity: *in situ* ('on site', 'in place') and *ex situ* ('off-site', 'out of place'). *In situ* conservation is carried out within their natural habitats such as national parks, wildlife sanctuaries, biosphere reserves, sacred forests, and lakes. Method of *in situ* conservation includes a description, organization, and monitoring of biodiversity within the ecosystem in which it is found (Ferraro and Agnes, 2002). CBD described that "*ex situ* measures should predominantly employ for the purpose of harmonize *in situ* measures." As of 1994, nearly 37,000 protected areas (PA) are there in the world. Among these around 531 wildlife sanctuaries (Table 10.1), 105 national parks (Table 10.2), 18 biosphere reserves (Table 10.3) exist in India. *Ex situ* conservation is the conservation of biodiversity outside their natural habitats such as gene banks, botanic gardens, zoos, and captive breeding programs. Preserving the genetic materials of target species and reintroducing the breeds which are threatened in their natural habitat is a type of *ex situ* conservation strategy. *Ex situ* techniques adopted for conserving plants include storing seeds, conserving pollens and *in vitro* conservation of plant shoots. Storage of embryos, ovule/semen/DNA or captive breeding techniques is used for conserving animals.

Throughout the world about 13.2 million sq km of protected area involves 1.3 million sq km of reserved marine area and around 149,811,204 sq km of land area which is legally conserved by the government. According to IUCN some of these areas which are under categories I–III where direct human intervention and modification of the environment have been limited and categories IV-VI significantly greater intervention and modification will be found (James et al., 2001). Based on the estimation of habitat loss and species diversity within an ecosystem around 36,500 to 50,000 species are eliminated in every year. Today, thousands of animal and plant species are endangered and threatened. If proper conservation measures are not implemented many of the threatened species will become extinct. Natural extinction of species is an evolutionary fact, but accelerated extinction is only because of human activities. According to Subsidiary Body on Scientific, Technological, and Technical Advice (SBSTTA), about 85% of species comes under IUCN Red list due to their habitat loss.

TABLE 10.1   List of Wildlife Sanctuaries in India

State/Union Territory	Name of the sanctuary	Year of establishment	Area (in Km²)
Andaman and Nicobar Islands	Landfall Island Wildlife Sanctuary	1987	0.05
	Bamboo Island Wildlife Sanctuary	1987	0.05
	Barren Island Wildlife Sanctuary	1987	8.10
	Battimalv Island Wildlife Sanctuary	1987	2.23
	Belle Island Wildlife Sanctuary	1987	0.08
	Benett Island Wildlife Sanctuary	1987	3.46
	Bingham Island Wildlife Sanctuary	1987	0.08
	Blister Island Wildlife Sanctuary	1987	0.26
	Bluff Island Wildlife Sanctuary	1987	1.14
	Bondoville Island Wildlife Sanctuary	1987	2.55
	Brush Island Wildlife Sanctuary	1987	0.23
	Buchanan Island Wildlife Sanctuary	1987	9.33
	Chanel Island Wildlife Sanctuary	1987	0.13
	Cinque Islands Wildlife Sanctuary	1987	9.51
	Clyde Island Wildlife Sanctuary	1987	0.54

TABLE 10.1 *(Continued)*

State/Union Territory	Name of the sanctuary	Year of establishment	Area (in Km²)
	Cone Island Wildlife Sanctuary	1987	0.65
	Curlew (BP) Island Wildlife Sanctuary	1987	0.16
	Curlew Island Wildlife Sanctuary	1987	0.03
	Cuthbert Bay Wildlife Sanctuary	1997	5.82
	Defence Island Wildlife Sanctuary	1987	10.49
	Dot Island Wildlife Sanctuary	1987	0.13
	Dottrell Island Wildlife Sanctuary	1987	0.13
	Duncan Island Wildlife Sanctuary	1987	0.73
	East Island Wildlife Sanctuary	1987	6.11
	Inglis Island Wildlife Sanctuary	1987	3.55
	Egg Island Wildlife Sanctuary	1987	0.05
	Elat Island Wildlife Sanctuary	1987	9.36
	Entrance Island Wildlife Sanctuary	1987	0.96
	Gander Island Wildlife Sanctuary	1987	0.05
	Galathea Bay Wildlife Sanctuary	1987	11.44
	Galathea Bay Wildlife Sanctuary	1987	0.16
	Galathea Bay Wildlife Sanctuary	1987	0.01
	Hump Island Wildlife Sanctuary	1987	0.47
	James Island Wildlife Sanctuary	1987	2.10
	Jungle Island Wildlife Sanctuary	1987	2.10
	Kwangtung Island Wildlife Sanctuary	1987	0.57
	Kyd Island Wildlife Sanctuary	1987	8.00
	Lohabarrack Saltwater Crocodile Wildlife Sanctuary	1987	22.21
	Mangrove Island Wildlife Sanctuary	1987	0.39
	Mask Island Wildlife Sanctuary	1987	0.78
	Mayo Island Wildlife Sanctuary	1987	0.10
	Mask Island Wildlife Sanctuary	1987	0.12
	Montogemery Island Wildlife Sanctuary	1987	0.21
	Narcondam Island Wildlife Sanctuary	1987	6.81

TABLE 10.1   *(Continued)*

State/Union Territory	Name of the sanctuary	Year of establishment	Area (in Km²)
	North Brother Island Wildlife Sanctuary	1987	0.75
	North Brother Island Wildlife Sanctuary	1987	0.49
	North Reef Island Wildlife Sanctuary	1987	3.48
	Oliver Island Wildlife Sanctuary	1987	0.16
	Orchid Island Wildlife Sanctuary	1987	0.10
	Ox Island Wildlife Sanctuary	1987	0.13
	Oyster Island-I Wildlife Sanctuary	1987	0.08
	Oyster Island-II Wildlife Sanctuary	1987	0.21
	Paget Island Wildlife Sanctuary	1987	7.36
	Parkinson Island Wildlife Sanctuary	1987	0.34
	Passage Island Wildlife Sanctuary	1987	0.62
	Patric Island Wildlife Sanctuary	1987	0.13
	Pitman Island Wildlife Sanctuary	1987	1.37
	Point Island Wildlife Sanctuary	1987	3.07
	Potanma Island Wildlife Sanctuary	1987	0.16
	Ranger Island Wildlife Sanctuary	1987	4.26
	Reef Island Wildlife Sanctuary	1987	1.74
	Roper Island Wildlife Sanctuary	1987	1.46
	Ross Island Wildlife Sanctuary	1987	1.01
	Rowe Island Wildlife Sanctuary	1987	0.01
	Sandy Island Wildlife Sanctuary	1987	1.58
	Sea Serpent Island Wildlife Sanctuary	1987	0.78
	Shark Island Wildlife Sanctuary	1987	0.60
	Shearme Island Wildlife Sanctuary	1987	7.85
	Sir Hugh Rose Island Wildlife Sanctuary	1987	1.06
	Sir Hugh Rose Island Wildlife Sanctuary	1987	0.36
	Snake Island-I Wildlife Sanctuary	1987	0.73
	Snake Island-II Wildlife Sanctuary	1987	0.03

TABLE 10.1 *(Continued)*

State/Union Territory	Name of the sanctuary	Year of establishment	Area (in Km²)
	South Brother Island Wildlife Sanctuary	1987	1.24
	South Reef Island Wildlife Sanctuary	1987	1.17
	South Sentinel Island Wildlife Sanctuary	1987	1.61
	Spike Island-I Wildlife Sanctuary	1987	0.42
	Spike Island-II Wildlife Sanctuary	1987	11.70
	Stoat Island Wildlife Sanctuary	1987	0.44
	Surat Island Wildlife Sanctuary	1987	0.31
	Swamp Island Wildlife Sanctuary	1987	4.09
	Table (Delgarno) Island Wildlife Sanctuary	1987	2.29
	Table (Excelsior) Island Wildlife Sanctuary	1987	1.69
	Talabaicha Island Wildlife Sanctuary	1987	3.21
	Temple Island Wildlife Sanctuary	1987	1.04
	Tillongchang Island Wildlife Sanctuary	1985	16.83
	Tree Island Wildlife Sanctuary	1987	0.03
	Trilby Island Wildlife Sanctuary	1987	0.96
	Tuft Island Wildlife Sanctuary	1987	0.29
	Turtle Island Wildlife Sanctuary	1987	0.39
	West Island Wildlife Sanctuary	1987	6.40
	West Island Wildlife Sanctuary	1987	0.11
	White Cliff Island Wildlife Sanctuary	1987	0.47
	Interview Island Wildlife Sanctuary	1987	133.87
	Peacock Island Wildlife Sanctuary	1987	0.62
Andhra Pradesh	Coringa Wildlife Sanctuary	1978	235.70
	Gundla Brahmeswaram Wildlife Sanctuary	1990	1194.00
	Kambalakonda Wildlife Sanctuary	2002	71.39
	Koundinya Wildlife Sanctuary	1990	357.60
	Kolleru Wildlife Sanctuary	1953	308.55

TABLE 10.1   *(Continued)*

State/Union Territory	Name of the sanctuary	Year of establishment	Area (in Km²)
	Krishna Wildlife Sanctuary	1989	194.81
	Nagarjuna Sagar-Srisailam Wildlife Sanctuary	1978	3568.09
	Nellapattu Wildlife Sanctuary	1976	4.59
	Pulicat Lake Wildlife Sanctuary	1976	500.00
	Rollapadu Wildlife Sanctuary	1988	6.14
	Sri Lankamalleswara Wildlife Sanctuary	1988	464.42
	Sri Penusila Narasimha Wildlife Sanctuary	1997	1030.85
	Sri Venkateswara Wildlife Sanctuary	1985	172.35
Arunachal Pradesh	Yordi Rabe Supse Wildlife Sanctuary	1996	397.00
	Tale Wildlife Sanctuary	1995	337.00
	Sessa Orchid Wildlife Sanctuary	1989	100.00
	Pakhui Wildlife Sanctuary	1977	861.95
	Mahao Wildlife Sanctuary	1980	281.50
	Kane Wildlife Sanctuary	1991	31.00
	Kamlang Wildlife Sanctuary	1989	783.00
	Itanagar Wildlife Sanctuary	1978	140.30
	Eaglenest Wildlife Sanctuary	1989	217.00
	Dibang Wildlife Sanctuary	1991	4149.00
	Daying Ering Memorial Wildlife Sanctuary	1978	190.00
Assam	Amchang Wildlife Sanctuary	2004	78.64
	Barail Wildlife Sanctuary	2004	326.24
	Barnadi Wildlife Sanctuary	1980	26.22
	Bherjan-Borajan-Padumoni Wildlife Sanctuary	1999	7.22
	Burachapari Wildlife Sanctuary	1995	44.06
	Chakrasila Wildlife Sanctuary	1994	45.57
	Deepor Beel Wildlife Sanctuary	1989	4.14
	Dihing Patkai Wildlife Sanctuary	2004	111.19
	East Karbi Anglong Wildlife Sanctuary	2000	221.81

TABLE 10.1 *(Continued)*

State/Union Territory	Name of the sanctuary	Year of establishment	Area (in Km²)
	Garampani Wildlife Sanctuary	1952	6.05
	Hoollongapar Gibbon Wildlife Sanctuary	1997	20.98
	Lawkhowa Wildlife Sanctuary	1972	70.13
	Marat Longri Wildlife Sanctuary	2003	451.00
	Nambor Wildlife Sanctuary	2000	37.00
	Nambor-Doigrung Wildlife Sanctuary	2003	97.15
	Pabitora Wildlife Sanctuary	1987	38.81
	Pani Dihing Bird Wildlife Sanctuary	1995	33.93
	Sonai Rupai Wildlife Sanctuary	1998	220.00
Bihar	Barela Jheel Salim Ali Bird Wildlife Sanctuary	1997	1.96
	Bhimbandh Wildlife Sanctuary	1976	681.99
	Gautam Budha Wildlife Sanctuary	1976	138.34
	Kaimur Wildlife Sanctuary	1982	1342.00
	Kanwarjheel Wildlife Sanctuary	1989	29.17
	Kusheshwar Asthan Bird Wildlife Sanctuary	1994	29.17
	Nagi Dam Wildlife Sanctuary	1987	1.92
	Nakti Dam Wildlife Sanctuary	1987	3.33
	Rajgir Wildlife Sanctuary	1978	35.84
	Udaipur Wildlife Sanctuary	1978	8.87
	Valmiki Wildlife Sanctuary	1978	545.15
	Vikramshila Gangetic Dolphin Wildlife Sanctuary	1990	50.00
Chandigarh	City Bird Wildlife Sanctuary	1998	0.03
	Sukhna Wildlife Sanctuary	1986	25.98
Chhattisgarh	Achanakmar Wildlife Sanctuary	1975	551.55
	Badalkhol Wildlife Sanctuary	1975	104.45
	Barnawapara Wildlife Sanctuary	1976	244.66
	Bhairamgarh Wildlife Sanctuary	1983	138.95
	Bhoramdev Wildlife Sanctuary	2001	351.24

TABLE 10.1 *(Continued)*

State/Union Territory	Name of the sanctuary	Year of establishment	Area (in Km²)
	Sarangarh-Gomardha Wildlife Sanctuary	1975	277.82
	Pamed Wildlife Sanctuary	1983	262.12
	Semarsot Wildlife Sanctuary	1978	430.35
	Sitanadi Wildlife Sanctuary	1974	553.36
	Tamor Pingla Wildlife Sanctuary	1978	608.51
	Udanti Wildlife Sanctuary	1985	237.27
Dadra and Nagar Haveli	Dadra and Nagar Haveli Wildlife Sanctuary	2000	92.16
Daman and Diu	Fudam Wildlife Sanctuary	1991	2.18
Delhi	Asola Bhatti Wildlife Sanctuary	1992	27.82
Goa	Bondla Wildlife Sanctuary	1969	7.95
	Salim Ali Wildlife Sanctuary	1988	1.78
	Cotigaon Wildlife Sanctuary	1968	85.65
	Madei Wildlife Sanctuary	1999	208.48
	Bhagwan Mahavir Wildlife Sanctuary	1967	133.00
	Netravali Wildlife Sanctuary	1999	211.05
Gujarat	Balaram Ambaji Wildlife Sanctuary	1989	542.08
	Barda Wildlife Sanctuary	1979	192.31
	Gaga Wildlife Sanctuary	1988	3.33
	Gir Wildlife Sanctuary	1965	1153.42
	Girnar Wildlife Sanctuary	2008	178.80
	Hingolgadh Wildlife Sanctuary	1980	6.54
	Jambughoda Wildlife Sanctuary	1990	130.38
	Jessore Sloth Bear Wildlife Sanctuary	1978	180.66
	Kachchh Great Indian Bustard Wildlife Sanctuary	1995	2.03
	Kutch Desert Wildlife Sanctuary	1986	7506.22
	Khijadiya Bird Wildlife Sanctuary	1981	6.05
	Marine (Gulf of Kachchh) Wildlife Sanctuary	1980	295.03

TABLE 10.1 *(Continued)*

State/Union Territory	Name of the sanctuary	Year of establishment	Area (in Km²)
	Mitiyala Wildlife Sanctuary	2004	18.22
	Nal Sarovar Bird Wildlife Sanctuary	1969	120.82
	Narayan Sarovar Chinkara Wildlife Sanctuary	1995	442.91
	Paniya Wildlife Sanctuary	1989	39.63
	Porbandar Bird Wildlife Sanctuary	1988	0.09
	Purna Wildlife Sanctuary	1990	160.84
	Rampara Wildlife Sanctuary	1988	15.01
	Ratanmahal Sloth Bear Wildlife Sanctuary	1982	55.65
	Shoolpaneshwar Wildlife Sanctuary	1982	607.70
	Thol Wildlife Sanctuary	1988	6.99
	Wild Ass Wildlife Sanctuary	1973	4953.71
Haryana	Abubshehar Wildlife Sanctuary	1987	115.30
	Bhindawas Wildlife Sanctuary	1986	4.12
	Bir Shikargarh Wildlife Sanctuary	1987	7.67
	Chhilchhila Wildlife Sanctuary	1986	0.29
	Kalesar Wildlife Sanctuary	1996	54.06
	Khaparwas Wildlife Sanctuary	1991	0.83
	Khol-Hi-Raitan Wildlife Sanctuary	2004	48.83
	Nahar Wildlife Sanctuary	1987	2.11
Himachal Pradesh	Bandli Wildlife Sanctuary	1962	32.11
	Chail Wildlife Sanctuary	1976	16.00
	Chandratal Wildlife Sanctuary	2007	38.56
	Churdhar Wildlife Sanctuary	1985	55.52
	Daranghati Wildlife Sanctuary	1962	171.50
	Dhauladhar Wildlife Sanctuary	1994	982.86
	Gamgul Siyabehi Wildlife Sanctuary	1962	108.40
	Kais Wildlife Sanctuary	1954	12.61
	Kalatop-Khajjiar Wildlife Sanctuary	1958	17.17
	Kanawar Wildlife Sanctuary	1954	107.29
	Khokhan Wildlife Sanctuary	1954	14.94
	Kibber Wildlife Sanctuary	1992	2220.12

TABLE 10.1 *(Continued)*

State/Union Territory	Name of the sanctuary	Year of establishment	Area (in Km²)
	Kugti Wildlife Sanctuary	1962	405.49
	Lippa Asrang Wildlife Sanctuary	1962	31.00
	Majathal Wildlife Sanctuary	1954	30.86
	Manali Wildlife Sanctuary	1954	29.00
	Nargu Wildlife Sanctuary	1962	132.3731
	Pong Dam Lake Wildlife Sanctuary	1982	207.59
	Renuka Wildlife Sanctuary	2013	4.00
	Rupi Bhaba Wildlife Sanctuary	1982	503.00
	Sainj Wildlife Sanctuary	1994	90.00
	Rakcham Chitkul Wildlife Sanctuary	1989	304.00
	Rupi Bhaba Wildlife Sanctuary	1982	503.00
	Sainj Wildlife Sanctuary	1994	90.00
	Sechu Tuan Nala Wildlife Sanctuary	1962	390.29
	Shikari Devi Wildlife Sanctuary	1962	29.94
	Shimla Water Catchment Wildlife Sanctuary	1958	10.00
	Talra Wildlife Sanctuary	1962	46.48
	Tirthan Wildlife Sanctuary	1992	61.00
	Tundah Wildlife Sanctuary	1962	64.00
Jammu and Kashmir	Baltal-Thajwas Wildlife Sanctuary	1987	210.50
	Changthang Wildlife Sanctuary	1987	4000.00
	Gulmarg Wildlife Sanctuary	1987	180.00
	Hirapora Wildlife Sanctuary	1987	110.00
	Hokersar Wildlife Sanctuary	1992	13.75
	Jasrota Wildlife Sanctuary	1987	25.75
	Karakoram Wildlife Sanctuary	1987	5000.00
	Lachipora Wildlife Sanctuary	1987	80.00
	Limber Wildlife Sanctuary	1987	26.00
	Nandni Wildlife Sanctuary	1981	33.34
	Overa-Aru Wildlife Sanctuary	1987	425.00
	Rajparian Wildlife Sanctuary	2002	20.00
	Ramnagar Wildlife Sanctuary	1981	31.50

TABLE 10.1   *(Continued)*

State/Union Territory	Name of the sanctuary	Year of establishment	Area (in Km²)
	Surinsar Mansar Wildlife Sanctuary	1981	55.50
	Trikuta Wildlife Sanctuary	1981	31.77
Jharkhand	Udhwa Lake Bird Wildlife Sanctuary	1991	5.65
	Topchanchi Wildlife Sanctuary	1978	12.82
	Parasnath Wildlife Sanctuary	1984	49.33
	Palkot Wildlife Sanctuary	1990	182.83
	Palamau Wildlife Sanctuary	1976	752.94
	Mahuadanr Wolf Wildlife Sanctuary	1976	63.26
	Lawalong Wildlife Sanctuary	1978	211.03
	Kodarma Wildlife Sanctuary	1985	177.35
	Hazaribagh Wildlife Sanctuary	1976	186.25
	Gautam Budha Wildlife Sanctuary	1976	121.14
	Dalma Wildlife Sanctuary	1976	193.22
Karnataka	Adichunchanagiri Wildlife Sanctuary	1981	0.84
	Arabithittu Wildlife Sanctuary	1985	13.50
	Attiveri Bird Wildlife Sanctuary	1994	2.22
	Bhadra Wildlife Sanctuary	1974	492.46
	Bhimgad Wildlife Sanctuary	2010	190.42
	Biligiri Rangaswamy Temple Wildlife Sanctuary	1987	539.52
	Brahmagiri Wildlife Sanctuary	1974	181.29
	Cauvery Wildlife Sanctuary	1987	1027.53
	Chincholi Wildlife Sanctuary	2012	134.88
	Dandeli Wildlife Sanctuary	1987	886.41
	Daroji Bear Wildlife Sanctuary	1992	82.72
	Ghataprabha Wildlife Sanctuary	1974	29.79
	Gudavi Bird Wildlife Sanctuary	1989	0.73
	Gudekote Sloth Bear Wildlife Sanctuary	2013	38.48
	Malai Mahadeshwara Wildlife Sanctuary	2013	906.19
	Melkote Temple Wildlife Sanctuary	1974	49.82
	Mookambika Wildlife Sanctuary	1974	370.37
	Nugu Wildlife Sanctuary	1974	30.32

TABLE 10.1 *(Continued)*

State/Union Territory	Name of the sanctuary	Year of establishment	Area (in Km²)
	Pushpagiri Wildlife Sanctuary	1987	102.96
	Ranebennur Blackbuck Wildlife Sanctuary	1974	119.00
	Ranganathittu Wildlife Sanctuary	1940	0.67
	Vulture Wildlife Sanctuary	2012	3.46
	Rangayyanadurga Wildlife Sanctuary	2011	77.24
	Sharavathi Wildlife Sanctuary	1974	431.23
	Shettihalli Wildlife Sanctuary	1974	395.60
	Someshwara Wildlife Sanctuary	1974	314.25
	Talakaveri Wildlife Sanctuary	1987	105.01
Kerala	Aralam Wildlife Sanctuary	1984	55.00
	Chimmony Wildlife Sanctuary	1984	85.00
	Chinnar Wildlife Sanctuary	1984	90.44
	Chulannur Peafowl Wildlife Sanctuary	2007	3.42
	Idukki Wildlife Sanctuary	1976	70.00
	Kottiyoor Wildlife Sanctuary	2011	30.38
	Kurinjimala Wildlife Sanctuary	2006	32.00
	Malabar Wildlife Sanctuary	2010	74.22
	Mangalavanam Bird Wildlife Sanctuary	2004	0.03
	Neyyar Wildlife Sanctuary	1958	128.00
	Parambikulam Wildlife Sanctuary	1973	285.00
	Peechi-Vazhani Wildlife Sanctuary	1958	125.00
	Peppara Wildlife Sanctuary	1983	53.00
	Periyar Wildlife Sanctuary	1950	427.00
	Shendurney Wildlife Sanctuary	1984	100.32
	Thattekad Bird Wildlife Sanctuary	1983	25.00
	Wayanad Wildlife Sanctuary	1973	344.44
Lakshadweep	Pitti Bird Wildlife Sanctuary	1995	0.01
Madhya Pradesh	Bagdara Wildlife Sanctuary	1978	478.00
	Bori Wildlife Sanctuary	1977	485.72
	Gandhi Sagar Wildlife Sanctuary	1981	368.62

TABLE 10.1   *(Continued)*

State/Union Territory	Name of the sanctuary	Year of establishment	Area (in Km²)
	Ghatigaon Wildlife Sanctuary	1981	368.62
	Karera Wildlife Sanctuary	1981	202.21
	Ken Gharial Wildlife Sanctuary	1981	45.20
	Kheoni Wildlife Sanctuary	1982	122.70
	Narsighgarh Wildlife Sanctuary	1978	59.19
	National Chambal Wildlife Sanctuary	1978	435.00
	Noradehi Wildlife Sanctuary	1984	1194.67
	Orcha Wildlife Sanctuary	1994	44.91
	Pachmarhi Wildlife Sanctuary	1977	417.78
	Kuno Wildlife Sanctuary	1981	344.68
	Gangau Wildlife Sanctuary	1979	68.14
	Panpatha Wildlife Sanctuary	1983	245.84
	Pench Wildlife Sanctuary	1975	118.47
	Phen Wildlife Sanctuary	1983	110.74
	Ralamandal Wildlife Sanctuary	1989	2.35
	Ratapani Wildlife Sanctuary	1978	823.84
	Sailana Wildlife Sanctuary	1983	12.96
	Sanjay Dubri Wildlife Sanctuary	1975	364.59
	Sardarpur Wildlife Sanctuary	1983	348.12
	Singhori Wildlife Sanctuary	1976	287.91
	Son Gharial Wildlife Sanctuary	1981	41.80
	Veerangna Durgawati Wildlife Sanctuary	1997	23.97
Maharashtra	Amba Barwa Wildlife Sanctuary	1997	127.11
	Andhari Wildlife Sanctuary	1986	509.27
	Aner Dam Wildlife Sanctuary	1986	82.94
	Bhamragarh Wildlife Sanctuary	1997	104.38
	Bhimashankar Wildlife Sanctuary	1985	130.78
	Bor Wildlife Sanctuary	1970	61.10
	Chaprala Wildlife Sanctuary	1986	134.78
	Deulgaon-Rehekuri Wildlife Sanctuary	1980	2.17

TABLE 10.1 *(Continued)*

State/Union Territory	Name of the sanctuary	Year of establishment	Area (in Km²)
	Dhyanganga Wildlife Sanctuary	1997	205.23
	Gautala-Autramghat Wildlife Sanctuary	1986	260.61
	Great Indian Bustard Wildlife Sanctuary	1986	1222.61
	Jaikwadi Wildlife Sanctuary	1986	341.05
	Kalsubai Harishchandragad Wildlife Sanctuary	1986	361.71
	Karnala Fort Wildlife Sanctuary	1968	4.48
	Karanja Lad-Sohal Blackbuck Wildlife Sanctuary	2000	18.32
	Katepurna Wildlife Sanctuary	1988	73.63
	Koyana Wildlife Sanctuary	1985	423.55
	Lonar Wildlife Sanctuary	2000	1.17
	Malvan Marine Wildlife Sanctuary	1987	29.12
	Mansingdeo Wildlife Sanctuary	2010	182.59
	Mayureswar Wildlife Sanctuary	1997	5.15
	Melghat Wildlife Sanctuary	1985	778.75
	Nagzira Wildlife Sanctuary	1970	152.81
	Naigaon Mayur Wildlife Sanctuary	1994	29.89
	Nandur Madhmeshwar Wildlife Sanctuary	1986	100.12
	Karnala Bird Wildlife Sanctuary	1997	12.35
	Nawegaon Wildlife Sanctuary	2012	122.76
	Bor Wildlife Sanctuary	2012	60.70
	Nagzira Wildlife Sanctuary	2012	151.33
	Painganga Wildlife Sanctuary	1986	324.62
	Phansad Wildlife Sanctuary	1986	69.79
	Radhanagari Wildlife Sanctuary	1958	351.16
	Sagareshwar Wildlife Sanctuary	1985	10.87
	Tansa Wildlife Sanctuary	1970	304.81
	Thane Creek Flamingo Wildlife Sanctuary	2015	16.91
	Tipeshwar Wildlife Sanctuary	1997	148.63

TABLE 10.1 *(Continued)*

State/Union Territory	Name of the sanctuary	Year of establishment	Area (in Km²)
	Tungareshwar Wildlife Sanctuary	2003	85.00
	Yawal Wildlife Sanctuary	1969	177.52
	Yedshi Ramling Wildlife Sanctuary	1997	22.38
	Umred-Kharngla Wildlife Sanctuary	2012	189.30
	Wan Wildlife Sanctuary	1997	211.00
Manipur	Yangoupokpi Lokchao Wildlife Sanctuary	1989	184.40
Meghalaya	Baghmara Pitcher Plant Wildlife Sanctuary	1984	0.02
	Nongkhyllem Wildlife Sanctuary	1981	29.00
	Siju Wildlife Sanctuary	1979	5.18
Mizoram	Dampa Wildlife Sanctuary	1985	500.00
	Khawnglung Wildlife Sanctuary	1992	35.00
	Lengteng Wildlife Sanctuary	1999	60.00
	Ngengpui Wildlife Sanctuary	1991	110.00
	Pualreng Wildlife Sanctuary	2004	50.00
	Tawi Wildlife Sanctuary	1978	35.75
	Thorangtlang Wildlife Sanctuary	2002	50.00
	Tokalo Wildlife Sanctuary	2007	250.00
Nagaland	Fakim Wildlife Sanctuary	1980	6.41
	Puliebadze Wildlife Sanctuary	1980	9.23
	Rangapahar Wildlife Sanctuary	1986	4.70
Odisha	Badrama Wildlife Sanctuary	1962	304.03
	Baisipalli Wildlife Sanctuary	1981	168.35
	Balukhand-Konark Wildlife Sanctuary	1984	71.72
	Bhitarkanika Wildlife Sanctuary	1975	525.00
	Chandaka Dampara Wildlife Sanctuary	1982	175.79
	Chilika Wildlife Sanctuary	1987	15.53
	Debrigarh Wildlife Sanctuary	1985	346.91
	Gahirmatha Wildlife Sanctuary	1997	1435.00
	Hadgarh Wildlife Sanctuary	1978	191.06

TABLE 10.1 *(Continued)*

State/Union Territory	Name of the sanctuary	Year of establishment	Area (in Km²)
	Kapilash Wildlife Sanctuary	2011	125.50
	Karlapat Wildlife Sanctuary	1992	147.66
	Khalasuni Wildlife Sanctuary	1982	116.00
	Kothagarh Wildlife Sanctuary	1981	399.50
	Kuldiha Wildlife Sanctuary	1984	272.75
	Lakhari Valley Wildlife Sanctuary	1985	185.87
	Nandankanan Wildlife Sanctuary	1979	14.16
	Satkosia Gorge Wildlife Sanctuary	1976	745.52
	Simlipal Wildlife Sanctuary	1979	1354.30
	Sunabeda Wildlife Sanctuary	1988	500.00
Puducherry	Oussudu Wildlife Sanctuary	2008	3.90
Punjab	Abohar Wildlife Sanctuary	1988	186.50
	Bir Aishwan Wildlife Sanctuary	1952	2.64
	Bir Bhadson Wildlife Sanctuary	1952	10.23
	Bir Bhunerheri Wildlife Sanctuary	1952	6.62
	Bir Dosanjh Wildlife Sanctuary	1952	5.18
	Bir Gurdialpura Wildlife Sanctuary	1977	6.20
	Bir Mehas Wildlife Sanctuary	1952	1.23
	Bir Moti Bagh Wildlife Sanctuary	1952	6.54
	Harike Lake Wildlife Sanctuary	1982	86.00
	Jhajjar Bacholi Wildlife Sanctuary	1998	1.16
	Kathlaur Kushlian Wildlife Sanctuary	2007	7.58
	Takhni-Rehampur Wildlife Sanctuary	1992	3.82
	Nangal Wildlife Sanctuary	2009	2.90
Rajasthan	Bandh Baratha Wildlife Sanctuary	1985	199.50
	Bassi Wildlife Sanctuary	1988	138.69
	Bhensrodgarh Wildlife Sanctuary	1983	229.14
	Darrah Wildlife Sanctuary	1955	80.75
	Jaisamand Wildlife Sanctuary	1955	52.00
	Jamwa Ramgarh Wildlife Sanctuary	1982	300.00
	Jawahar Sagar Wildlife Sanctuary	1975	153.41

TABLE 10.1 *(Continued)*

State/Union Territory	Name of the sanctuary	Year of establishment	Area (in Km²)
	Kailadevi Wildlife Sanctuary	1983	676.38
	Kesarbagh Wildlife Sanctuary	1955	14.76
	Kumbhalgarh Wildlife Sanctuary	1971	608.58
	Mount Abu Wildlife Sanctuary	1960	326.10
	Nahargarh Wildlife Sanctuary	1980	50.00
	National Chambal Wildlife Sanctuary	1979	274.75
	Phulwari Wildlife Sanctuary	1983	692.68
	Ramgarh Vishdhari Wildlife Sanctuary	1982	252.79
	Ramsagar Wildlife Sanctuary	1955	34.40
	Sajjangarh Wildlife Sanctuary	1987	5.19
	Sariska Wildlife Sanctuary	1955	219.00
	Sawaimadhopur Wildlife Sanctuary	1955	131.30
	Sawai Mansingh Wildlife Sanctuary	1984	103.25
	Shergarh Wildlife Sanctuary	1983	98.71
	Sitamata Wildlife Sanctuary	1979	422.94
	Tal Chhapar Wildlife Sanctuary	1971	7.19
	Todgarh Raoli Wildlife Sanctuary	1983	495.27
	Van Vihar Wildlife Sanctuary	1955	25.60
Sikkim	Barsey Rhododendron Wildlife Sanctuary	1998	104.00
	Fambong Lho Wildlife Sanctuary	1984	51.76
	Kitam Bird Wildlife Sanctuary	2005	6.00
	Kyongnosla Alpine Wildlife Sanctuary	1977	31.00
	Maenam Wildlife Sanctuary	1987	35.34
	Pangolakha Wildlife Sanctuary	2002	128.00
	Shingba Rhododendron Wildlife Sanctuary	2002	43.00
Tamil Nadu	Cauvery Wildlife Sanctuary	2014	504.33
	Chitrangudi Bird Wildlife Sanctuary	1989	0.48
	Gangaikondan Spotted Deer Wildlife Sanctuary	2013	2.88
	Annamalai Wildlife Sanctuary	1976	841.49

TABLE 10.1 *(Continued)*

State/Union Territory	Name of the sanctuary	Year of establishment	Area (in Km²)
	Kalakad Wildlife Sanctuary	1976	223.58
	Kanjirankulam Bird Wildlife Sanctuary	1989	1.04
	Kanyakumari Wildlife Sanctuary	2002	457.78
	Karaivetti Bird Wildlife Sanctuary	1999	4.54
	Karikili Bird Wildlife Sanctuary	1989	0.61
	Kodaikanal Wildlife Sanctuary	2013	608.95
	Koonthankulam-Kadankulam Wildlife Sanctuary	1994	1.29
	Mudumalai Wildlife Sanctuary	1942	217.76
	Mundanthurai Wildlife Sanctuary	1977	567.38
	Nellai Wildlife Sanctuary	2015	356.73
	Oussudu Wildlife Sanctuary	2015	3.32
	Point Calimere Wildlife Sanctuary	1967	17.26
	Sathyamangalam Wildlife Sanctuary	2008	1411.61
	Grizzled Squirrel Wildlife Sanctuary	1988	485.20
	Udayamarthandapuram Lake Wildlife Sanctuary	1991	0.45
	Vaduvoor Bird Wildlife Sanctuary	1991	1.28
	Vedanthangal Lake Bird Wildlife Sanctuary	1936	0.30
	Vallanadu Wildlife Sanctuary	1987	16.41
	Vellode Bird Wildlife Sanctuary	1997	0.77
	Vettangudi Bird Wildlife Sanctuary	1977	0.38
	Melaselvanoor-Keelaselvanoor Wildlife Sanctuary	1998	5.93
Telengana	Eturnagaram Wildlife Sanctuary	1953	806.15
	Kawal Wildlife Sanctuary	1965	892.23
	Kinnersani Wildlife Sanctuary	1977	635.41
	Lanja Madugu Siwaram Wildlife Sanctuary	1978	29.81
	Manjira Wildlife Sanctuary	1978	20.00
	Pakhal Wildlife Sanctuary	1952	860.00
	Pocharam Wildlife Sanctuary	1952	130.00
	Pranahita Wildlife Sanctuary	1980	136.03

TABLE 10.1   *(Continued)*

State/Union Territory	Name of the sanctuary	Year of establishment	Area (in Km²)
Tripura	Gumti Wildlife Sanctuary	1988	389.54
	Rowa Wildlife Sanctuary	1988	0.86
	Sepahijala Wildlife Sanctuary	1987	13.45
	Trishna Wildlife Sanctuary	1988	163.08
Uttar Pradesh	Bakhira Wildlife Sanctuary	1990	28.94
	Chandraprabha Wildlife Sanctuary	1957	78.00
	Dr. Bhimrao Ambedkar Bird Wildlife Sanctuary	2003	4.27
	Hastinapur Wildlife Sanctuary	1986	2073.00
	Kaimur Wildlife Sanctuary	1982	500.73
	Katerniaghat Wildlife Sanctuary	1976	400.09
	Kishanpur Wildlife Sanctuary	1972	227.00
	Lakh Bahosi Wildlife Sanctuary	1988	80.24
	Mahavir Swami Wildlife Sanctuary	1977	5.41
	National Chambal Wildlife Sanctuary	1979	635.00
	Nawabganj Bird Wildlife Sanctuary	1984	2.25
	Okhla Wildlife Sanctuary	1990	4.00
	Parvati Aranga Wildlife Sanctuary	1990	10.84
	Patna Wildlife Sanctuary	1990	1.09
	Ranipur Wildlife Sanctuary	1977	230.31
	Saman Bird Wildlife Sanctuary	1990	5.26
	Samaspur Wildlife Sanctuary	1987	7.99
	Sandi Bird Wildlife Sanctuary	1990	3.09
	Sohagibarwa Wildlife Sanctuary	1987	428.20
	Sohelwa Wildlife Sanctuary	1988	452.47
	Sur Sarovar Wildlife Sanctuary	1991	4.03
	Jai Prakash Narayan Bird Wildlife Sanctuary	1991	34.32
	Turtle Wildlife Sanctuary	1989	7.00
	Vijai Sagar Wildlife Sanctuary	1990	7.00
Uttarakhand	Askot Wildlife Sanctuary	1986	600.00
	Binsar Wildlife Sanctuary	1988	47.07

TABLE 10.1   *(Continued)*

State/Union Territory	Name of the sanctuary	Year of establishment	Area (in Km²)
	Govind Pashu Vihar Wildlife Sanctuary	1955	485.89
	Kedarnath Wildlife Sanctuary	1972	975.20
	Mussoorie Wildlife Sanctuary	1993	10.82
	Nandhaur Wildlife Sanctuary	2012	269.96
	Sonanadi Wildlife Sanctuary	1987	301.18
West Bengal	Ballavpur Wildlife Sanctuary	1977	2.02
	Bethuadahari Wildlife Sanctuary	1980	0.67
	Bibhutibhusan Wildlife Sanctuary	1980	0.64
	Buxa Wildlife Sanctuary	1986	267.92
	Chapramari Wildlife Sanctuary	1976	9.60
	Chintamani Kar Bird Wildlife Sanctuary	1986	0.07
	Haliday Island Wildlife Sanctuary	1976	5.95
	Jore Pokhri Salamander Wildlife Sanctuary	1982	0.04
	Lothian Island Wildlife Sanctuary	1976	38.00
	Mahananda Wildlife Sanctuary	1976	158.04
	Raiganj Wildlife Sanctuary	1985	1.30
	Ramnabagan Wildlife Sanctuary	1981	0.14
	Sajnakhali Wildlife Sanctuary	1976	362.40
	Senchal Wildlife Sanctuary	1976	38.88
	West Sunderban Wildlife Sanctuary	2013	556.45

TABLE 10.2   List of National Parks in India

Name of national park	State/Union Territory	Year of establishment	Area (in Km²)
Anamudi Shola National Park	Kerala	2003	7.50
Anshi National Park	Karnataka	1987	417.34
Balphakram National Park	Meghalaya	1986	220.00
Bandhavgarh National Park	Madhya Pradesh	1968	446.00
Bannerghatta National Park	Karnataka	1986	104.30
Bandipur National Park	Karnataka	1974	874.20

Name of national park	State/Union Territory	Year of establishment	Area (in Km²)
Balphakram National Park	Meghalaya	1986	220.00
Bandhavgarh National Park	Madhya Pradesh	1968	446.00
Bhitarkanika National Park	Odisha	1988	145.00
Blackbuck National Park	Gujarat	1976	34.08
Buxa National Park	West Bengal	1992	117.10
Campbell Bay National Park	Andaman and Nicobar Islands	1992	426.30
Chandoli National Park	Maharashtra	2004	317.67
Clouded Leopard National Park	Tripura	2003	5.08
Darrah National Park	Rajasthan	2006	200.54
Dachigam National Park	Jammu and Kashmir	1981	141.00
Desert National Park	Rajasthan	1980	3162.00
Dibru-Saikhowa National Park	Assam	1999	340.00
Dudhwa National Park	Uttar Pradesh	1977	490.29
Eravikulam National Park	Kerala	1978	97.00
Galathea National Park	Andaman and Nicobar Islands	1992	110.00
Gangotri National Park	Uttarkhand	1989	2390.00
Gir Forest National Park	Gujarat	1965	1412.00
Gorumara National Park	West Bengal	1994	79.45
Govind Pashu Vihar National Park	Uttarkhand	1990	472.08
Great Himalayan National Park	Himachal Pradesh	1984	754.40
Gugamal National Park	Maharashtra	1987	361.28
Guindy National Park	Tamil Nadu	1976	2.82
Gulf of Mannar Marine National Park	Tamil Nadu	1980	6.23
Guru GhasidasNational Park	Chhattisgarh	1981	1440.71
Hemis National Park	Jammu and Kashmir	1981	4400.00
Hazaribagh National Park	Jharkhand	1954	183.89
Inderkilla National Park	Himachal Pradesh	2010	104.00
Indira Gandhi National Park	Tamil Nadu	1989	117.10

TABLE 10.2   *(Continued)*

Name of national park	State/Union Territory	Year of establishment	Area (in Km²)
Indravati National Park	Chhattisgarh	1981	1258.70
Jaldapara National Park	West Bengal	2012	216.00
Jim Corbett National Park	Uttarkhand	1936	1318.70
Kalesar National Park	Haryana	2003	100.88
Kanha National Park	Madhya Pradesh	1995	940.00
Kanger Ghati National Park	Chhattisgarh	1982	200.00
Kasu Brahmananda Reddy National Park	Telangana	1994	1.42
Kaziranga National Park	Assam	1974	858.98
Keibul Lamjao National Park	Manipur	1977	40.00
Keoladeo National Park	Rajasthan	1981	28.73
Khangchendzonga National Park	Sikkim	1977	1784.00
Khirganga National Park	Himachal Pradesh	2010	710.00
Kishtwar National Park	Jammu and Kashmir	1981	400.00
Kudremukh National Park	Karnataka	1987	600.32
Madhav National Park	Madhya Pradesh	1959	375.22
Mahatma Gandhi Marine National Park	Andaman and Nicobar Islands	1983	281.50
Mahavir Harina Vanasthali National Park	Telangana	1994	14.59
Manas National Park	Assam	1990	500.00
Mandla Plant Fossils National Park	Madhya Pradesh	1983	0.27
Marine National Park	Gujarat	1980	162.89
Mathikettan Shola National Park	Kerala	2003	12.82
Middle Button Island National Park	Andaman and Nicobar Islands	1987	0.44
Mollem National Park	Goa	1978	107.00
Mouling National Park	Arunachal Pradesh	1986	483.00
Mount Harriet National Park	Andaman and Nicobar Islands	1987	46.62
Mrugavani National Park	Telangana	1994	3.60

TABLE 10.2   *(Continued)*

Name of national park	State/Union Territory	Year of establishment	Area (in Km²)
Mudumalai National Park	Tamil Nadu	1940	321.55
Mukurthi National Park	Tamil Nadu	2001	78.46
Murlen National Park	Mizoram	1991	100.00
Namdapha National Park	Arunachal Pradesh	1974	1985.24
Nameri National Park	Assam	1978	137.07
Nanda Devi National Park	Uttarkhand	1982	630.33
Navegaon National Park	Maharashtra	1975	133.88
Neora Valley National Park	West Bengal	1986	88.00
Nokrek National Park	Meghalaya	1986	47.48
North Button Island National Park	Andaman and Nicobar Islands	1979	0.44
Ntangki National Park	Nagaland	1993	202.02
Omkareshwar National Park	Madhya Pradesh	2004	293.56
Orang National Park	Assam	1999	78.81
Pambadum Shola National Park	Kerala	2003	1.32
Panna National Park	Madhya Pradesh	1981	542.67
Papikonda National Park	Andhra Pradesh	2008	1012.85
Pench National Park	Madhya Pradesh	1977	758.00
Periyar National Park	Kerala	1982	305.00
Phawngpui National Park	Mizoram	1992	50.00
Pin Valley National Park	Himachal Pradesh	1987	807.36
Rajbari National Park	Tripura	2007	31.63
Rajaji National Park	Uttarakhand	1983	820.00
Nagarhole National Park	Karnataka	1988	643.39
Rajiv Gandhi National Park	Karnataka	2005	2.40
Rani Jhansi Marine National Park	Andaman and Nicobar Islands	1996	256.14
Ranthambore National Park	Rajasthan	1981	392.00
Saddle Peak National Park	Andaman and Nicobar Islands	1979	32.54
Salim Ali National Park	Jammu and Kashmir	1992	9.07
Sanjay National Park	Madhya Pradesh	1981	466.70
Sanjay Gandhi National Park	Maharashtra	1969	104.00
Sariska National Park	Rajasthan	1955	866.00

TABLE 10.2 *(Continued)*

Name of national park	State/Union Territory	Year of establishment	Area (in Km²)
Satpura National Park	Madhya Pradesh	1981	524.00
Silent Valley National Park	Kerala	1980	237.00
Simbalbara National Park	Himachal Pradesh	2010	27.88
Sirohi National Park	Manipur	1982	41.30
Simlipal National Park	Odisha	1980	845.70
Singalila National Park	West Bengal	1986	78.60
South Button Island National Park	Andaman and Nicobar Islands	1987	0.03
Sri Venkateswara National Park	Andhra Pradesh	1989	353.00
Sultanpur National Park	Haryana	1989	1.43
Sundarbans National Park	West Bengal	1984	1330.12
Tadoba National Park	Maharashtra	1955	625.00
Valley of Flowers National Park	Uttarakhand	1982	87.50
Valmiki National Park	Bihar	1976	898.45
Vansda National Park	Gujarat	1979	23.99

TABLE 10.3 List of Biosphere Reserves in India

Name of the biosphere reserve	State/Union Territory	Year of establishment	Area (in km²)
Achanakamar–Amarkantak Biosphere Reserve	Madhya Pradesh, Chhattisgarh	2005	3835.51
Agasthyamalai Biosphere Reserve	Kerala, Tamil Nadu	2001	3500.08
Cold Desert Biosphere Reserve	Himachal Pradesh	2009	7770.00
Dehang-Dibang Biosphere Reserve	Arunachal Pradesh	1998	5111.50
Dibru-Saikhowa Biosphere Reserve	Assam	1997	765.00
Great Nicobar Biosphere Reserve	Andaman and Nicobar Islands	1989	885.00
Gulf of Mannar Biosphere Reserve	Tamil Nadu	1989	10500.00

TABLE 10.3   *(Continued)*

Name of the biosphere reserve	State/Union Territory	Year of establishment	Area (in km²)
Kachchh Biosphere Reserve	Gujarat	2008	12454.00
Khangchendzonga Biosphere Reserve	Sikkim	2000	2619.92
Manas Biosphere Reserve	Assam	1989	2837.00
Nanda Devi Biosphere Reserve	Uttarakhand	1988	5860.69
Nilgiri Biosphere Reserve	Karnataka, Kerala, Tamil Nadu	1986	5520.00
Nokrek Biosphere Reserve	Meghalaya	1988	820.00
Pachmarhi Biosphere Reserve	Madhya Pradesh	1999	4926.00
Panna Biosphere Reserve	Madhya Pradesh	2011	2998.98
Seshachalam Hills Biosphere Reserve	Andhra Pradesh	2010	4755.99
Simlipal Biosphere Reserve	Orissa	1994	4374.00
Sundarbans Biosphere Reserve	West Bengal	1989	9630.00

10.2   VALUES OF BIODIVERSITY

10.2.1   BIODIVERSITY RESOURCES

Biological resources are living components of the environment that are either helpful or harmful for human beings. Plants, animals, microorganisms, forests, agricultural crops are all examples for biological resources. Biological resources provide food, timber, fuel, oil, medicines, and many other useful materials. Timber is obtained from plant species such as Australian teak (*Acacia mangium*), Catachu (*Acacia catechu*), Mahogany (*Swietenia mahogani*), Teak (*Tectona grandis*) and Arjuna (*Terminalia arjuna*) (Kohli et al., 2001; Alka, 2010). Jatropha (*Jatropha curcas*), Karanjia (*Pongamia pinnata*), Neem (*Azadirachta indica*) and Mahua (*Madhuca indica*) are sources of biofuels (Sahoo et al., 2008). Some plants are sources of many drugs. Galantamine used for treating Alzheimer's disease is obtained from *Galanthus nivalis*. Other drugs extracted from plants include titropium extracted from *Atropa belladonna* used to cure pulmonary diseases, apomorphine extracted from *Papaver somniferum* is used to treat Parkinson's disease, betulinic acid extracted from *Betula alba* is an anticancerous drug, quinine used in the treatment of malaria extracted from *Cinchona ledgeriana*, vinblastine, and vincristine extracted from *Catharanthus roseus* is used in the treatment of blood cancer (Saravanan and Karthi, 2014). Essential oils and cooking oils are extracted from plants like *Olea europaea* (olive oil), *Prunus amygdalus* (almond oil), *Syzygium aromaticum* (clove oil), *Eucalyptus globulus* (eucalyptus oil), *Cocos nucifera* (coconut oil), *Helianthus annuus* (sunflower oil), *Sesamum indicum* (sesame oil) and *Brassica nigra* (mustard oil) (Chanchal and Swarnlata, 2010).

Plants are also an excellent food source for animals and humans. Leaves (cabbage, spinach, and lettuce), stem (potato and ginger), roots (carrot, radish, and turnip), flower (cauliflower and broccoli) are used as food. Cereals such as rice, wheat, and maize, pulses such as beans, grams, and peas, spices, and condiments such as tea, coffee, cardamom, cloves, and cinnamon, fruits such as orange, grapes, pineapple are food resources that are obtained from plants (Prasad and Chiranjib, 1995). Animals are also used as food resources. Fishes such as sardine, mackerel, sole fish, pomfret, seer fish, tuna, tilapia, and snapper, crustaceans such as crabs, prawns, and lobsters, mollusks such as clams, squid, and mussels are used as food by many other species (Shinoj et al., 2008). Oil is extracted from fishes such as cod, tuna, salmon, and shark.

Biological resources can also have a negative impact on the ecosystem. Certain organisms are known to pose a threat for agricultural crops. Black point (*Alternaria alternata*), Armyworm (*Mythimna separata*), Pink borer (*Sesamia inferens*) are some of the pests that infect agricultural crops (Singh, 2012). Certain organisms are known to cause diseases in human beings. Microorganisms such as fungi cause psoriasis, candidiasis, aspergillosis, fungal meningitis and Athlete's foot, bacteria causes cholera, anthrax, tetanus, and leprosy, and virus causes rabies, mumps, polio, and dengue fever (Doyle and Bryan, 2000; Pfaller and Diekema, 2007). Fungi such as *Fusarium* spp., *Verticillium* spp. and *Puccinia* spp., bacteria such as *Xanthomonas* spp. and *Pseudomonas* spp. and virus such as tobacco mosaic virus cause diseases in plants. Invasive species or exotic species are known to pose a threat to native species. Example include *Nile perch* (this fish was introduced to Lake Victoria in Africa which posed a threat to native fish species), Eucalyptus, Parthenium, and Casurina (Dara, 1993; Pimental et al., 2005; Pringle, 2005).

10.2.2 ECOSYSTEM SERVICES

An ecosystem consists of biotic components (living organisms such as plants, animals, microorganisms) which interact with themselves and also with the abiotic components (non-living such as sunlight, carbon dioxide, air, water, soil) surrounding them. The benefits that human beings get from ecosystem are known as ecosystem services. The services provided by ecosystem can be supporting, provisional, cultural, and regulating. Supporting services include cycling of nutrients, the formation of soil, primary production, providing habitat for organisms, atmospheric oxygen production and cycling of water. Other types of ecosystem services require supporting services for their production. Supporting services do not have a direct impact on humans. Provisioning services are the products which human beings get from the ecosystem. Food, water, and medicinal resources come under provisioning services of the ecosystem. Cultural services are services that are non-material. Tourism, cultural diversity, aesthetic value, and educational value come under cultural services. Regulation of climate, maintenance of soil fertility, carbon sequestration, maintaining air quality, treatment of wastewater, biological control of pests, pollination, and prevention of soil erosion comes under regulatory functions of the ecosystem. For providing regulating services, ecosystem acts themselves as regulators (Boyd and Banzhaf, 2007).

10.2.3 SOCIAL BENEFITS

Recreational and aesthetic come under the social benefits of biodiversity. Bird watching, visiting wildlife sanctuaries and national parks, butterfly watching, hiking, and visiting tourist places is some of the recreational activities. People always prefer places with beautiful scenery and a good climate for spending their leisure time; for instance, hill stations are preferred by many for school excursions and study trips. Areas which have varieties of species of plants, animals, and birds are worth to visit. Places which are biologically rich have always been important tourist attractions: example includes national parks, zoos, botanical gardens, and wildlife sanctuaries. Since biologically rich areas are important tourist spots, this can increase the local economy and could provide various job opportunities for local people. Biodiversity also provide platform for research activities, which are carried out on different aspects of biodiversity such as diversity of certain species in a particular area, effect of certain factors on certain species, extraction of some compounds from plants and animals, extraction, and production of medicines, impact of pests on agricultural crops, and, impact of climate change on certain species (Anjaneyalu, 2004).

10.2.4 SPIRITUAL VALUES

Spiritual services are related to culture and religion. Certain plants and animals have religious importance. Plants such as tulsi (*Ocimum sanctum*), neem (*Azadirachta indica*), lotus (*Nelumbo nucifera*), peepal (*Ficus religiosa*) are considered as sacred. Their flowers, leaves, and fruits are used for performing certain rituals during prayers. Cow, house crow, vulture, monkey, black buck and snake are considered holy in Hindu mythology, whereas yak, elephant, horse, and the peacock is considered as sacred by Buddhists. Sacred grove, a type of *in situ* conservation is also related to religion. Sacred groves are a way in which biodiversity is protected culturally and religiously. Sacred groves are forest fragments that vary in size from several kilometers to several acres of land. Sacred groves are protected as a part of belief (Anjaneyalu, 2004; Laxman et al., 2014).

10.3 THREATS TO BIODIVERSITY

10.3.1 HABITAT ALTERATION

Habitat alterations devastate the animal and plant species at a high rate. MacKinnon and MacKinnon (1986, 1991) reported that in Africa during 1986 about 65% of animal and plant species habitats were destroyed. During 1950 and 1985 the deforestation rate in the eastern region of Madagascar is 1.5%, if the same alteration continues mostly the forests of slopes will survive up to 35 years (Sussman et al., 1996). The situation is similar in regions of Neotropics and certain regions in Asia (Peres et al., 2010; Sodhi et al., 2010). In Indomalayan regions during 1990, the original habitat of 42 primate species was lost up to 31 to 96% (MacKinnon and

MacKinnon, 1991). Habitat loss is closely allied with the changes in the remaining habitat which is ruined by selective logging, disturbances for seedling growth by cattle, pigs, and goats. If the animals left in fragmented or degraded forests, causes the discontinuous supply of food, shelter, and also some pressures from introduced species. Populations also play an important role in a natural disaster such as storms, climatic change and seasonal droughts (Malhi et al., 2008). Habitat alteration causes easier access for predators and simultaneously increases the predators pressure (Wilcove, 1985; Wilcove et al., 1986; Andrén and Angelstam, 1988; Estrada and Coates-Estrada, 1996; Onderdonk and Chapman, 2000; Irwin et al., 2009). Due to improved visibility and ease of use resulted in the classification and disintegration of species, which is the best reason for the increase in the raptor population (Karpanty, 2003; Colquhoun, 2006). Hunters are the novel threats for the wild species as the forest were exposed to hunting. The forest area was fragmented, and the continuing fragmentation in the forest do not remain structurally integral. Edge effect determines the variation in a community that presents at the edge of two different habitats; due to human activities, it creates pressure to the biotic and abiotic environment (Ries et al., 2004; Lehman et al., 2006a,b). Edge effect leads to the changes in forest dynamics, affects the forest climate, mortality of plant species, and the animal exists in that habitat (Laurance et al., 2011). Habitat destruction may show effects even on intimately related animals. Irwin et al. (2010) demonstrated that responses of species to an anthropogenic disturbance in Madagascar resulted that response of species may be negative, the adjacent factor may differ in taxonomic groups, the responses for taxonomic group alters between ecoregions and the anthropogenic activity leads to the decline of native or endemic species.

10.3.2 OVERHARVESTING OF BIORESOURCES

Overexploitation is the major loss for habitat-based individuals since from last few decades, due to which the species no longer uphold itself in the forest dynamics. Due to a reduction in individuals, it may also reduce the ability of exploited species to abide them from technological aspects which is indirect drivers of biodiversity loss. The extinction of individual species may cause fluctuation in the composition of other species habitat (David and Clarence, 2001). For instances, during the 1800s in the United States million tons of fish especially Sturgeon (Acipenseridae) was exported, but unfortunately, Sturgeon has become threatened species due to overharvesting and pollution (Serap and Ibrahim, 2004). Also the numerous tree species, for example, Mahogany (*Swietenia mahagoni*) was well known for its hardwood used for a decorative purpose has become endangered due to overharvesting. According to Millennium Ecosystem Assessment (MEA) majority of overexploited species belongs to marine organisms, and animals hunted for food, certain plants and animals are grazed for medicine and trade purpose. Each year about 8% of total aquatic ecosystems were destructed due to human activities from last few years; commercial fish stocks are almost exploited (Bennett et al., 2000; Andrew and Marc, 2011). In Central Africa, tropical species are at risk, about 1–5 million tons of meat were exploited annually (Sagan et al., 2015). The Convention on International Trade in Endangered Species of Wild Fauna and Flora (CITES), aims to avoid over-trading of wild animals and plant species which cause trouble for their endurance, approximately millions of species and its derivatives were traded annually. International Standard for Sustainable Wild Collection of Medicinal and Aromatic Plants (ISSC-MAP) aims to endorse the sustainable utilization of plants in medicine and cosmetics field, but the conservation of species is complicated has the trade law may alter from regional and national levels.

10.3.3   ENVIRONMENTAL POLLUTION

Industrial revolution leads to the addition of natural and inert components into the ecosystem which has become a growing threat to biodiversity. Pollution can be profound either with a single confrontation or along with some other components to the environment over a long period of time (Wilson, 1989; Golovko and Izhevsky, 1996; Grodzinsky, 1999). Some toxic substances were released, and they are resisting to biodegradation and reside in the environment for a long duration, such form of organic substances are known as persistent organic pollutants (POPs); toxins were accumulated in tissues, and thus the pollutants were scattered through the food chain. Due to such pollutants, the species of tropic levels especially marine mammal *Delphinapterus leucas* weakens the reproductive and immunological functions (Ehrlich and Wilson, 1991; Monisha et al., 2014). Organisms show resistance towards certain chemicals which causes the reduction in species diversity. Carson (1962) reported that utilization of restricted chemicals such as DDT is still used as pesticides in many countries. Pollution often affects the higher level organisms by changing the structure and function of the ecosystem. Chemical components manufactured by humans are also the major pollutants causing ecological problems such as eutrophication, acid deposition and ozone depletion (Anita and Madhoolika, 2008). Eutrophication resulted in the chronic accumulation of waste products such as nitrogen and phosphorous released from the agriculture sector, and also the mineral deposition from industry. The ozone layer is depleted due to chlorofluorohydrocarbons and some other chemicals released into the atmosphere, which permits penetration of ultraviolet light that can be detrimental for biological organisms such as marine plankton communities (Sivasakthivel and Reddy, 2011).

10.3.4 INVASIVE ALIEN SPECIES

Globally, invasive alien species is the major threats to biological diversity. To realize the impact of biological diversity one should know the characterization of biodiversity. Biodiversity is composed of three major features: functional, compositional, and structural diversity (Noss, 1990). Compositional diversity is most widely accepted to determine the number of species in the ecosystem. Though, functional, and structural diversity plays a vital role to know the ecosystem dynamics, which are repeatedly as alerted by biological invasions. Biological invasions are an important component of changes in the ecological system (Vitousek et al., 1996). Introducing *Bromus tectorum* in American rangelands increases the incidence of fires, which convert the shrublands into grasslands. In California chaparral, the biological invasion of grasses causes a difference in species components (Zedler et al., 1983). Plant invasions are best illustrated by *Mimosa pigra* of northern Australia, which collapses the plant and animal population (Braithwaitte et al., 1989). Fynbos is the plant invasion of South Africa which has threatened some plant species, in which few are endemic (Musil, 1993). The introduction of alien fish directly in Lake Victoria leads to a loss of more than 200 cichlids species, along with the addition of hyacinth collapsed the whole lake (Bright, 1999). The utilization of invasive alien species eliminate the keystone species which may alter the system dynamics, collapses the ecosystem, and invasive species revealed the changes in the ecosystems from heterogeneity to homogeneity (Baskin, 2002).

10.3.5 CLIMATE CHANGE

Changes in climatic condition will probably affect all the levels of biological diversity, from organism to biotic community (Parmesan, 2006). Due to directional selection and rapid migration, climatic conditions reduce the genetic diversity of species which in turn affect functions of the ecological system (Meyers and Bull, 2002; Botkin et al., 2007). On the other hand, studies are concentrated on impacts of higher biomes, whereas the genetic effects of climatic condition were studied for the small group of species. Apart from this, different effects on population will alter the web of interactions at the biome level (Gilman et al., 2010; Walther, 2010). Variation in climate leads to the alteration in cyclic and seasonal phenomena of flowering and insect pollinators, leads to conflicts in plant and pollinator inhabitants which eliminates both the plant and insect pollinator (Kiers et al., 2010; Rafferty and Ives, 2010). Various adaptations of interspecific competition such as commensalism, mutualism, parasitism, predation, amensalism, neutralism, and protocooperation also adapt community structure and ecosystem functions (Lafferty, 2009; Walther, 2010; Yang and Rudolf, 2010). In higher organisms, climate may change the vegetative community, affecting biome integrity as well. According to the study conducted by MEA, there is a shift of 5–20% terrestrial ecosystem, especially coniferous forest, tundra, scrubland, savanna, and boreal forest biome (Sala et al., 2005). Predominantly in tipping points, the ecosystem thresholds led to permanent shifts in biome community (Leadley et al., 2010). In Africa, due to reduced rainfall and increasing temperature may dry the lake (Campbell et al., 2009). Due to raising in temperature ocean will become acidic and causes degradation of tropical coral reefs (Hoegh-Guldberg et al., 2007). The change in climatic conditions for genetic and species diversity possess potentiality for ecosystem services. Species extinction is the most common and permanent causes for declining the system dynamics. To overcome such issues biodiversity can acquire certain ways, by adopting various types of mechanisms.

10.3.6 HUMAN POPULATION

The increase in human population dimension, intensification, density, and relocation of species is the fundamental causes for loss of biodiversity

(Liu et al., 2003). World population was supposed to exceed 8–10 billion during the middle of the century, where a large number of growths will take place in humid tropics which is having the richest form of biodiversity. Population plays a key role in the loss of biodiversity, to meet the needs of growing population more pressure has created on the components of biodiversity (ecosystems, genes, and species) (Myers, 1994). Urbanization leads to the loss in the biodiversity, where the household demographic aspect is an important cause for ecological resource consumption (Liu et al., 2003; Richard and Gorenflo, 2011). Rapid species growing areas possess a high number of threatened and vulnerable plant species due to over-exploitation, and habitat loss of population and some other external pressures pose a high risk for the extinction of plant and animal species, it takes place where the humans are entirely dependent on biological diversity for their basic occupation (Cincotta and Engelman, 2000). Many parts of the world especially in Asia and Africa the threatened species and people habitually present on the same localities (Craig et al., 2009). The place where human growth rates are high, the number of threatened species are more likely to increase (Mittermeier, 1999). Habitat loss is the predominant loss of biodiversity, current tendency and protuberance indicate the excess use of land for human occupations is the important driver of biological diversity and ecosystem deterioration. Based on MEA, the major groups including forests, grasslands, and coastal zones were destroyed by human activities led to deprivation to the ecosystem. Land expansion for agriculture is the major factor for habitat loss; a combination of unsustainable forest management along with agricultural land contributes furthermore cause for the extinction of species (Myers, 1990).

10.4   CONSERVATION OF BIODIVERSITY

Conservation of biodiversity includes two fundamental concepts composed of different techniques as *in situ* and *ex situ*. CBD provides definitions for these strategies: *ex situ* conservation means preserving the biological components outside their natural habitat, whereas *in situ* conservation means the ecosystem conservation as well as its habitat maintained for the recovery of viable populations of an organism. Both the methods have basic differences: *ex situ* conservation includes sampling, transformation, storage of taxa from the collection region, whereas, *in situ* conservation includes designation, management, and monitoring of target species by which they are encountered (Maxted et al., 1997). Conservation criteria are further subdivided to precise techniques such as seed storage, gene banks, cryopreservation, zoological parks and arboreta for conservation by *ex situ*, and protected areas, biosphere reserves, sanctuaries for conservation of species by *in situ* method; both the techniques represents its own advantages and certain limitations (Engels and Wood, 1999). During 1950s and 1960s, great advancement in the field of plant breeding brought up the "green revolution" resulted in adopting high-yield varieties and genetically modified crops, particularly wheat and rice. Due to loss of genetic diversity among these crops, farmers replaced those local and conventional varieties with the genetically modified crops. The International Agricultural Research Centers (IARC) of the Consultative Group on International Agricultural Research (CGIAR) was established to collect the germplasm of major crop plants. To collect and conserve the threatened plant genetic resources the International Board for Plant Genetic Resources was started during 1974. There are about 1,300 genebanks and germplasm collection centers throughout the world (FAO, 1996). Collection and preserving is mainly concerned to the food crops, such as cereals and legumes, i.e., species which are conserved easily as seed, which resulted to over-representation of such species in the world's major storage banks. Field genebanks were established with new technologies including cryopreservation; *in vitro* storing to conserve species for which preservation of seed is impossible or inappropriate, was given attention by various international communities.

10.4.1   IN SITU BIODIVERSITY CONSERVATION

The *in situ* conservation involves description, maintenance, and monitoring of biological diversity in the equivalent region where it comes across. CBD describes that *in situ* conservation is the primitive conservation criteria, and suggest that *ex situ* methods act as a supporting factor to reach the conservation strategy. The main strategy of *in situ* conservation is to establish protected areas with suitable environmental conditions for the biologically diversified organisms. The main purpose is to protect the endangered species in its natural environment, either by shielding them in sanctuaries and national parks or protecting it from predators. To conserve agricultural biodiversity the unconventional farming method is used, e.g., Nilgiri biosphere reserve.

10.4.1.1   PROTECTED AREAS AND BIOSPHERE RESERVES

For wild species conservation in India, National Forest Policy in 1952 planned for national parks and sanctuaries (Singh and Kushwaha, 2008). The Wildlife Act (1972) also showed potential involvement for natural biosphere conservation and preservation of biodiversity in protected areas. In India, there are about 28 tiger reserving zones launched during 1973 with an area of 37,762 km^2 (Jimmy et al., 2010) and for elephant during 1992 with 25 reserved zones of 61,200 km^2 areas considered as protected areas (Ceballos et al., 2005). Biosphere reserves are mandated approach of Wildlife Act to conserve the life forms and overall the supporting system and later permitting them to monitor or evaluate the changes occur in nature (McCool, 2006). Therefore, the core area of the biosphere reserve is a national park or wildlife sanctuaries. About 4.78% area of India is covered by PAs, i.e., 105 national parks, 531 wildlife sanctuaries, and 18 biosphere reserve areas. It is a never-ending process, and more PAs are supposed to be launched for conserving biological diversity. PAs are very small, and few are fragmented, which leads to the isolation of species populations. Hence the size needs to be optimized (Blom, 2004). Increase in population leads to the use and abuse of protected areas causes destruction to biological diversity (Mackinnon and Mackinnon, 1986). The Wildlife (Protection) Act 1972 divides two new kinds of PA, namely, conservation, and community reserves implemented to cover all biogeographical zones and different types of forests; the communal were also involved in the establishment and management of such protected areas. National Environment Policy visualized a legal approach for an environmentally sensitive zone of India by various environmental criteria required for conservation. Those entities are regulated by the Environment (Protection) Act, 1986. Both were evaluated for environmentally susceptible zones for the well-being of biodiversity-rich areas of PA system (Singh and Kushwaha, 2008).

10.4.1.2   SACRED GROVES

Many Indian cultures have deep astonishment for nature and natural objects, which are not only confined but also worshiped. The conventional practice for nature worship, certain communities, was established in forest zone for ancestral spirits and deities. Those zones are called as sacred groves (Ramakrishnan, 1998). There are about 5,000 small-large sized sacred groves in India. Thus the number of sacred groves may be more than 1,00,000 covering 2% of the countries geographical area (Malhotra, 1998). Malhotra et al. (1997) have reported 322 sacred groves in Koraput district of Odisha. Likewise, 953 groves are recorded in Maharashtra (Deshmukh et al., 1998). The place around Mount Kanchenjunga in West Sikkim, widely called as Demojong, most of the sacred groves are of Buddhists. Meghalaya, Kerala, Maharashtra, Tamil Nadu, Uttarakhand, and Himachal Pradesh, few other states sacred groves are rich in floral and faunal diversity (Pushpangadan et al., 1998). The maintenance of sacred groves varies from region to region. For instance, in Rajasthan the Orans groves are maintained by Regional Panchayats; Meghalaya sacred groves are managed by certain communities; whereas, in santhals, munda, kharia, and various tribes of Northeastern region of India were protected by the clan-based system. Sacred groves possess vegetation in natural habitat at its peaks or near peak. Sacred groves of India is at risk due to the eradication of conventional values among the local populations, degradation due to the open canopy, overgrazing, assault by exotic weeds, infringement for agriculture and natural disaster. Criteria for conserving sacred groves include: (i) investigation of sacred groves and other lands of all bio-geographical areas and list out the organic, environmental, and socio-cultural values related with each; (ii) educating about the cultural values to the local populations; (iii) conservation and restoration programs in consultation with local people to implement by public; and (iv) to provide incentives to associations who are controlling and managing of sacred groves for posterity (Singh and Kushwaha, 2008).

10.4.2 EX SITU BIODIVERSITY CONSERVATION

In *ex situ* conservation, a sample of organisms, variety, and breeds are collected and preserved as living material of plants in seed and gene banks, botanical gardens and arboreta, captive breeding, animal trans-location, cryopreservation, and zoological parks were maintained under artificial conditions (Engelmann and Engels, 2002). National Bureau of Plant Genetic Resources (NBPGR) for the conservation of crop wild relatives established has India's first national gene preserving bank. The National Facility for Plant Tissue Culture Repository (NFPTCR) aims at seed storage, pollen, and embryos; also *in vitro* culture of plants is cost-effective and possesses scientific value (Chandel, 1995). Botanical gardens are meant for conserving native and wild plants. National Bioresource Development Board (NBDB) launched in 1999 by the Indian Government to evolve conservation by an *ex situ* method for several biological assets through potential cost-effective research.

10.4.2.1   SEED BANK AND GENE BANK

Seed banking is the most suitable practice to preserve plant genetic resources (Kaviani, 2011; Borokini, 2013; Ramya et al., 2015). Usually, seeds were dried to lessen the moisture level and kept at subzero temperatures in cold stores or deep freezers. Thus it helps to preserve seeds for many years (Adebooye and Opabode, 2005). Throughout the globe, the FAO has recorded 90% of 6 million accessions (FAO, 1996; Engelmann and Engels, 2002). Seed genebanks are widely spread throughout the world when compared to the field genebanks. For example, in Ethiopia 60,000 accessions of a seed gene bank is available and approximately 9,000 field genebanks were situated. In Burkina Faso, Niger, and Zambia have numerous seed genebank than field genebank. In Nigeria, seed genebank is used for conserving the orthodox seeds, such as vegetables and cereals. National Centre for Genetic Resources and Biotechnology (NACGRAB) is a newly established seed genebank for the preservation of orthodox seeds. According to the survey on 2010, the NACGRAB has 1,757 accessions of cereals, 585 accessions of legumes and 549 accessions of vegetables. In spite of this, International Institute of Tropical Agriculture (IITA) having 15,122 cowpea accessions, 1,827 groundnut accessions, 1,615 wild relatives of *Vigna*, 1,754 accessions of soybean, nearly 600 accessions of legumes and 876 accessions of maize. Farmers have their own associations for seed preservation, in which the low-tech system helps to preserve farm products and the conserved materials were used for planting in next season. Indigenous methods such as smearing seeds with kerosene, preservation in bottles and sun drying, and such methods for conserving seeds were maintained individually by farmers. Various other methods, tying of maize cobs and fixing it above the cooking fire commonly in the kitchen may drive out pests.

10.4.2.2 CAPTIVE BREEDING

Captive breeding techniques are the process of breeding animals within the controlled environment such as protected areas, biosphere reserves, and various commercial and noncommercial conserved reservoirs. It also includes the releasing of the individual organism to the wild, when there is an adequate natural resource to prop-up the new organism and for recovering endangered species. Captive breeding aims to assist biological diversity and to save species from extinction. For instance, *Gymnogyps californianus*, *Falco punictatus*, *Mustela nigripes*, and *Rallus owstoni* showed divergence in survival and extinction of species in short tenure (Snyder and Snyder, 1989; Derrickson and Snyder, 1992; Jones et al., 1995). Captive Breeding Specialist Group (CBSG) recently introduced Conservation Assessment and Management Plans (CAMPs) for breeding of several taxa. Engebretson (2006) reported that in parrots, the CAMP document shows long-term captive breeding of 330 species globally. The main criteria are the recovery of the organism, i.e., reintroduction to the wild and not for other research and conservation aspects. Although it is having conservation value, it also includes various characteristics and different measures. Captive breeding is employed to recover only for a limited number of endangered species, and it should be implicated only when other feasible alternatives are unavailable (Snyder et al., 1996).

10.4.2.3 ANIMAL TRANSLOCATION

Animal translocation is deliberation of organisms to the wild to establish, reestablish or to increase the population rate (Griffith et al., 1989). Translocation is used to establish the organisms of non-native species and to re-establish native species eradicated by hunting. Translocation has created its own perception for the conservation of biological diversity of rare native species. The main aim includes reinforcing of genetic heterogeneity of diminutive community, minimizing the risk of species loss caused by catastrophes by establishing population; it also increases the recovery rate of species soon after rehabilitating from negative effects of environmental toxicants (Minckley, 1995; Leberg, 1999). There are several types of translocations: (a) translocation by introduction - in this method, the non-endemic species were moved to the place where they did not previously exist, and it includes hunting, fishing, economic development and also biological control agents; (b) translocation by reintroduction - the process in which the introduction of organism into a place where it exists previously before eradication by human activities or by natural calamities; and, (c) translocation by restocking - restocking refers to the liberation of species to place in which it already exists, and it is useful for the small isolated population where it is becoming hazardously inbred or population biogeographical islands.

10.4.2.4   CRYOPRESERVATION

About 100,000 of plant species out of the one-third of the entire world plant species is in the face of the global plant extinction crisis (Benson, 1999). Preserving the plant species is crucial for conventional and contemporary plant breeding technique. Since the 1970s, the native breeds and primitive relatives of cultivated crops were sampled and preserved in *ex situ* gene banks. Approximately 6 million accessions of plant resources were stored in provincial, national, international, and some other gene bank in and around the world (Hitmi et al., 1997). The dried seeds were stored at low temperature, and it is one of the most convincible techniques to conserve plant genetic resources. It has some limitation, wherein it is not used for certain plant crops which do not produce seed (e.g., bananas), non-orthodox seeds, fruits, and various timber yielding and ornamental trees. Maintaining the germplasm only in the field is unsafe as it may cause genetic erosion due to pests, disease, and other natural calamities. Further, to maintain clonal orchards requires a huge number of labor and also the high-cost price for processing, and also *in vitro* method is always dangerous as it may cause contamination, somaclonal variation or some human errors which may lead to germplasm loss. Thus cryopreservation or freeze-preservation is one of the phenomenal tool for long-term conservation of plant genetic resources at the ultra-low temperature ($-196°C$). At this condition, the biochemical, physicochemical reactions are completely detained, and it can be stored for several years as such. In spite of conservation of genetic resources, cryopreservation paved the way for conservation of medicinal and secondary metabolites (Elleuch, 1998); hairy root meristem, suspension culture, phenomics, and transformation-competent culture line (Gordon-Kamm et al., 1990). Recent research revealed that cryotherapy is used to remove viruses from plum, banana, and grape completely (Brison et al., 1997; Helliot et al., 2002; Wang et al., 2003).

10.4.2.5   BOTANICAL GARDEN

Conservation of plant diversity is the major contribution through the world facing changes in the environment which causes an unprecedented loss of biodiversity. Approximately 60,000 species of plants out of 2,87,655 species around the globe are facing a threat of extinction. Oldfield et al. (1998) reported that 7300 tree species are globally threatened. Hence, several threatened tree species were cultivated and preserved in various botanic gardens (Jackson, 2002). Throughout the world, about 1800 botanic gardens are situated in 148 countries altogether preserving almost 4 million plant species belongs to 80,000 species of higher plants. Among the world, India is one of the richest diversified countries having about 140 botanical gardens located in various parts of the country. In 1890 the Botanical Survey of India was started with an objective to carry out the floristic survey basically supported by Ministry of Environment and Forests, which maintains the botanical gardens and vigorously engages in conservational aspects of plants throughout the country. The major role of botanical gardens, arboreta or any other conservation center is documenting the regional, local, and national flora, to assess threats and to know the risk level of species and its population to employ suitable course of action, recovering, and restoring of rare and endangered plant species and also other plant genetic resources, to educate the professionals and communal about the importance of conserving the plant resources (Bazai et al., 2013). It also plays a major role by training in some areas like horticulture, gardening, landscaping, and environmental awareness.

10.4.2.6   ZOOLOGICAL PARKS

Zoos or zoological parks where wild animals are restrained within open areas or partially like the forest, which is flaunted to the public. It was established by environmentalists for conserving biological diversity (Ratledge, 2001; Balcombe, 2009; Melfi, 2012). It is not only meant to observe the animals activities and an entertainment object but also to as study material to various research laboratories, institutions, and to gain information of rare and extinct species (Carrizo et al., 2013; Lacy, 2013). As such, per year more than 450 million people visit the zoological parks which help to enhance the knowledge as well as to improve economic values. Hence from last few decades, zoological parks have made considerable changes in the maintenance of the *ex situ* population of various biologically diversified species (Leus, 2011). Captive breeding in zoological parks helps to put aside the endangered and extinct species by increasing in its number (Melfi, 2012). Throughout India, around 350 animal collection centers are situated in which about 50 million people visit every year. In 1972, the Government of India organized a committee for maintenance of zoological parks and came into existence during 1973. The main aim of the zoological parks is to maintain the acquisition of animals, animal housing, maintenance of animal collections, health care, research, and training, breeding programs for species, education, and outreach activity (Ratledge, 2001).

10.5   CONCLUSION

Global biodiversity loss is increasing at a rapid pace of rate. Biodiversity loss can be measured by the declining number of individual species, a group of species or number of individual organisms of an ecosystem. Ecosystem functioning and dynamics are regulated by biodiversity, but population growth, unplanned developmental activities and modern cultivation have increasingly threatened biodiversity. Both *ex situ* and *in situ* conservations are different conservation strategies which are being used since the ancient times. Except for some conditions *ex situ* conservation are less effective than *in situ* and they are extremely costly. *In situ* measures are the more holistic approach for conservation process than *ex situ*. An important disadvantage of *in situ* conservation is that it requires enough space to protect the full component of biotic diversity and human interference completely barred that locality, but it is too complicated for growing human population demand for space. *Ex situ* conservation methods offer 'insurance policy' against extinction and endangered species, thereby supposed to support the *in situ* conservation. Thus, both conservation strategies are essential and complementary to each other for proper management of biodiversity. For the existence of humans, biodiversity is necessary. No matter people are rich or poor, everyone utilizes the planet and all kind of benefits from biodiversity. Hence it is necessary to work for the conservation of biodiversity so that the future generation could experience the cultural, spiritual, and economic benefits from nature.

KEYWORDS

- **animal translocation**
- **biodiversity**
- **bioresources**
- **biosphere reserve**
- **captive breeding**
- **cryopreservation**
- **ecosystem**
- **environmental pollution**
- **ex situ conservation**
- **gene bank**
- **habitat alteration**
- **in situ conservation**
- **invasive alien species**
- **protected areas**
- **sacred groves**
- **seed bank**
- **spiritual values**

REFERENCES

Adebooye, O. C., & Opabode, J. T., (2005). Status of conservation of the indigenous leaf vegetables and fruits of Africa. *Afr. J. Biotechnol., 3*(12), 700–705.

Alka, G., (2010). *Environmental Geography* (pp. 163–164). Sharda Pustak Bhawan, Allahabad, India.

Andrén, H., & Angelstam, P., (1988). Elevated predation rates as an edge effect in habitat islands: Experimental evidence. *Ecology, 69*, 544–547.

Andrew, O. S., & Marc, M., (2011). Fluctuations of fish populations and the magnifying effects of fishing. *Proc. Natl. Sci., 108*(17), 7075–7080.

Anita, S., & Madhoolika, A., (2008). Acid rain and its ecological consequences. *J. Env. Biol., 29*(1), 15–24.

Anjaneyalu, Y., (2004). *An Introduction to Environmental Sciences* (pp. 285–293). BS Publications, India.

Balcombe, J., (2009). Animal pleasure and its moral significance. *Appl. Anim. Behav. Sci., 118*, 208–216.

Baskin, Y., (2002). *A Plague of Rats and Rubbervines: The Growing Threats of Species Invasions.* Shearwater Books/Island Press, Washington, D. C., USA.

Bazai, Z. A., Tareen, R. B., Achakzai, A. K. K., & Batool, H., (2013). Application of the participatory rural appraisal (PRA) to assess the ethnobotany and forest conservation status of the Zarghoon Juniper ecosystem, Balochistan, Pakistan. *Int. J. Exp. Bot., 82*(1), 69–74.

Bennett, E. L., Nyaoi, A., & Sompud, J., (2000). Saving Borneo's bacon: the sustainability of hunting in Sarawak and Sabah. In: Robinson, J. G., & Bennett, E. L., (eds.), *Hunting for Sustainability* (pp. 305–324). Columbia Press, New York.

Benson, E. E., (1999). Cryopreservation. In: Benson, E. E., (ed.), *Plant Conservation Biotechnology* (pp. 83–95). Taylor, and Francis, London.

Blom, A., (2004). An estimate of the costs of an effective system of protected areas in the Niger Delta–Congo Basin Forest Region. *Biodiv. Conserv., 13*, 2661–2678.

Borokini, T. I., (2013). The state of *ex situ* conservation in Nigeria. *Int. J. Conserv. Sci., 4*(2), 197–212.

Botkin, D. B., Saxe, H., Araujo, M. B., Betts, R., Bradshaw, R. H. W., & Cedhagen, T., (2007). Forecasting the effects of global warming on biodiversity. *Bioscience, 57*, 227–236.

Boyd, J., & Banzhaf, S., (2007). What are ecosystem services? The need for standardized environmental accounting units. *Ecol. Econom., 63*(1), 1–11.

Braithwaitte, R. W., Lonsdale, W. M., & Estbergs, J. A., (1989). Alien vegetation and native biota in tropical Australia: The impact of *Mimisa pigra. Biol. Conserv., 48*, 189–210.

Bright, C., (1999). *Life Out of Bounds: Bio-Invasions in a Borderless World* (pp. 1–288). Earthscan Publications Ltd, London.

Brison, M., De Boucaud, M. T., Pierronnet, A., & Dosba, F., (1997). Effect of cryopreservation on the sanitary state of a prunus rootstock experimentally contaminated with plum pox potyvirus. *Plant Sci., 123*, 189–196.

Campbell, A., Kapos, V., Scharlemann, J. P. W., Bubb, P., Chenery, A., Coad, L., Dickson, B., Doswald, N., Khan, M. S. I., Kershaw, F., & Rashid, M., (2009). *Review of the Literature on the Links Between Biodiversity and Climate Change: Impacts, Adaptation, and Mitigation* (pp. 1–124). Secretariat of the Convention on Biological Diversity, Montreal, Canada.

Carrizo, S. F., Smith, K. G., & Darwall, W. R. T., (2013). Progress towards a global assessment of the status of freshwater fishes (Pisces) for the IUCN Red List: Application to conservation programmes in zoos and aquariums. *Int. Zoo Yearbook, 47*, 46–64.

Carson, R., (1962). *Silent Spring* (pp. 1–368). Houghton Mifflin Company, Boston, MA.

Ceballos, G., Ehrlich, P. R., Soberon, J., Salazar, I., & Fay, J. P., (2005). Global mammal conservation: What must we manage? *Science, 309*, 603–607.

Chanchal, D., & Swarnlata, S., (2009). Herbal photoprotective formulations and their evaluation. *Open Nat. Prod. J., 2*, 71–76.

Chandel, K. P. S., (1995). Biodiversity conservation and use through *in vitro* strategies. In: Rana, R. S., Chandel, K. P. S., Bhat, S. R., Mandai, B. B., Karihaloo, J. L., Bhat, K. V., & Pandey, R., (eds.), *Plant Germplasm Conservation: Biotechnological Approaches* (pp. 23–36). NBPGR Publication, New Delhi.

Chiras, D. D., (2012). *Environmental Science* (pp. 194–213). Jones and Bartlett, Burlington, USA.

Cincotta, R. P., & Engelman, R., (2000). *Nature's Place: Human Population and the Future of Biological Diversity* (pp. 1–81). Population Action International, Washington, DC.

Colquhoun, I. C., (2006). Predation and cathemerality: Comparing the impact of predators on the activity patterns of lemurids and ceboids. *Folia Primatol.*, *77*, 143–165.

Craig, H. T., Caroline, M. P., Janice, S. C., Stuart, H. M. B., Thomasina, E. E. O., & Vineet, K., (2009). State of the world's species. In: Christophe, J. V., Craig, H. T., & Simon, N. S., (eds.), *Wildlife in a Changing World - An Analysis of the 2008 IUCN Red List of Threatened Species* (pp. 15–43). IUCN, Gland, Switzerland.

Dara, S. S. A., (1993). *Textbook of Environmental Chemistry and Pollution Control* (pp. 313–320). S. Chand and Company, New Delhi, India.

David, T., & Clarence, L., (2001). Human-caused environmental change: Impacts on plant diversity and evolution. *Proc. Natl. Acad. Sci. USA*, *98*(10), 5433–5440.

Derrickson, S. R., & Synder, N. F. R., (1992). Potentials and limits of captive breeding in parrot conservation. In: Beisinger, S. R., & Synder, N. F. R., (eds.), *New World Parrots in Crisis: Solutions From Conservation Biology* (pp. 133–166). Smithsonian Institution Press, Washington, DC.

Deshmukh, S., Gogate, M. G., & Gupta, A. K., (1998). Sacred groves and biological diversity: Providing new dimensions to conservation issue. In: Ramakrishnan, P. S., Saxena, K. G., & Chandrashekhara, U. M., (eds.), *Conserving the Sacred for Biodiversity Management* (pp. 397–414). Oxford, and IBH Publishing Co., New Delhi, India.

Dobson, A. P., Bradshaw, A. D., & Baker, A. J. M., (1997). Hopes for the future: Restoration ecology and conservation biology. *Science*, *277*, 515–521.

Doyle, T. J., & Bryan, R. T., (2000). Infectious disease morbidity in the US region bordering Mexico. *J. Infect. Dis.*, *182*(5), 1503.

Ehrlich, P. R., & Wilson, E. O., (1991). Biodiversity studies: Science and policy. *Science*, *253*(5021), 758–762.

Elleuch, H., Gazeau, C., David, H., & David, A., (1998). Cryopreservation does not affect the expression of a foreign Sam gene in transgenic *Papaver somniferum* cells. *Plant Cell Rep.*, *18*, 94–98.

Engebretson, M., (2006). The welfare and suitability of parrots as companion animals: A review. *Anim. Welfare*, *15*, 263–276.

Engelmann, F., & Engels, J. M. M., (2002). Technologies and strategies for *ex situ* conservation. In: Engels, J., Rao, V. R., Brown, A. H. D., & Jackson, M. T., (eds.), *Managing Plant Genetic Diversity* (pp. 89–104). CABI Publishing, New York.

Engels, J. M. M., & Wood, D., (1999). Conservation of agrobiodiversity. In: Wood, D., & Lenné, J. M., (eds.), *Agrobiodiversity: Characterization, Utilization, and Management* (pp. 355–386). CAB International, Wellingford.

Estrada, A., & Coates, E. R., (1996). Tropical rain forest fragmentation and wild populations of primates at Los Tuxtlas, Mexico. *Int. J. Primatol.*, *17*, 759–783.

FAO, (1996). *Report on the State of the World's Plant Genetic Resources for Food and Agriculture* (pp. 1–75). FAO, Rome.

Ferraro, P. J., & Agnes, K., (2002). Direct payment to conserve biodiversity. *Science*, *298*, 1718–1719.

Fisher, B., & Christopher, T., (2007). Poverty and biodiversity: Measuring the overlap of human poverty and the biodiversity hotspots. *Ecol. Econom.*, *62*, 94–96.

Gilman, S. E., Urban, M. C., Tewksbury, J., Gilchrist, G. W., & Holt, R. D., (2010). A framework for community interactions under climate change. *Trends Ecol. Evol.*, *25*, 325–331.

Golovko, O. V., & Izhevsky, P. V., (1996). Studies of the reproductive behavior in Russian and Belorussian populations, under impact of the Chernobyl ionizing irradiation. *Rad. Biol. Radioecol.*, *36*(1), 3–8.

Gordon-Kamm, W. J., Spencer, T. M., Mangano, M. L., Adams, T. R., Daines, R. J., Start, W. G., et al., (1990). Transformation of maize cells and regeneration of fertile transgenic plants. *Plant Cell*, *2*, 603–618.

Griffith, B., Scott, J. M., Carpenter, J. W., & Reed, C., (1989). Translocation as a species conservation tool: status and strategy. *Science*, *245*(4917), 477–480.

Grodzinsky, D. M., (1999). General situation of the radiological consequences of the Chernobyl accident in Ukraine. In: Imanaka, T., (ed.), *Recent Research Activities About the Chernobyl NPP Accident in Belarus, Ukraine, and Russia* (pp. 18–28). Research Reactor Institute, Kyoto University, Japan.

Helliot, B., Panis, B., Poumay, Y., Swennen, R., Lepoivre, P., & Frison, E., (2002). Cryopreservation for the elimination of cucumber mosaic and banana streak viruses from banana (*Musa* spp.). *Plant Cell Rep.*, *20*, 1117–1122.

Hitmi, A., Sallanon, H., & Barthomeuf, C., (1997). Cryopreservation of *Chrysanthemum cinerariaefolium* Vis. cells and its impact on their pyrethrin biosynthesis ability. *Plant Cell Rep.*, *17*, 60–64.

Hoegh-Guldberg, O., Mumby, P. J., Hooten, A. J., Steneck, R. S., Greenfield, P., Gomez, E., et al., (2007). Coral reefs under rapid climate change and ocean acidification. *Science*, *318*, 1737–1742.

Irwin, M. T., Raharison, J. L., & Wright, P. C., (2009). Spatial and temporal variability in predation on rainforest primates: Do forest fragmentation and predation act synergistically? *Anim. Conserv.*, *12*, 220–230.

Irwin, M. T., Wright, P. C., Birkinshaw, C., & Fisher, B., (2010). Patterns of species change in anthropogenically disturbed forests of Madagascar. *Biol. Conserv.*, *143*, 2351–2362.

Jackson, W. P., (2002). Development and adoption of the global strategy for plant conservation by the convention on biological diversity: An NGO's perspective. *BGC News*, *3*(8), 25–32.

James, A., Kevin, J. G., & Andrew, B., (2001). Can we afford to conserve biodiversity? *BioScience*, *51*(1), 43–51.

Jimmy, B. M., Firoz, A., & Pranjit, K. S., (2010). Brahmaputra river islands as potential corridors for dispersing tigers: A case study from Assam, India. *Int. J. Biodiv. Conserv.*, *2*(11), 350–358.

Jones, C. G., Heck, W., Lewis, R. E., Mungroo, Y., Slade, G., & Cade, T., (1995). The restoration of the Mauritius kestrel *Falco punctatus* population. *Ibis*, *137*(1), 173–180.

Karpanty, S. M., (2003). Rates of predation by diurnal raptors on the lemur community of Ranomafana National Park, Madagascar. *Am. J. Phys. Anthropol.*, *36*, 126–127.

Kaviani, B., (2011). Conservation of plant genetic resources by cryopreservation. *Aust. J. Crop. Sci.*, *5*(6), 778–800.

Kiers, E. T., Palmer, T. M., Ives, A. R., Bruno, J. F., & Bronstein, J. L., (2010). Mutualisms in a changing world: An evolutionary perspective. *Ecol. Lett.*, *13*, 1459–1474.

Kohli, R. K., Batish, D. R., & Singh, H. P., (2001). Important tree species. *Forest and Forest Plants*, *2*, 2–12.

Kumar, U., & Asija, M. J., (2007). *Biodiversity Principles and Conservation* (pp. 43–61). Agrobios, Jodhpur, India.

Lacy, R. C., (2013). Achieving true sustainability of zoo populations. *Zoo Biol., 32*, 19–26.

Lafferty, K. D., (2009). The ecology of climate change and infectious diseases. *Ecology, 90*, 888–900.

Laurance, W. F., Camargo, J. L. C., Luizao, R. C. C., Laurance, S. G., Pimm, S. L., Bruna, E. M., et al., (2011). The fate of Amazonian forest fragments: a 32-year investigation. *Biol. Conserv., 144*, 56–67.

Laxman, S. K., Vinod, K. B., Meenakshi, B., & Ashok, K. T., (2014). Conservation and management of sacred groves, myths, and beliefs of tribal communities: A case study from north-India. *Env. Syst. Res., 3*(16), 5–6.

Leadley, P., Pereira, H. M., Alkemade, R., Fernandez-Manjarres, J. F., Proenca, V., Scharlemann, J. P. W., & Walpole, M. J., (2010). *Biodiversity Scenarios: Projections of 21st Century Change in Biodiversity and Associated Ecosystem Services* (pp. 1–132). Secretariat of the Convention on Biological Diversity, Montreal.

Leberg, P. L., (1999). Using genetic markers to assess the success of translocation programs. *Proceedings of the 64th Transnational North American Wildlife and Natural Resources Conference* (pp. 174–188).

Lehman, S. M., Rajaonson, A., & Day, S., (2006a). Edge effects and their influence on lemur density and distribution in southeast Madagascar. *Am. J. Phys. Anthropol., 129*, 232–241.

Lehman, S. M., Rajaonson, A., & Day, S., (2006b). Edge effects on the density of *Cheirogaleus major. Int. J. Primatol., 27*, 1569–1588.

Leus, K., (2011). Captive breeding and conservation. *Zool. Middl. East, 54*(3), 151–158.

Liu, J., Daily, G. C., Ehrlich, P. R., & Luck, G. W., (2003). Effects of household dynamics on resource consumption and biodiversity. *Nature, 421*(6922), 530–533.

Loreau, M., Naeem, S., Inchausti, P., Bengtsson, J., Grime, J. P., Hector, A., et al., (2001). Biodiversity and ecosystem functioning: Current knowledge and future challenges. *Science, 294*, 804–807.

Mackinnon, J., & Mackinnon, K., (1986). *Review of the Protected Areas System in the Indo-Malayan Realm* (pp. 1–284). International Union for the Conservation of Nature and Natural Resources, Switzerland.

Mackinnon, K., & MacKinnon, J., (1991). Habitat protection and reintroduction programmes. In: Gipps, J. H. W., (ed.), *Beyond Captive Breeding - Re-Introducing Endangered Mammals to the Wild* (pp. 173–198). Clarendon Press, London.

Malhi, Y., Roberts, J. T., Betts, R. A., Killeen, T. J., Li, W., & Nobre, C. A., (2008). Climate change, deforestation, and the fate of the Amazon. *Science, 319*, 169–172.

Malhotra, K. C., (1998). Anthropological dimensions of sacred groves in India: An overview. In: Ramakrishnan, P. S., Saxena, K. G., & Chandrashekhara, U. M., (eds.), *Conserving the Sacred for Biodiversity Management* (pp. 423–428). Oxford and IBH Publishing Co., New Delhi, India.

Malhotra, K. C., Stanley, S., Hemam, N. S., & Das, K., (1997). Biodiversity conservation and ethics: Sacred groves and pools. In: Fujiki, N., & Macer, D. R. J., (eds.), *Bioethics in Asia* (pp. 338–345). Eubois Ethics Institute, Japan.

Maxted, N., Ford-Lloyd, B. V., & Hawkes, J. G., (1997). Complementary conservation strategies. In: Maxted, N., Ford-Lloyd, B. V., & Hawkes, J. C., (eds.), *Plant Genetic Resources Conservation* (pp. 15–39). Chapman, and Hall, London.

McCool, S., (2006). Managing for visitor experiences in protected areas: Promising opportunities and fundamental challenges. *Parks*, *16*(2), 3–9.

Melfi, V. A., (2012). *Ex situ* gibbon conservation: Status, management, and birth sex ratios. *Int. Zoo. Yearbook*, *46*(1), 241–251.

Meyers, L. A., & Bull, J. J., (2002). Fighting change with change: Adaptive variation in an uncertain world. *Trends Ecol. Evol.*, *17*, 551–557.

Minckley, W. L., (1995). Translocation as tool for conserving imperiled fishes: Experience in western United States. *Biol. Conserv.*, *72*, 297–309.

Mittermeier, R. A., Myers, N., Robles, G. P., & Mittermeier, C. G., (1999). *Hotspots: Earth's Biologically Richest and Most Threatened Ecosystems* (pp. 1–430). CEMEX/ Agrupación Sierra Madre, Mexico City.

Monisha, J., Tenzin, T., Naresh, A., Blessy, B. M., & Krishnamurthy, N. B., (2014). Toxicity, mechanism, and health effects of some heavy metals. *Interdisciplinary Toxicology*, *7*(2), 60–72.

Musil, C. F., (1993). Effect of invasive Australian acacias on the regeneration growth and nutrient chemistry of South African lowland fynbos. *J. Appl. Ecol.*, *30*, 361–372.

Mutia, T. M., (2009). *Biodiversity Conservation* (pp. 1–6). Geothermal Development Company Limited, Lake Naivasha, Kenya.

Myers, N., (1990). The biodiversity challenge: Expanded hot-spots analysis. *Environmentalist*, *10*(4), 243–256.

Myers, N., (1994). In: Graham-Smith, F., (ed.), *Population - the Complex Reality* (pp. 117–135). Royal Society, London.

Noss, R. F., (1990). Indicators for monitoring biodiversity: A hierarchical approach. *Conserv. Biol.*, *4*, 355–364.

Oldfield, S., Lusty, C., & MacKinven, A., (1998). *The World List of Threatened Trees*. World Conservation Press, Cambridge.

Onderdonk, D. A., & Chapman, C. A., (2000). Coping with forest fragmentation: The primates of Kibale National Park, Uganda. *Int. J. Primatol.*, *21*, 587–611.

Parmesan, C., (2006). Ecological and evolutionary responses to recent climate change. *Ecol. Evol.*, *37*, 637–669.

Peres, C. A., Gardner, T. A., Barlow, J., & Zuanon, J., (2010). Biodiversity conservation in human modified Amazonian forest landscapes. *Biol. Cons.*, *143*, 2314–2327.

Pfaller, M. A., & Diekema, D. J., (2007). Epidemiology of invasive candidiasis: A persistent public health problem. *Clinical Microbiology Reviews*, *20*(1), 133.

Pimental, D., Zuniga, R., & Morrison, D., (2005). Update on the environmental and economic costs associated with alien-invasive species in the United States. *Ecological Economics*, *58*(3), 273–288.

Prasad, S. V., & Chiranjib, B., (1995). Estimation of phenolic acids in cinnamon, clove, cardamom, nutmeg, and mace by high-performance liquid chromatography. *J. Spice Aroma Crop*, *4*(2), 129–134.

Pringle, R. M., (2005). The origins of the Nile Perch in Lake Victoria. *Biosciences*, *55*(9), 780.

Pushpangadan, P., Rajendraprasad, M., & Krishnan, P. N., (1998). Sacred groves of Kerala- a synthesis on the state of art of knowledge. In: Ramakrishnan, P. S., Saxena, K. G., & Chandrashekhara, U. M., (eds.), *Conserving the Sacred for Biodiversity Management* (pp. 193–210). Oxford and IBH Publishing Co., New Delhi, India.

Rafferty, N. E., & Ives, A. R., (2010). Effects of experimental shifts in flowering phenology on plant-pollinator interactions. *Ecol. Lett.*, *14*, 69–74.

Ramakrishnan, P. S., (1998). Conserving the sacred for biodiversity: The conceptual framework. In: Ramakrishnan, P. S., Saxena, K. G., & Chandrashekhara, U. M., (eds.), *Conserving the Sacred for Biodiversity Management* (pp. 3–16). Oxford, and IBH Publishing Co., New Delhi, India.

Ramya, A. R., Rajesh, V., Jyothi, P., & Swathi, G., (2015). A review on *in situ* and *ex situ* conservation strategies for crop germplasm. *Int. J. App. Biol. Pharma. Tech.*, *5*(1), 267–273.

Ratledge, C., (2001). Towards conceptual framework for wildlife tourism. *Tourism Management*, *22*, 31–40.

Richard, P. C., & Gorenflo, L. J., (2011). Influences of human population on biological diversity. In: Cincotta, R. P., & Gorenflo, L. J., (eds.), *Human Population its Influence on Biological Diversity* (pp. 1–8). Springer, New York.

Ries, L., Fletcher, R. J., Battin, J., & Sisk, T. D., (2004). Ecological responses to habitat edges: Mechanisms, models, and variability explained. *Annu. Rev. Ecol. Evol. Syst.*, *35*, 491–522.

Sagan, F., Sarah, B. P., & Tony, L. G., (2015). Drivers of bushmeat hunting and perceptions of zoonoses in Nigerian hunting communities. *PLoS Negl. Trop. Dis.*, *9*(5), doi: 10.1371/journal.pntd.

Sahoo, P. K., & Das, L. M., (2009). Combustion analysis of Jatropha, Karanja, and Polanga based biodiesel as fuel in a diesel engine. *Fuel*, *88*(6), 994–995.

Sala, O. E., Detlef van Vuuren, D., Pereira, H. M., Lodge, D., Alder, J., Cumming, G., Dobson, A., Wolters, V., & Xenopoulos, M. A., (2005). Biodiversity across scenarios. In: *Ecosystems and Human Well-Being Scenarios* (Vol. 2, pp. 375–408). Island Press, Washington, DC.

Saravanan, S., & Karthi, S., (2014). HPLC analysis for methanolic extract of *Catharanthus roseus* under elevated CO_2. *J. Pharm. Pharm. Sci.*, *3*(10), 683–693.

Serap, U., & Ibrahim, O., (2004). The sturgeons: Fragile species need conservation. *Turk J. Fish Aquat. Sc.*, *4*, 49–57.

Shinoj, P., Kumar, B. G., Sathiadas, R., Datta, K. K., Menon, M., & Singh, S. K., (2008). Spatial price integration and price transmission among major fish markets in India. *Agricultural Economics Research Review*, *21*, 327–335.

Singh, B., (2012). Incidence of the pink noctuid stem borer, *Sesamia inferens* (Walker), on wheat under two tillage conditions and three sowing dates in north-western plains of India. *J. Entmol.*, doi: 10.3923/je.2012.

Singh, J. S., & Kushwaha, S. P. S., (2008). Forest biodiversity and its conservation in India. *Int. Forest. Rev.*, *10*(2), 292–304.

Sivasakthivel, T., & Siva, K. R. K. K., (2011). Ozone layer depletion and its effects: A review. *International Journal of Environmental Science and Development*, *2*(1), 30–37.

Snyder, N. F. R., Scott, R. D., Steven, R. B., James, W. W., Thomas, B. S., William, D. T., & Brian, M., (1996). Limitations of captive breeding in endangered species recovery. *Conserv. Biol.*, *10*(2), 338–348.

Sodhi, N., Koh, L. P., Clements, R., & Wanger, T., (2010). Conserving Southeast Asian forest biodiversity in human-modified landscapes. *Biol. Cons.*, *143*, 2375–2384.

Sussman, R. W., Green, G. M., & Sussman, L. K., (1996). The use of satellite imagery and anthropology to assess the causes of deforestation in Madagascar. In: Sponsel, L. E.,

Headland, T. N., & Bailey, R. C., (eds.), *Tropical Deforestation: The Human Dimension* (pp. 296–315). Columbia University Press, New York.

Synder, N. R. F., & Synder, H. A., (1989). Biology and conservation of the *California condor. Current Ornithology, 6*, 175–263.

Vitousek, P. M., D'Antonio, C. M., Loope, L. L., & Westborrkes, R., (1996). Biological invasions as global environmental change. *American Scientist, 4*, 468–478.

Walther, G. R., (2010). Community and ecosystem responses to recent climate change. *Philos. Trans. R. Soc. B-Biol. Sci., 365*, 2019–2024.

Wang, Q. C., Mawassi, M., Li, P., Gafny, R., Sela, I., & Tanne, E., (2003). Elimination of grapevine virus A (GVA) by cryopreservation of *in vitro*-grown shoot tips of *Vitis vinifera, L., Plant Sci., 165*, 321–327.

Wilcove, D. S., (1985). Nest predation in forest tracts and the decline of migratory songbirds. *Ecology, 66*, 1211–1214.

Wilcove, D. S., McLellan, C. H., & Dobson, A. P., (1986). Habitat fragmentation in the temperate zone. In: Soulé, M. E., (ed.), *Conservation Biology: The Science of Scarcity and Diversity* (pp. 237–256). Sinauer Associates, Sunderland, MA.

Wilson, E. O., (1989). Threats to biodiversity. *Sci. Am., 261*, 108–116.

Yang, L. H., & Rudolf, V. H. W., (2010). Phenology, ontogeny, and the effects of climate change on the timing of species interactions. *Ecol. Lett., 13*, 1–10.

Zedler, P. H., Gautier, C. R., & McMaster, G. S., (1983). Vegetation change in response to extreme events: The effect of a short interval between fires in *California chaparral* and coastal scrub. *Ecology, 64*, 809–818.

CHAPTER 11

UNDERSTANDING THE PROCESS OF EVOLUTION AND THE FUTURE OF BIODIVERSITY UNDER A CHANGING CLIMATE WITH SPECIAL REFERENCE TO INFECTIOUS DISEASES

TAPAN KUMAR BARIK and JAYA KISHOR SETH

Post Graduate Department of Zoology, Berhampur University, Berhampur 760007, Odisha, India

11.1 INTRODUCTION

Biodiversity encompasses fluctuation in the living beings that surviving in all different kinds of the ecosystem and ecological complex to which they belong (Ali, 1999). There may be around 100 million species on earth, and about 1.75 million species have been identified so far. However, it is very much difficult to describe the sustenance of all these different kinds of diversified species on the globe. All species tend to increase their fitness to survive, but the world has no space to contain all these organisms. The most successful species, recognized by the large size of their populations, pose the premises for their own collapse, because their use of resources limits the turn-over of the resources themselves, hence, famine, war, and disease (Boero, 2010). Charles Darwin's Origin of Species suggested that the struggle for existence was the chief driving force for the process of evolution of new species and the survival of the strongest indicated as evidence of adaptation by individuals to the environment. The rapid growth of human population and urbanization in the last ten decades cause a disturbance in terms of habitat loss, fragmentation, overexploitation of natural resources, pollution, the introduction of exotic species and change in climatic condition cause an evolutionary imbalance between the surroundings to which the species are adapted and the condition now they survives. The aftermath of which, lead to biodiversity loss. As indicated by the Millennium Ecosystem Assessment (2005), biodiversity supports at regulating diseases, with disease regulation as one of the services provided by ecosystems (Walpole et al., 2009). On the other side, new infectious diseases have emerged out at a rapid pace. Therefore, by studying the biodiversity in relation to their evolution and ecology, we can comprehend the importance of past theories and give a more extensive comprehension of life existence on Earth.

11.2   BIODIVERSITY: UNDERSTANDING ITS VALUE AND LOSS

As we all know biodiversity is a result of evolution and continuously it is trying to adapt new circumstances and evolving. Hence, biodiversity ought not to be regarded as a static phenomenon rather to be exist as a dynamic process. On the opposite side, competition, and mutation lead to preserve and adapt species and not their diversification. Biodiversity provides nine major benefits to humans: utilitarian (direct economic importance of products of nature and their services), scientific (biological knowledge), aesthetic (motivation from nature's beauty), humanistic (sentiments profoundly established in our intrinsic connection to different species), dominionistic (physical and mental prosperity advanced by a few sorts of associations with nature), moralistic (with motivational elevating), naturalistic (interest driven fulfillment from the living scene), symbolic (nature encouraged imagination, correspondence, and thought), and negativistic (fears and tensions about nature, which can really enhance individuals' beneficial experience) (Kellert, 2005). Biodiversity loss may cause weak and fragile ecosystem. According to Wilson (1993), Earth at present losing 0.25% of its residual species every year (approximately 12,000 species may be in the verge of extinction annually). This biodiversity loss will shrink the available natural resources. The current disappearance of species not only affects the megafauna but also affects the microorganism in general. For example, around 10–15% of parasitic helminths like Trematoda, Cestoda, Acanthocephala, and Nematoda are now at risk of extinction because of their dependency activity only one on threatened or endangered species of vertebrate host. Further, the diversity of parasites does not make a chart of host diversity and about three-quarters of entire links in food webs involve a parasitic species. Thus, the extinction process has a more unexpected consequence for ecosystem operation (Avise et al., 2008).

11.3   EVOLUTION: INTERPLAY BETWEEN SPECIES ORIGIN AND ENVIRONMENTAL CHANGE

Evolution, as we all know, is the way how a population will change in due course of time, and the speciation is the way of origin of new species from the existing one. Speciation can be sympatric or allopatric. The organism has to change according to their changed environment. Environmental change can be slow like the formation of mountain or desert or can be sudden like the floods, earthquakes, and volcanoes. Environmental dynamics force a particular organism to change in order to adapt; which ultimately causing it to evolve into a new species. Further, if it fails, it may also lead to the extinction of that particular species.

The great extinctions of the species during the past, like in the Permian-Triassic and the Cretaceous-Tertiary, were all caused by climatic and physical-chemical changes (Courtillot and Gaudemer, 1996; Purvis and Hector, 2000) and not by competition for space and source between the species.

Anthropogenic activities have fastened the environmental changes that have at least three major repercussions on the process of speciation (Myers and Knoll, 2001): (1) *Outbursts of speciation:* As large numbers of niches are vacated, there could be explosive adaptive radiations within certain taxa - notably small mammals, insects, and terrestrial plants - capable to increase their population in human-dominated landscapes. (2) *Reduced speciation rates:* Biogeography theory suggests that the rate of speciation correlate with the area (Rosenzweig, 1995; Losos, 1996; Losos and Schluter, 2000; Rosenzweig, 2001). Therefore, even the largest protected areas and nature reserves may prove far very small to help the speciation of large vertebrates. Even for smaller species, fragmentation of habitat may severely curb the rate of speciation. (3) *Depletion of evolutionary powerhouses:* The unrelenting depletion and destruction of tropical biomes that have served in the past as pre-eminent powerhouses of evolution and speciation (Jablonski, 1993) could entail grave consequences for the long-lasting resurgence of the biosphere. Dieckmann et al. (1995) considered a case of prey-predator coevolution where predator's extinction occurs due to the prey's adaptation. The phenotype of a predator needs to remain adequately near to that of its prey for the predator's harvesting efficiency to remain high enough to ensure predator survival. This may reflect the necessity for a match between the prey size and the dimensions of the feeding apparatus of the predator. Thus, whenever evolutionary process in the prey takes its phenotype too far different from the predator's matching phenotype, harvesting efficiency drops below a critical level, and so causes the predator to become extinct (Dieckmann and Ferrière, 2004).

Our understanding on the role of genes and their mutations responsible for evolutionary changes in phenotype is increasing at an accelerating pace. It has been reported that, more than 400 genes and mutations responsible for domesticated, intraspecific changes in morphology, physiology or behavior. After a relatively long period of time, it is conceivable that some integration of genome parts coming from successively evolved cell non-living parasites (virus and bacteriophages). These latter are just a mixture of genes moving among cells and using them to replicate. Then the genetic sequences of the initial species transcribe, giving rise to the origin of some organisms that are slightly or hugely different from their ancestors. This highlights to understand how the organism with changed genetic composition will adopt to a changed environment.

11.4   EXTINCTION: FAILURE OF ADAPTATION

Usually, extinction of organism occurs when environmental conditions change rapidly or drastically enough that a species cannot be genetically able to adapt to the change. The crucial point is that the evolutionary process constantly produces novelties which eventually go extinct, more or less rapidly in more cases. This process is healthy if the novelties are potentially more numerous as compared to the extinctions. The problem thus is not that same species are going extinct, which could be corrected by saving some of them, but that the process itself is not working properly anymore. Each and every species have three choices to avoid extinction: (i) react plastically, (ii) develop adaptively in a particular place (Gienapp et al., 2007; Williams et al., 2008) and (iii) migration to more suitable habitats (Smith et al., 2014). Species extinction may not be a bad thing because the smallpox, for example, is thought to be extinct in the wild and poliovirus is now only restricted to a small patch of regions. Rinderpest, the most serious cattle disease was almost wiped out in the year 2000 making it the first non-human animal virus to be eradicated. It has been proposed that the planned 30 mosquito vector species extinction would be justified both morally and economically.

11.5   CLIMATE: IN THE CHANGING SCENARIO

Climate change can be defined as a long-term change of the weather condition, including variations in the average weather or in the dissemination of weather conditions around the average (i.e., excessive weather events) (Wu et al., 2016). Mid–1800s onward, the global temperature increased of about 0.6°C, influencing the whole world (McCarthy, 2001), which is significantly affecting the sea level, the overall volume of the glacier and also on the total volume of water. The biodiversity on the earth needed to manage a changing climate and need to adjust to increased temperature and precipitation, which has been a noteworthy impact on evolutionary changes that created the plant and animal species. Variation in the climatic condition is perfectly compatible with the existence of the ecosystems and their function on which the life exist. There are various reasons why living organisms are less able to adapt to the present global warming. One is the extremely quick pace of progress: it is expected that over the next century, increase in average world temperatures will be rapid than anything practiced by the globe for at least 10,000 years. Most of the species will be unable to adapt the changed climatic condition rapidly or to shift to regions more favorable for their survival (CBD, 2007). Humans made a massive change to the landscape, river basins and ocean of the world that have shut off survival options to species available previously under pressure due to climate change. There are some other human-induced factors as well. Pollution derived from nutrients such as nitrogen, the addition of alien invasive species and excess harvesting of wild animals through hunting or fishing can minimize the elasticity of ecosystems, and thus the likelihood that they will adjust naturally to the changed climatic condition (CBD, 2007). Various mechanism has been projected for how high range of temperature might give rise to species diversity by impacting the rate of speciation, species' physiological tolerance and rate of extinction (Dowle et al., 2013).

11.6   INFECTIOUS DISEASES: AN ALARMING PUBLIC HEALTH ISSUE

Infectious agents obtain their essential nutrients and energy by parasitization of higher life forms. Most of such kinds of infections are benign, and some are even useful to both host and microorganism. Only a minority of infections that adversely affect the biology of the host are called as infectious disease. Infectious agents differ in their size, type, and method of infection. There are some unicellular viruses, bacteria, protozoa, and multicellular parasites. Those microbial organisms that cause "anthroponoses" have adapted, by means of evolution, to human beings as their essential or primary, particularly exclusive host. Usually, most pathogens have a very short lifetime with large population sizes that favor and act as selective agent in natural populations to survive because they can be able to spread rapidly and cause significant negative impacts on host fitness. Parasites account for more than half of all living species. Then, no free-living species are free from infection. However, at an interspecific level, a great difference is found among hosts, with very few host species that harbor high parasitic species diversity and several other harbors with less parasitic species diversity (Morand, 2011). Drastic vast scale environmental change due to anthropogenic activities could directly influence the life cycle and transmission of pathogen giving rise to an evolutionary shift in other parasitic traits (Table 11.1). Further, investigation on various types of pathogens that cause infection particularly to the endangered host populations along with broader genomic surveys that inspect a wide range of immune genes in living organisms are needed to give further light on the benefits and feasibility of maintenance of resistance traits in captive breeding programs. Several studies reveal that pathogen can cause a decline in the healthy populations and can cause many threats to declining species, but host extinction due to a pathogen is very rare (Lafferty and Gerber, 2002; Anderson et al., 2004). Examples include a decline in populations of African apes population due to Ebola virus, amphibian due to chytridiomycosis and several Hawaiian forest birds due to avian malaria.

TABLE 11.1 Examples of Environmental Changes and Possible Effects of Infectious Diseases (Wilson, 2001)

Environmental changes	Example diseases	Pathway of effect
Dams, canals, irrigation	Schistosomiasis	↑Snail host habitat, human contact
	Malaria	↑Breeding sites for mosquitoes
	Helminthiasies	↑Larval contact due to moist soil
	River blindness	↓Blackfly breeding, ↓disease
Agricultural intensification	Malaria	Crop insecticides and↑ vector resistance
	Venezuelan haemorraghic fever	↑Rodent abundance, contact
Urbanization, urban crowding	Cholera	↓ Sanitation, hygiene, ↑water contamination
	Dengue	Water collecting trash, ↑*Aedes aegypti* mosquito breeding sites
	Cutaneous leishmaniasis	↑Proximity, sandfly vectors
Deforestation and a new habitation	Malaria	↑Breeding sites and vectors, immigration of susceptible people
	Oropouche	↑Contact, breeding of vectors
	Visceral leishmaniasis	↑Contact with sandfly vectors
Reforestation	Lyme disease	↑Tick hosts, outdoor exposure
Ocean warming	Red ride	↑Toxic algal blooms
Elevated precipitation	Rift valley fever	↑Pools for mosquito breeding
	Hantavirus pulmonary syndrome	↑Rodent food, habitat, abundance

11.7   CLIMATE CHANGE: ITS IMPACT ON INFECTIOUS DISEASES

Effect of climatic variation on infectious diseases is resolved to a great extent by the transmission cycle of each and every pathogen (Table 11.1). Transmission cycle of each and every pathogen require a vector or non-human host that are more prone to the influences of the external environment than those diseases which include only the pathogen and human. Most significant environmental factors are temperature, precipitation, and humidity. Various probable transmission components are a pathogen (virus and bacteria), vector (mosquito and snail), human host, the non-human reservoir (mice and deer) and non-biological physical vehicle (water and soil). Important ecological variables integrate temperature, moistness, and precipitation. Transmission cycle of such diseases basically includes two elements such as pathogen and human host. Usually, these diseases are not much influenced by climatic factors since these agents spend less or no time outside the human host. However, these diseases are highly vulnerable to change in human behavior like crowding and inadequate sanitation that may cause due to altered land use occurred by climatic changes. Malaria disease is highly sensitive to long-term change in climatic condition and also varies seasonally in a certain highly endemic region.

The most important climatic factors for most vector-borne diseases are temperature and precipitation; however, the rise in sea level, wind action and duration of daylight are some additional important considerations. A small increase in temperature may affect the pathogen. Increase in temperature may cause an increase in the developmental period, incubation, and replication of the pathogen (Bradley, 1993; Lindsay and Birley, 1996). The temperature may alter the disease-carrying vectors by modifying their biting rates, as well as vector population dynamics and change the rate at which they come into contact with humans. At last, a change in temperature regime can change the duration of the transmission season (Gubler et al., 2001). If global temperatures increase by 2–3°C, it is estimated that the population at risk for malaria will increase by 3–5%, which means that millions of additional individuals would likely to suffer from malaria each year (Shuman, 2011). Hales et al. (2002) assessed that 5–6 billion individuals (around half of the populace) would be in danger for dengue fever by 2085, utilizing projections in light of the normal impact of environmental change on humidity. Raghavendra et al. (2010) reported that thermal stress also play a vital role in the development of adaptive cross-tolerance to other forms of stress such as insecticide treatment in mosquito vector species and also oocyst rate was more in adult mosquitoes that derived from larvae pre-adapted at 40°C in the laboratory condition. The expected increase in sea level in association with climate change is likely to decrease or remove breeding sites for salt-marsh mosquitoes. Bird and mammalian hosts of the same ecological niche may be threatened by extinction that would also help the elimination of viruses endemic to this particular habitat (Reeves et al., 1994). On the other hand, inland invasion of salt water may transform former freshwater habitats into salt-march territories which could give support for the vector and host species displaced from previous salt-marsh habitats (Reeves et al., 1994). Similarly, expanding sea surface temperature can directly affect the viability of enteric pathogens such as *Vibrio cholera* by increasing the food supply of the reservoir (Colwell, 1996). Climate change will alter the relationship between pathogens and hosts directly by changing the development and life histories timings of the pathogen, changing seasonal patterns of pathogen survival and altering susceptibility to pathogens (Gubler et al., 2001; Harvell et al., 2002) (Table 11.2). Change of the climatic condition may cause a change in the level of precipitation that affects the distribution of waterborne pathogens. Similarly, change of the climate may also cause variations in range, period, and intensity of infectious diseases through its impacts on disease vectors (Wu et al., 2016) (Figure 11.1). The pattern of precipitation can affect the transport and distribution of infectious agents, but temperature can influence their growth and survival (Rose et al., 2001). Rainfall also plays a vital role in the growth and development of water-borne disease pathogens. Rainy season is related to the increase of fecal pathogens as heavy rain may stir up sediments in the water, giving rise to the gathering of fecal microorganisms (Jofre et al., 2010). Unusual rainfall fall after a long drought can cause an increase of the pathogen population leading to a disease outbreak (Wilby et al., 2005). Droughts or low rainfall lead to Low River flows, resulting in the concentration of water-borne effluent pathogens (Semenza and Menne, 2009; Hofstra, 2011). Humidity change also impacts the pathogens of infectious diseases (Wu et al., 2016). The pathogens of air-borne infectious disease such as influenza tend to be responsive to humidity condition. Humidity change also affects the viruses of water-borne diseases. The virus of vector-borne diseases may also affected by the change in the humidity (Wu et al., 2016).

Sunshine is one more important climate variable that may affect the pathogens of infectious diseases. For instance, sunshine hours and temperature accelerate at the time of cholera periods to develop a suitable condition for the proliferation of *Vibrio cholera* in aquatic environments (Islam et al., 2009). Wind is a key factor affecting the pathogens of air-borne diseases. Literature indicates the positive correlation between dust particle association and virus survival (Chung and Sobsey, 1993; Chen et al., 2010). According to the Institute of Medicine, the influences are complex and inter-related, "climate interacts with a range of factors that shape the course of infectious disease emergence, including the host, vector, and pathogen population dynamics; land use, trade, and transportation; human and animal migration," and these connections confuse the attribution of impacts (EASAC, 2010). Fundamental influences of climatic change on the infectious disease can already be able to perceive, and it is likely that new vectors and pathogens will develop and become established within coming few years. Further, expansion of pathogen to a new locality and their interaction with new varieties of hosts may offer new evolutionary opportunities and give rise to the appearance of pathogens with unique virulence (Pallen and Wren, 2007).

TABLE 11.2 Biodiversity Loss and Increase Disease Transmission (Keesing et al., 2010)

Diseases	Mechanism	References
Bacteriophage of *Pseudomonas syringae*	Host/vector/parasite behavior	Dennehy et al. (2007)
Coral diseases	Host/vector abundance	Raymundo et al. (2009)
Fungal disease of *Daphnia*	Host/vector/parasite behavior	Hall et al. (2009)
Hantavirus disease	Host/vector abundance, host/vector/parasite behavior	Tersago et al. (2008); Clay et al. (2009); Dizney and Ruedas (2009); Suzan et al. (2009)
Helminthic parasite of fish	Host/vector abundance	Kelly et al. (2009)
Lyme disease	Host/vector abundance, host/vector/parasite behavior	Brunner and Ostfeld (2008); LoGiudice et al. (2008); Keesing et al. (2009)
Malaria	Host/vector abundance	Carlson et al. (2009)
Schistosomiasis	Host/vector/parasite behavior	Johnson et al. (2009)

TABLE 11.2 *(Continued)*

Diseases	Mechanism	References
Trematode diseases of snails and birds	Host/vector/parasite behavior	Kopp and Jokela (2007); Thieltges et al. (2008, 2009)
West Nile fever	Host/vector abundance, host/vector/parasite behavior	Ezenwa et al. (2006); Swaddle and Calos (2008); Allan et al. (2009); Koenig et al. (2010)

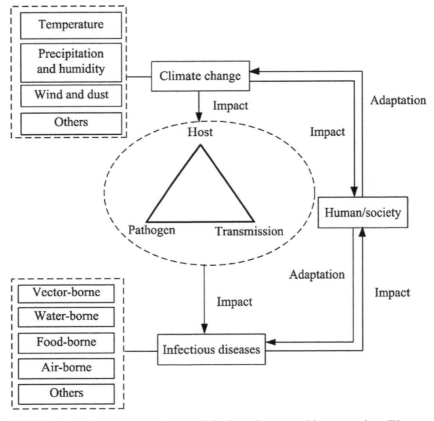

FIGURE 11.1 Climate change, human infectious diseases and human society (Wu et al., 2016).

11.8   FUTURE PERSPECTIVES

Evolution gives the principal context and structure for comparative analyses of biodiversity, and its implications and utilization of overall levels of biological organization. Further, genomic studies and phylogenetic analysis will help us to understand the process of evolution and the future trend of biodiversity. Extracting genome-wide information from the vast amount of sequence data enables one to follow a genome-scanning approach, where signs of collection and selection can be investigated in the genome, candidate genes that play a role in adaptation to changing environments can be detected and screening of variation in gene sequence and activity can be promising at the population level. Genome analyses help us to survey a large amount of hereditary loci underlying phenotypic traits. But paradoxically in such extensive whole-genome scan approaches, the pitfall of reductionism is lurking, simplifying the processes underlying the complexity of gene expression regulation and determining evolutionary potential.

Phylogeographical studies demonstrated that genetic diversity is often unevenly distributed over a species' geographical distribution. Conservation priorities, however, are usually based on ecosystem and/or species diversity. Given that intraspecific genetic diversity is most relevant for evolution, conservation strategies should also include a measure of such genetic diversity. Most common rationales for not including evolutionary approaches in management are that: (a) evolution acts very slowly as compared to ecological threats, (b) management of evolution is beyond our ability and, (c) manipulating evolution runs oppose to the integrity of natural phenomena (Smith and Bernatchez, 2008).

Advanced technology like next-generation sequencing help to give more accurate genetic diversity information and ultimately might be used to standardize the evolutionary rescue of a declining population (Smith et al., 2014). Palaeontological studies reveal that species have large geographical ranges and with high dispersal activity are best at surviving at large scale environmental setup and mass extinctions (Stigall and Lieberman, 2006). Fortunately, new technology is helping to elaborate the rate of biodiversity discovery ranging from rapid biodiversity assessments of particular critical areas (www.biosurvey.conservation.org) to metagenomic surveys of microbial communities (Venter et al., 2004; Lambais et al., 2006). Molecular analyses are not appropriate to examine the rates of extinction because extinct taxa are not available for sampling. Molecular phylogenetic studies indicate indirect ancestral relationships among extant taxa and do not explain how extinction affect observed branch lengths. Varying degree of rates of speciation and extinction can leave same patterns of molecular phylogeny (Rabosky and Lovette, 2008; Crisp and Cook, 2009), so techniques like lineage through time plots may be confusing. Therefore, it is difficult to unravel the relative involvement to the diversity of speciation and extinction without a good quality fossil record (Quental and Marshall, 2010). In response to environmental change, if natural selection is changing species and ecosystem then what should be our approach for the management and conservation of biodiversity and how vital in the long term is the extinction of species?

11.9   CONCLUSION

We cannot immediately stop climate change, pollution, and invasive species, so we should find out how to make the best of a bad situation, and try to find out how evolutionary processes have a role in this. If humankind does not stop its "unnatural" competitive spirit in the huge elimination of species, it will take a long time before the various set of life forms that we now call biodiversity will be regenerated. Ehrlich and Pringle (2008) reported that "the fate of biological diversity for the next ten million years will be determined during the next 50–100 years by the activities of a single species (*Homo sapiens*)." With the anticipated increase by mid-century of 2.6 billion individuals to an already overloaded planet, the projection for preserving significant biodiversity are dim, unless societal mindsets and comportments vary dramatically and quickly (Avise et al., 2008). Humankind has had a tendency to respond to emerging diseases as they occur, using our understanding of epidemiology in an attempt to diminish the damage done. Neither we can stop climate change, nor we can stop these emerging diseases.

At last, formally linking phylogenetic data to explicit geographical information can also be made. For instance, linking phylogenetic trees to geographic maps available in the Google Earth can help to distinguish the origin and pathway of spread of invasive species and diseases. Current examples include avian influenza (Janies et al., 2007) and SARS (Janies et al., 2008; www.supramap.osu.edu/cov/janiesetal2008covsars.kmz). In conclusion, the scientific personnel should help policymakers to identify that climate change could have considerable influence on human and animal infectious diseases as well. There is an urgent need to strengthen the evidence base and to realize that most of the adaptation that is needed to respond to the influence of climate change is basic preventive public health.

KEYWORDS

- adaptation
- biodiversity
- climate change
- ecology
- ecosystem
- evolution
- extinction
- global warming
- infectious diseases
- interplay
- public health
- species origin

REFERENCES

Ali, S. S., (1999). *Fresh Water Fisheries Biology*. Naseem Book Depot, Hyderabad, India.

Allan, B. F., Langerhans, R. B., Ryberg, W. A., Landesman, W. J., Griffin, N. W., Katz, R. S., et al., (2009). Ecological correlates of risk and incidence of West Nile virus in the United States. *Oecologia, 155*, 699–708.

Anderson, P. K., Cunningham, A. A., Patel, N. G., Morales, F. J., Epstein, P. R., & Daszak, P., (2004). Emerging infectious diseases of plants: Pathogen pollution, climate change, and agrotechnology drivers. *Trends Ecol. Evol., 19*, 535–544.

Avise, J. C., Hubbell, S. P., & Ayala, F. J., (2008). In the light of evolution II: Biodiversity and extinction. *Proc. Natl. Acad. Sci. USA, 105*(1), 11453–11457.

Boero, F., (2010). Changing organisms in changing anthropogenic landscapes, evolution, and biodiversity: The evolutionary basis of biodiversity and its potential for adaptation to global change. *Report of an Electronic Conference, 37*.

Bradley, D. J., (1993). Human tropical diseases in a changing environment. *Ciba Foundation Symposium, 175*, 146–170.

Brunner, J., & Ostfeld, R. S., (2008). Multiple causes of variable tick burdens on small mammal hosts. *Ecology, 89*, 2259–2272.

Carlson, J. C., Dyer, L. A., Omlin, F. X., & Beier, J. C., (2009). Diversity cascades and malaria vectors. *Journal of Medical Entomology, 46*, 460–464.

CBD, (2007). *Biodiversity and Climate Change*, https: //www.cbd.int/doc/bioday/2007/ ibd–2007-booklet–01-en.pdf.

Chen, P. S., Tsai, F. T., Lin, C. K., Yang, C. Y., Chan, C. C., Young, C. Y., & Lee, C. H., (2010). Ambient influenza and avian influenza virus during dust storm days and background days. *Environ Health Perspect, 118*, 1211–1216.

Chung, H., & Sobsey, M. D., (1993). Comparative survival of indicator viruses and enteric viruses in seawater and sediment. *Water Sci. Technol., 27*, 425–428.

Clay, C., Lehmer, E. M., St Jeor, S., & Dearing, M. D., (2009). Sin Nombre virus and rodent species diversity: A test of the dilution and amplification hypotheses. *PLoS One, 4*, e6467.

Colwell, R. R., (1996). Global climate and infectious disease: The cholera paradigm. *Science, 274*(5295), 2025–2031.

Courtillot, V., & Gaudemer, Y., (1996). Effects of mass extinctions on biodiversity. *Nature, 381*, 146–148.

Crisp, M. D., & Cook, L. G., (2009). Explosive radiation or cryptic mass extinction? Interpreting signatures in molecular phylogenies. *Evolution, 63*, 2257–2265.

Dennehy, J. J., Friedenberg, N. A., Yang, Y. W., & Turner, P. E., (2007). Virus population extinction via ecological traps. *Ecology Letters, 10*, 230–240.

Dieckmann, U., & Ferrière, R., (2004). Adaptive dynamics and evolving biodiversity. In: Ferrière, R., Dieckmann, U., & Couvet, D., (eds.), *Evolutionary Conservation Biology* (pp. 188–224). Cambridge University Press, Cambridge.

Dieckmann, U., Marrow, P., & Law, R., (1995). Evolutionary cycling of predator-prey interactions: Population dynamics and the Red Queen. *Journal of Theoretical Biology, 176*, 91–102.

Dizney, L. J., & Ruedas, L. A., (2009). Increased host species diversity and decreased prevalence of Sin Nombre virus. *Emerging Infectious Diseases, 15*, 1012–1018.

Dowle, E. J., Morgan-Richards, M., & Trewick, S. A., (2013). Molecular evolution and the latitudinal biodiversity gradient. *Heredity, 110*, 501–510.

EASAC, (2010). *Climate Change and Infectious Diseases: The View From EASAC* (pp. 1–3). European Academies Science Advisory Council.

Ehrlich, P. R., & Pringle, R. M., (2008). Where does biodiversity go from here? A grim business-as-usual forecast and a hopeful portfolio of partial solutions. *Proc. Natl. Acad. Sci. USA, 105*(1), 11579–11586.

Ezenwa, V. O., Godsey, M. S., King, R. J., & Guptill, S. C., (2006). Avian diversity and West Nile virus: Testing associations between biodiversity and infectious disease risk. *Proceedings of the Royal Society B - Biological Sciences, 273*, 109–117.

Gienapp, P., Leimu, R., & Merila, J., (2007). Responses to climate change in avian migration time-microevolution versus phenotypic plasticity. *Climate Res., 35*, 25–35.

Gubler, D. J., Reiter, P., Ebi, K. L., Yap, W., Nasci, R., & Patz, J. A., (2001). Climate variability and change in the United States: Potential impacts on vector- and rodent-borne diseases. *Environmental Health Perspectives, 109*(S2), 223–233.

Hales, S., De Wet, N., Maindonald, J., & Woodward, A., (2002). Potential effect of population and climate changes on global distribution of dengue fever: An empirical model. *Lancet, 360*, 830–834.

Hall, S. R., Becker, C. R., Simonis, J. L., Duffy, M. A., Tessier, A. J., & Cáceres, C. E., (2009). Friendly competition: Evidence for a dilution effect among competitors in a planktonic host-parasite system. *Ecology, 90*, 791–801.

Harvell, C. D., Mitchell, C. E., Ward, J. R., Altizer, S., Dobson, A. P., Ostfeld, R. S., & Samuel, M. D., (2002). Climate warming and disease risks for terrestrial and marine biota. *Science, 296*(5576), 2158–2162.

Hofstra, N., (2011). Quantifying the impact of climate change on enteric waterborne pathogen concentrations in surface water. *Curr. Opin. Environ. Sustain., 3*, 471–479.

Islam, M. S., Sharker, M. A. Y., Rheman, S., Hossain, S., Mahmud, Z. H., Islam, M. S., et al., (2009). Effects of local climate variability on the transmission dynamics of cholera in Matlab, Bangladesh. *Trans. R. Soc. Trop. Med. Hyg., 103*, 1165–1170.

Jablonski, D., (1993). The tropics as a source of evolutionary novelty through geological time. *Nature, 364*, 142–144.

Janies, D., Habib, F., Alexandrov, B., Hill, A., & Pol, D., (2008). Evolution of genomes, host shifts and the geographic spread of SARS-CoV and related coronaviruses. *Cladistics, 24*, 111–130.

Janies, D., Hill, A. W., Guralnick, R., Habib, F., Waltari, E., & Wheeler, W. C., (2007). Genomic analysis and geographic visualization of the spread of avian influenza (H5N1). *Syst. Biol., 56*, 321–329.

Jofre, J., Blanch, A. R., & Lucena, F., (2010). Water-borne infectious disease outbreaks associated with water scarcity and rainfall events. In: Sabater, S., & Barcelo, D., (eds.), *Water Scarcity in the Mediterranean: Perspectives Under Global Change* (pp. 147–159). Springer, Berlin.

Johnson, P. T. J., Lund, P., Hartson, R. B., & Yoshino, T., (2009). Community diversity reduces *Schistosoma mansoni* transmission and human infection risk. *Proceedings of the Royal Society B - Biological Sciences, 276*, 1657–1663.

Jones, K. E., Patel, N. G., Levy, M. A., Storeygard, A., Balk, D., Gittleman, J. L., & Daszak, P., (2008). Global trends in emerging infectious diseases. *Nature, 451*, 990–993.

Keesing, F., Belden, L. K., Daszak, P., Dobson, A., Harvell, C. D., Holt, R. D., et al., (2010). Impacts of biodiversity on the emergence and transmission of infectious diseases. *Nature, 468*, 647–652.

Keesing, F., Brunner, J., Duerr, S., Killilea, M., LoGiudice, K., Vuong, S. H., & Ostfeld, R. S., (2009). Hosts as ecological traps for the vector of Lyme disease. *Proceedings of the Royal Society B - Biological Sciences, 276*, 3911–3919.

Kellert, S. R., (2005). Perspectives on an ethic toward the sea. In: Barnes, P. W., & Thomas, J. P., (eds.), *Benthic Habitats and the Effects of Fishing* (pp. 703–712). American Fisheries Society, Bethesda, MD.

Kelly, D. W., Paterson, R. A., Townsend, C. R., Poulin, R., & Tompkins, D. M., (2009). Has the introduction of brown trout altered disease patterns in native New Zealand fish? *Freshwater Biology, 54*, 1805–1818.

Koenig, W. D., Hochachka, W. M., Zuckerberg, B., & Dickinson, J. L., (2010). Ecological determinants of American crow mortality due to West Nile virus during its North American sweep. *Oecologia, 163,* 903–909.

Kopp, K., & Jokela, J., (2007). Resistant invaders can convey benefits to native species. *Oikos, 116,* 295–301.

Lafferty, K., & Gerber, L., (2002). Good medicine for conservation biology: The intersection of epidemiology and conservation theory. *Conserv. Biol., 16,* 593–604.

Lambais, M. R., Crowley, D. E., Cury, J. C., Bull, R. C., & Rodrigues, R. R., (2006). Bacterial diversity in tree canopies of the Atlantic Forest. *Science, 312,* 1917.

Lindsay, S. W., & Birley, M. H., (1996). Climate change and malaria transmission. *Annals of Tropical Medicine and Parasitology, 90*(6), 573–588.

LoGiudice, K., Duerr, S. T., Newhouse, M. J., Schmidt, K. A., Killilea, M. E., & Ostfeld, R. S., (2008). Impact of host community composition on Lyme disease risk. *Ecology, 89,* 2841–2849.

Losos, J. B., & Schluter, D., (2000). Analysis of an evolutionary species-area relationship. *Nature, 408,* 847–850.

Losos, J. B., (1996). Ecological and evolutionary determinants of the species-area relation in Caribbean anoline lizards. *Philosophical Transactions of the Royal Society of London, B., 351,* 847–854.

McCarthy, J. J., Canziani, O. F., Leary, N. A., Dokken, D. J., & White, K. S., (2001). *Climate Change: Impacts, Adaptation, and Vulnerability* (pp. 1–1032). Cambridge University Press, Cambridge.

Millenium Ecosystem Assessment, (2005). *Ecosystems and Human Well–Being: Biodiversity Synthesis.* Island Press, Washington, DC.

Morand, S., (2011). Infectious diseases, biodiversity, and global changes: How the biodiversity sciences may help. In: López-Pujol, J., (ed.), *The Importance of Biological Interactions in the Study of Biodiversity* (pp. 231–254). InTech publisher, Croatia.

Myers, N., & Knoll, A. H., (2001). The biotic crisis and the future of evolution. *Proceedings of the National Academy of Sciences of the USA, 98,* 5389–5392.

Pallen, M. J., & Wren, B. W., (2007). Bacterial pathogenomics. *Nature, 449,* 835–842.

Purvis, A., & Hector, A., (2000). Getting the measure of biodiversity. *Nature, 405*(6783), 212–219.

Quental, T. B., & Marshall, C. R., (2010). Diversity dynamics: Molecular phylogenies need the fossil record. *Trends Ecol. Evol., 25,* 434–441.

Rabosky, D. L., & Lovette, I. J., (2008). Explosive evolutionary radiations: Decreasing speciation or increasing extinction through time? *Evolution, 62,* 1866–1875.

Raghavendra, K., Barik, T. K., & Adak, T., (2010). Development of larval thermotolerance and its impact on adult susceptibility to malathion insecticide and *Plasmodium vivax* infection in *Anopheles stephensi. Parasitol Res., 107,* 1291–1297.

Raymundo, L. J., Halforda, A. R., Maypab, A. P., & Kerr, A. M., (2009). Functionally diverse reef-fish communities ameliorate coral disease. *Proc. Natl. Acad. Sci. USA, 106,* 17067–17070.

Reeves, W. C., Hardy, J. L., Reisen, W. K., & Milby, M. M., (1994). Potential effect of global warming on mosquito-borne arboviruses. *Journal of Medical Entomology, 31*(3), 323–332.

Rose, J. B., Epstein, P. R., Lipp, E. K., Sherman, B. H., Bernard, S. M., & Patz, J. A., (2001). Climate variability and change in the United States: Potential impacts on water- and foodborne diseases caused by microbiologic agents. *Environmental Health Perspectives*, *109*(S2), 211–221.

Rosenzweig, M. L., (1995). *Species Diversity in Space and Time*. Cambridge University Press, Cambridge, UK.

Rosenzweig, M. L., (2001). Loss of speciation rate will impoverish future diversity. *Proceedings of the National Academy of Sciences of the USA*, *98*, 5404–5410.

Semenza, J. C., & Menne, B., (2009). Climate change and infectious diseases in Europe. *Lancet Infect. Dis.*, *9*, 365–375.

Shuman, E. K., (2011). Global climate change and infectious diseases. *Climate Change and Infectious Diseases*, *2*(1), 11–19.

Smith, T. B., & Bernatchez, L., (2008). Evolutionary change in human-altered environments. *Mol. Ecol.*, 171–178.

Smith, T. B., Kinnison, M. T., Strauss, S. Y., Fuller, T. L., & Carroll, S. P., (2014). Prescriptive evolution to conserve and manage biodiversity, *Annu. Rev. Ecol. Evol. Syst.*, *45*, 1–22.

Stigall, A. L., & Lieberman, B. S., (2006). Quantitative palaeobiogeography: GIS, phylogenetic biogeographical analysis, and conservation insights. *J. Biogeogr.*, *33*, 2051–2060.

Suzán, G., Marcé, E., Giermakowski, J. T., Mills, J. N., Ceballos, G., Ostfeld, R. S., et al., (2009). Experimental evidence for reduced rodent diversity causing increased Hantavirus prevalence. *PLoS One*, *4*, e5461.

Swaddle, J. P., & Calos, S. E., (2008). Increased avian diversity is associated with lower incidence of human West Nile infection: Observation of the dilution effect. *PLoS One*, *3*, e2488.

Tersago, K., Schreurs, A., Linard, C., Verhagen, R., Dongen, S. V., & Leirs, H., (2008). Population, environmental, and community effects on local bank vole (*Myodes glareolus*) Puumala virus infection in an area with low human incidence. *Vector-Borne and Zoonotic Diseases*, *8*, 235–244.

Thieltges, D. W., Bordalo, M. D., Caballero-Hernandez, A., Prinz, K., & Jensen, K. T., (2008). Ambient fauna impairs parasite transmission in a marine parasite-host system. *Parasitology*, *135*, 1111–1116.

Thieltges, D. W., Reise, K., Prinz, K., & Jensen, K. T., (2009). Invaders interfere with native parasite-host interactions. *Biological Invasions*, *11*, 1421–1429.

Venter, J. C., Remington, K., Heidelberg, J. F., Halpern, A. L., Rusch, D., Eisen, J. A., et al., (2004). Environmental genome shotgun sequencing of the Sargasso Sea. *Science*, *304*, 66–74.

Walpole, M., Almond, R. E., Besançon, C., Butchart, S. H., Campbell-Lendrum, D., Carr, G. M., et al., (2009). Tracking progress toward the 2010 biodiversity target and beyond. *Science*, *325*, 5947.

Wilby, R. L., Hedger, M., & Orr, H., (2005). Climate change impacts and adaptation: A science agenda for the Environment Agency of England and Wales. *Weather*, *60*, 206–211.

Wilcox, B. A., & Gubler, D. J., (2005). Disease ecology and the global emergence of zoonotic pathogens. *Environmental Health and Preventive Medicine*, *10*, 263–272.

Williams, S. E., Shoo, L. P., Isaac, J. L., Hoffmann, A. A., & Langham, G., (2008). Towards an integrated framework for assessing the vulnerability of species to climate change. *PLoS Biol.*, *6*, 2621–2626.

Wilson, E. O., (1993). *The Diversity of Life*. Harvard University Press, Cambridge, MA.

Wilson, M. L., (2001). Ecology and infectious disease. In: Aron, J. L., & Patz, J. A., (eds.), *Ecosystem Change and Public Health: A Global Perspective* (pp. 283–324). John Hopkins University Press, Baltimore, USA.

Wu, X., Lub, Y., Zhou, S., Chen, L., & Xu, B., (2016). Impact of climate change on human infectious diseases: Empirical evidence and human adaptation. *Environment International*, *86*, 14–23.

CHAPTER 12

EVOLVING PARK VALUES AMONG THE LOCAL PEOPLE IN TAMAN NEGARA, MALAYSIA'S OLDEST NATIONAL PARK

FEI ERN CHING[1], SZE-SIONG CHEW[2], and HONG CHING GOH[3]

[1]University of Malaya Spatial-Environmental Governance for Sustainability Research, University of Malaya, 50603 Kuala Lumpur, Malaysia

[2]Centre for Sustainable Urban Planning and Real Estate, Faculty of Built Environment, University of Malaya, 50603 Kuala Lumpur, Malaysia

[3]Department of Urban and Regional Planning, Faculty of Built Environment, University of Malaya, 50603 Kuala Lumpur, Malaysia

12.1 INTRODUCTION

From the conservation perspective, protected areas are important particularly in the efforts to preserve biodiversity, protect the ecosystem and ensure the quality of the environment (Eagles et al., 2000; Putney, 2003; Harmon, 2004). A protected area is "a clearly defined geographical space, recognized, dedicated, and managed through legal or other effective means, to achieve the long-term conservation of nature with associated ecosystem services and cultural values" (Juffe-Bignoli et al., 2014). According to Eagles et al. (2000), Stone and Wall (2004), Karanth et al. (2006) and Karanth and Nepal (2012), apart from conserving the nature, protected areas also provide economic support to the local communities with its existing natural resources through human livelihood activities such as hunting, agriculture, collecting fuelwood and other forest products. The growing tourism industry, might help improve of the local people's quality of life as more tourism-related employment opportunities are offered in protected areas (Stone and Wall, 2004; Karanth et al., 2006; Karanth and Nepal, 2012). Therefore, the conservation and economic values are essentially significant in protected areas (Harmon, 2004). Further to that, protected areas also act as a platform where environmental education takes place, offering educational values which could enhance the public awareness and understanding on the significance of protecting the environment (Eagles et al., 2000).

While these values are important, it is also essential to recognize that the connection between human and protected areas exists long before an area is established as a "protected area," particularly referring to the communities living surrounding and within the areas before they were gated and guarded. Putney (2003) for instance highlighted that the research on the importance of protected areas have always focused on economic benefits and *in situ* biodiversity conservation while its relation to human existence through the spiritual, cultural, aesthetic, and relational aspects are less explored and have received less attention. According to Eagles et al. (2000), any human activities in the parks will bring different effects to the people around. It has been challenging in determining the park values based on the perceived costs and benefits of these people and their different point of views. The local people are often affected when a park is established along with its policies and operations. For instance, designating the status of the national park to an area may limit the access of the local people thus interrupting their livelihood considering they rely mostly on natural resources or forest products as their source of income. Despite this, the decision of landscape protection is able to enhance human health and security by providing unspoilt or quality resources for their daily life.

Additionally, protecting the landscapes often lead to the growing appreciation of natural beauty through its aesthetic values. Harmon (2004) also suggested that it is important to increase the public's understanding and recognition of human values in order to gain their support and in turn improve the management protected area. Acknowledging the significance of human values especially among the local people, this chapter highlights the different dimensions of value perceived by the local people in Taman

Negara National Park and the evolution of these values throughout times (Figure 12.1).

FIGURE 12.1 **(See color insert.)** Local people being interviewed.

12.2 THE VALUES OF PROTECTED AREAS

A value of parks generally refers to the attribute or quality of physical features available within the area (Putney, 2003). Park values can be determined either objectively or subjectively depending on the desired outcome. The role of park managers is a very important role in protecting the physical characteristics if it comes to offering objective values. Otherwise public priorities will be the main concern if it is accessed based on subjective values. However, intrinsic values are indeed objectively present in nature (Rolston, 1988; Putney, 2003) followed by the act of appreciation or instrumental value once the intrinsic value is discovered. A park is more valuable if it can gain more instrumental value through deeper appreciation of its intrinsic values. Therefore, the value of national parks "is the sum of the interactions of the intrinsic values of the resource itself plus the myriad instrumental values assigned by humans" (Putney, 2003).

Putney (2003) classified the values present in protected areas based on their usefulness to humankind in terms of either tangible or intangible values. Different groups within the society may have differing views on the park values. For instance, indigenous, and traditional people may view protected areas differently than the urban populations in which the former perceive it as having a significant relationship between community, culture, spirituality, nature, and territory, in national parks while the latter see it as a natural area that provides recreational opportunities. Economic values are categorized as intangible values in which a community/country receives direct or indirect monetary, commercial, and employment benefits that are generated from activities including tourism, cottage industries, agricultural grazing and others (unpublished data). In protected areas, economic values are tangible because it helps sustain the livelihood of the local community around the park through the various employment opportunities protected areas provide. The opportunities for economic development that comes with the establishment of protected areas can occur in two ways for the local people. Firstly, recommending the attraction and resources to non-residents contributes to a rise in the rate of employment for the local people; and secondly, by showing the local communities how tourism practice occurring in their area can help them generate profit through businesses. Such economic benefits will help both the individuals and the community (Eagles and McCool, 2002; Lopoukhine, 2008). Tourists' presence in protected areas help generates the local economy through the money they spend on accommodation, food, and beverage, souvenirs, and recreational activities. Additionally, tourism activities also contribute to flourishing business opportunities such as boat operators, grocery shops, and tour agencies become rampant because of the increasing tourists' demand (Eagles and McCool, 2002). Subsequently, the tourism-related business also thrives and further helps enhance the local people's living standard as well as the reduction of their poverty level.

On the other hand, the World Commission on Protected Areas (WCPA) interprets intangible values of protected areas as "that which enriches the intellectual, psychological, emotional, spiritual, cultural, and/or creative aspects of human existence and well-being" (South African National Parks, 2016). According to SANParks (2016), recognizing the human values of parks is vital in obtaining the public's continuous support in the management of protected areas. However, the lack of interest in research related to the intangible value of parks of protected area management is still evident in the current literature. There are eleven types of intangible values in protected areas identified by the WCPA as summarized below:

1. *Recreational values*, which provide opportunities for a human to restore or refresh the mind, body, and soul through any consumptive or non-consumptive activities.
2. *Therapeutic values*, which create the potential for healing and enhancing physical and psychological well-being.

3. *Spiritual values*, which inspire humans' respects to the purity of nature.

4. *Cultural values*, which are reflected by a distinct group of people or society who were formed by different traditions, beliefs, and value systems. These values allow people to understand and connect in meaningful ways to the environment of its origin and the rest of nature.

5. *Identity values*, which link people to their landscape through myth, legend or history.

6. *Existence values*, which contain the satisfaction, symbolic importance and willingness to pay among people, knowing that natural and cultural landscapes are well protected so that they exist as physical and conceptual spaces where forms of life and culture are valued.

7. *Artistic values*, which inspire human imagination in creative expression.

8. *Aesthetic values*, which carry an appreciation of the beauty found in nature.

9. *Educational values* include the use of protected areas to create respect and understanding of the physical and natural environment and biodiversity between people and the natural environment.

10. *Scientific research and monitoring values* include the significance of protected areas in contributing to an understanding of natural areas and the consequences of human interference.

11. *Peace values*, which foster regional peace and stability through cooperative management across international land or sea boundaries, as "intercultural spaces" for the development of understanding between distinct cultures, or as places of "civic engagement" where difficult moral and political questions can be constructively addressed (adapted from Harmon and Putney, 2003; Pagan and Legare, 2004).

Despite the contextual significance of each value, most of these values are still present in each and every protected area. For instance, Borrini-Feyerabend et al. (2004) asserted that ecological, cultural, and spiritual values are used as the basis in the identification of most protected areas in Europe. Recognizing the presence of these values including social and religious values within protected areas is essential especially in protecting the indigenous people's right to practice their traditional way of life continuously. Vodouhê et al. (2010) revealed that the local people appreciate the existence of Pendjari National Park as they mostly depend on the park to hunt food and medicinal plants. The existence of protected areas further benefits the local people in terms of the ecosystem services provided namely pollution control, freshwater conservation, soil fertility and so on (Eagles and Hilled, 2008). Moreover, the local people are also hopeful that the conservation of wildlife in protected areas can benefit their future generation.

Further to that, it is important to acknowledge that different stakeholders may perceive the values differently due to the difference in their interests thus these values may be debated. Karanth and Nepal (2012) emphasized the significance of tourism in bringing up the economic values of a park to the local people. They indicated that most people living around protected areas in Nepal and India perceived economic values from tourism activities and the access to the available natural resources within the areas respectively. Generally, local communities value the existence of protected areas and support the establishment of protected areas by the government in the efforts to preserve wildlife, forests, stop poaching and promote tourism. Protected areas are valued because of their capacity to conserve community values through protection and the meaning, identity, and connection they give to the local people. Research indicated that the local people highly valued the existence of Ranthambore, Kanha, and Nagarahole in India, Chitwan National Park and Annapurna Conservation Area in Nepal despite the fact that establishing the parks had cost them their crops and caused damages to their properties. The local people's appreciation and cultural tolerance to the wildlife conservation program in both India and Nepal had helped contribute to the conservation goals in the region (Karanth and Nepal, 2012). Specifically in Nepal, Allendorf et al. (2007) reported that the residents living in the surrounding area of Royal Bardia National Park, acknowledged that the park is important because it offers them benefits in terms of economy, recreation, environment, aesthetics, and education and it also acts as a bequest for the future generations. However, the residents also held negative views toward the park especially with the implementation of conservation policies where their access to the resources and grazing areas became restricted.

In terms of recreational value, a park is of significant value for the local people as the recreational activities provided helped society to enhance self-esteem, increased social cohesiveness, cognitive development and more (Eagles and McCool, 2002). For instance, local residents around a coastal national park in Germany known as Hamburger Hallig mainly visited the park for recreational purposes like hiking or cycling (Kalisch, 2012). Through recreational activities, the interaction between local people occurs when they meet. Within protected areas and its surrounding environment, recreational, and educational values can simultaneously exist with the tangible values depending on the resources available. A large part of the local population living in the vicinity of the Eastern Macedonia and Thrace National Parks in Greece engaged in recreational and tourism activities such as camping, hunting, and fishing. At the same time, the parks have become a source of income for another part of the local population as they involved themselves in agricultural practices namely in producing crops and livestock as well as commercial fishing. Besides that, the parks' biological attractions along with the landscapes' aesthetic value offer valuable educational information (Pavlikakis and Tsihrintzis, 2006) that may benefit both the visitors and the local people. The locals can better learn about the protected areas especially in recognizing the endemic, unique, and threatened flora and fauna through their interactions with the environment and from the various scientific researchers conducted in the parks thus generating a sense of belonging and a need to safeguard the natural environment. Kideghesho et al. (2008) disclosed that the Robanda community in Western Serengeti, Tanzania possessed a high level of awareness towards the advantages of conserving a protected area resulting from the educational programs provided to the community.

Furthermore, the values found in protected areas may evolve over the times taking into account its changing roles which are constantly influenced by the policies implemented by the park management. Whenever a park is established, the access into the area and its natural resources are always restricted hence altering the economic foundation of the local community (Lindberg et al., 1996; Stone and Wall, 2004). According to Stone and Wall (2004), the establishment of Jianfengling National Forest Park and Diaoluoshan National Forest Park highly affected the livelihood of local communities from the nearby villages. Socioeconomic benefits were limited as ecotourism development at the parks was still at the early stage. However, as ecotourism grew and employment opportunities expanded, the local community was offered to fill positions or was able to start a tourism venture with provisions of business loans.

12.3 TAMAN NEGARA NATIONAL PARK

Taman Negara National Park is located in the heart of Peninsular Malaysia encompassing three neighboring states namely Pahang, Kelantan, and Terengganu. It is Malaysia's greatest national park covering an area of 4,343 km² and is dubbed as the oldest tropical rainforests in the world as it is approximately calculated to be more than 130 million years old (DWNP, 2015). The park first received protection as Gunung Tahan Game Reserve in 1925 where 1,300 km² of the area was gazetted. During the colonial era in 1938/39, the British Administrators declared it as King George V National Park under Enactment 1939. The name was later changed to Taman Negara meaning "national park" after the independence of Malaya in 1957.

In general, the Taman Negara National Park is divided into three different national parks with the respective enactments placed at state level namely Taman Negara Pahang, Taman Negara Kelantan, and Taman Negara Terengganu. Taman Negara Pahang is the largest out of all three covering 57% of the entire park's area. Taman Negara Kelantan is the second largest with an area coverage of 24% followed by Taman Negara Terengganu 19%. Due to nature if its establishment and geographical location, Taman Negara is accessible via four gateways namely Kuala Tahan, Sungai Relau, Kuala Koh and Tanjung Mentong (UNESCO World Heritage Centre, 2015). Despite being divided geographically, the park authority responsible for management of Taman Negara lies with the Department of Wildlife and National Parks (DWNP), a federal agency placed under the purview of the Ministry of Natural Resources and the Environment (DWNP, 2015). The park authority, DWNP is in charge of wildlife enforcement and biodiversity protection in the park, under the jurisdiction of the Wildlife Conservation Act 2010 [Act 716] securing the sustainability of natural resources.

Taman Negara National Park was established to "set aside and reserve in perpetuity and trust for the purpose of the propagation, protection, preservation of the indigenous fauna and flora of Malaya and of the preservation of objects and places of aesthetic, historical or scientific interest" (DWNP, 1987). In addition, it also offers educational and recreational opportunities for visitors' enjoyment (DWNP, 1987). Today, tourism has become a significant practice in the park. The tourist arrivals to the park has significantly increased over the last decade, from 79,269 in 2005 to

112,596 in 2014 (DWNP Kelantan, Pahang dan Terengganu, 2014). Due to the increasing number of tourists, DWNP eventually decided to privatize the accommodation and eating facilities particularly at Kuala Tahan, Kuala Trenggan and Kuala Keniam since 1991 (Ledesma et al., 2003) in order to improve the park management's effectiveness. In other words, this decision has benefited the local communities especially in terms of its contribution in diversifying the employment opportunities and sources of income.

As one of the world's oldest rainforests, Taman Negara National Park is well-known for its rich floral and faunal biodiversity where it hosts many threatened and rare species of plants and wildlife in Malaysia. In 1984, the park was declared as an ASEAN Heritage Park (UNESCO WHC, 2015). The park accommodates mammals such as the Asian elephant, wild boar, the Malayan tapir, mouse deer, and seladang. Wild dog, sun bear, white-handed gibbon, and Malayan tiger are also commonly spotted in the wild forest of the park. Besides, an estimated of 14,000 species of flora are found in the park including the *Rafflesia cantleyi*, one of the largest flowers in the world. Notably, the park is also home to at least 300 species of birds thus making it a popular bird watching spot especially among international tourists (DWNP, 2015). Geographically, the park's mountainous surface is formed by Gunung Teku, Gunung Tangga Dua Belas and Gunung Tahan. Standing at an altitude of 2,178 m, Gunung Tahan is the highest peak in Peninsular Malaysia (UNESCO WHC, 2015).

Other than being important biologically and ecologically, Taman Negara National Park especially Taman Negara Pahang is also home to the Bateq people, a group of indigenous people coming from a subgroup of the Negritos who happened to be the earliest inhabitants in Peninsular Malaysia. The Bateq people are hunters and gatherers, and they lived nomadically in the jungle. However, they are normally settled temporarily along the river basins (Fatanah et al., 2012; Yunus et al., 2013). There are several indigenous people villages within the park, namely Kampung Kuala Atok, Kampung Dedari, Kampung Teresek, Kampung Terenggan, Kampung Gua Telinga, Kampung Yong, Kampung Tabung and Kampung Yong. However, the existence of the villages may be different from time to time as the indigenous people practice the nomadic way of life. Apart from the indigenous people, many locals live in Kampung Kuala Tahan which is the main entrance to Taman Negara Pahang. It is located across Sungai

Tembeling which marks the park boundary from the village. It is estimated that there is a total of about 400 households in Kampung Kuala Tahan.

12.4   THE VALUES OF PARK AS PERCEIVED BY THE LOCAL PEOPLE

Questionnaire survey and interviews conducted targeted the local and indigenous people by visiting their villages. A self-administrated five-page questionnaire was designed and handed out to 111 respondents (one-third of the total population in Kampung Kuala Tahan, the largest settlement outside the park). The questionnaire used a five-point Likert scale measurement, and it aimed to evaluate the park values perceived by the local people of Taman Negara Pahang. The survey also divided the questions into two parts stratifying those who work in the tourism sector and those who do not. Overall, the analysis findings reveal that intangible values including recreational, education, artistic, aesthetic, and existence are among the most important values perceived by the local people towards Taman Negara Pahang.

According to Harmon and Putney (2003), recreational use of parks was defined as "visits by the local and regional residents" to fulfill their social need. Most local people agreed that the park was a place for them to engage in recreational activities and strengthen their relationships with other locals. Recreation was helpful for society to function and national parks could be the best place for it (Eagles and McCool, 2002). Furthermore, local people agreed that visiting the park enables them to refresh their minds as the park was deemed a quiet and beautiful place. The local people felt that their declining quality of life in the urban areas was improved by visiting the park through the enjoyment of natural areas. For this reason, they expressed better appreciation for nature conservation.

The local people also perceived educational values as abundant in Taman Negara Pahang. The local people could share their existing knowledge related to plants, herbs, medicines in the park not just with their children and other locals but also with the visitors. The Department of Wildlife and National Parks had organized programs such as the Wildlife Day in the year 2005 in the efforts to educate and increase the awareness of the local people. Educational values, in fact, included the use of the site to train or educate people so that they would be more aware of their physical and natural surroundings as well as the biodiversity (unpublished data).

Meanwhile, the local people also highly appreciated the park's artistic values because it allows them to enjoy the beauty of nature that inspired their sense of imagination. The respondents further described in the interviews that they felt serene during their walk in the park. They also felt the wind blow, heard the birds singing and saw the greenery which made them feel like being in a paradise on earth. Meanwhile, an aesthetic value which was defined as appreciation to the harmony, beauty, and profound meaning of nature (Harmon and Putney, 2003) was present among the local people of Taman Negara Pahang. The local people strongly agreed when it comes to maintaining the greenery in the park as they were concerned about nature around the park and held the strong positive opinion that nature conservation programs were requisite to maintain the natural settings.

The existence value was also highly rated by the local people of Taman Negara Pahang as the people have already witnessed the negative impact of modernization thus finding it important to protect the natural landscape and culture heritage from such development (Harmon and Putney, 2003). This was the highest intangible values rated by the local people of the park mainly due to the recognition that through biodiversity conservation, they could enjoy the natural landscape, protect the wildlife and sustain their culture for future generations. They also realized that by protecting Taman Negara Pahang, they could get fresher air, cleaner water, and their living environment would be free from major pollution issues.

Overall, local people rated that existence values as the most important values to them followed by recreational values that were deemed essential for their quality of life. Educational values were also important as the values provide an opportunity for the people to learn about flora and fauna in the park. Artistic and aesthetic values were rated slightly lower than other intangible values, but it was not because the local people considered both values as not important to them. Instead, it was because the local people were able to gain such values from the recreational activities offered. This proved that the local people could attain other values through existence values, thus making existence values the most important intangible values to existing in the park.

In terms of tangible values, economic values which were induced by the development of tourism in the park were deemed the most significant ones. Enjoyment of nature through recreational activities is among the leading objectives in establishing a national park (Amran et al., 2008). The recreational activities turn into tourism activities when tourists are involved. For many parks today, tourism has been one of the key activities offered in order to generate revenue which will not only ease the financial burden of the park authority but benefit the local people. Interviews with the local people outside Taman Negara Pahang reveal that the local people were able to gain economic benefits through their participation in tourism businesses. Tourism development in the park affects local communities in Kampung Kuala Tahan. Following the decision of accommodation priva-tization in the park, there were tourism facilities and services undertaken by private sectors or residents of the area (WWF, 1996). Nevertheless, the local people were given priority in running small enterprises such as the floating restaurants, accommodation facilities, souvenir, and convenience shops, boating within the park and guides (Ahmad et al., 2012). Therefore, Kuala Tahan has developed rapidly since the 1980s especially in terms of business involving services such as accommodation and food together with other small businesses to provide an easier access and assist in the park development.

This tourism-induced economic value of Taman Negara Pahang evolved from the earlier local dependency on gathering rattan and wood along with agricultural activities to support their livelihood. Today, the tourism-induced economic value of the park has become the priority among the local people because it is able to support their livelihood through the various opportunities for employment provided either on a part-time or full-time basis. Mukrimah et al. (2015) study found that each household received an average monthly income of RM 4035. Traditionally, they depended on the forest for their livelihood. Although they are restricted from entering the park limiting their source of income after the protected area was established, the development of tourism in the park has created alternative sources of income for the local people in Kuala Tahan. Activities related to forestry and ecotourism within and in the surrounding park area generated 47% of the local people's average monthly household income. Working as a tour guide enabled them to earn 13% of the average household income and 10% was generated from being employed in the operation of retail stores or restaurants. Hence ecotourism is an important part of Taman Negara Pahang as it helps to create incomes and jobs for the local community and eventually reduce their level of poverty (Mukrimah et al., 2015). Apart from operating individual lodging facilities, the only accommodation operator in the Park, for example, Mutiara Taman Negara also offered a significant number of employments to the local people. Many of the local people here worked as the assistant managers, division executives, private drivers as well as chalet and food service operators.

By involving in the tourism business, capacity building was observed among the local people where they gained experience and acquired hotel management and hospitality skills while they work. In fact, many of them started their own accommodation operation such as hostel, guesthouse, and chalets outside the park after gaining several years of experience working with Mutiara Taman Negara. Subsequently, there were several options of accommodation facilities made available by the local people outside the park. The upgraded road from Jerantut to Kuala Tahan since 2005 made the park more accessible and also contributed to the increase in tourists arrival to Taman Negara Pahang. Transportation services and van services were available to the park. The frequent boat service across the river to enter the park was provided by the local young men to gain extra income while providing convenience to the tourists entering the park. The economic values of the park also helped to upgrade the village condition as a result of the substantial income generated from tourism businesses, which in return, improved their living standards. An interview with the key personnel of the Village Development and Security Committee also reveals that various tourism development proposals submitted by the villagers have already been materialized. The proposed ideas include improving the overall landscape of the villages and enhancing the tourism facilities in the park such as constructing an observation deck on Bukit Teresek for sunset and sunrise viewing, building a "Garden in the Jungle" outside the park as well as upgrading Kampung Kuala Tahan into a mini town with better basic amenities, more floating restaurants, introducing water sport activities, mini museums to accommodate the tourist demand and including more villages to gain the economic benefits induced by tourism businesses. These facilities and activities have resulted in the increased length of stay among the tourists to Taman Negara Pahang (Figure 12.2).

FIGURE 12.2   (See color insert.) Night view of the floating restaurant.

Meanwhile, the economic values of Taman Negara Pahang were not solely contributed by tourism activities. In fact, after the park was officially gazetted, local people around TNP were offered employment opportunities to work with the park authority. Out of the 106 staff, more than half of them were local people, and few had worked in the park for longer than 20 years as general workers, maintenance crews and park rangers (Zaaba, 1999).

Similarly, the indigenous people also perceived significant economic value from the park ranging from the traditional lifestyle by collecting rattan, bamboos, and wildlife hunting to involving in tourism activities, working as guides, boat drivers or porters (MSTE, 1997; Zanisah et al., 2009; Zuriatunfadzliah et al., 2009). Notably, the traditional lifestyle of the indigenous people has always become among the leading attractions in Taman Negara Pahang thus reflecting the cultural value of the park. As tourism business grows, Kampung Dedari and Gua Telinga are among the Orang Asli settlements which have become semi-permanent and have started to accommodate tourist visits (Fatanah et al., 2012). Tourists learn about the forest survival skills such as using bamboo blow pipes and making fire using stones as well as learning the beliefs and customs of Orang Asli that form the basis of their way of life. Since practicing semi-nomadic lifestyle, the Orang Asli has now become less dependent on jungle resources. In fact, some work as guides for tourists in the park (Backhaus, 2005) (Figure 12.3).

FIGURE 12.3 **(See color insert.)** Indigenous people village and tourists are visiting the indigenous village.

12.5 CONCLUSION

The values of park vary, depending on the level of the interdependency of the park with the local communities. In Taman Negara Pahang, tourism development has enhanced the economic values of park perceived by the local and indigenous people. This, in fact, witnesses the evolving values perceived by both the local and indigenous people. How the local people perceived the park values may differ from how tourists perceived them because the latter may not establish a similar kind of dependency and intimate relationship with the park compared to what the local people have experienced. Meanwhile, recognizing the economic values of the protected area and its absolute dependency on the existence value is important as it forms the foundation for ecotourism development. The interdependency of these values must be acknowledged in order to avoid the intangible values from being overshadowed by the tangible values of the park, an important aspect that the park management must keep as the foundation for its tourism development management and park conservation efforts.

ACKNOWLEDGEMENTS

The findings discussed in this chapter are a part of the research by a master degree graduate and financed by the Ministry of Higher Education through Fundamental Research Grant Scheme (FRGS FP047–2013A). The authors would also like to express our appreciation for the technical and logistic aids given by the Department of Wildlife and National Park, the Department of Orang Asli Development (JAKOA), the local people of Kuala Tahan, and the indigenous people living within the vicinity of Taman Negara Pahang during the data collection stage.

KEYWORDS

- **aesthetic**
- **artistic**
- **biodiversity**
- **economical**
- **educational**
- ***in situ* conservation**
- **intangible values**
- **Kuala Tahan**
- **local people**
- **Malaysia**
- **national park**
- **protected areas**
- **recreational**
- **tangible values**
- **tourism**

REFERENCES

Ahmad, S., Chia, K. W., Ramachandran, S., Syamsul, H. M. A., Hafiza, A. K., & Kamaludin, M. H., (2012). *Indicators of Inclusive Business Model: Perception of Business and Government Agency Communities in Taman Negara Pahang, Kuala Tahan, Pahang,*

Malaysia. In Iskandar Malaysia National Ecotourism Summit 2012. Incorporating responsible rural tourism symposium, Johor Bahru.

Allendorf, T. D., Smith, J. L. D., & Anderson, D. H., (2007). Residents perceptions of Royal Bardia National Park, Nepal. *Landscape and Urban Planning, 82*(1 & 2), 33–40.

Amran, H., Zainab, K., Nor Azina, D., & Ahmad, T. K., (2008). Planning for ecotourism in protected areas of Malaysia: Some reflections on current approaches. In: Kalsom, K., Nurhazani, M. S., & Mohamad, K. M., (eds.), *Ecotouism in the IMT-GT Region: Issues and Challenges*. Universiti Utara Malaysia Press, Sintok, Kedah.

Backhaus, N., (2005). *Tourism and Nature Conservation in Malaysian National Parks*. Lit Verlag, Munster (pp. 1-277).

Borrini-Feyerabend, G., Kothari, A., & Oviedo, G., (2004). *Indigenous and Local Communities and Protected Areas: Towards Equity and Enhanced Conservation*. International Union of Conservation for Nature, Gland, Switzerland (pp. 1-111).

De Ledesma, C., Lewis, M., & Savage, P., (2003). *The Rough Guide to Malaysia, Singapore, and Brunei*. Rough Guides, London.

Department of Wildlife and National Parks (DWNP) Kelantan (2014).

Department of Wildlife and National Parks (DWNP) Pahang (2014).

Department of Wildlife and National Parks (DWNP) Terengganu (2014).

Department of Wildlife and National Parks (DWNP) (1987). *Taman Negara Master Plan*. Kuala Lumpur.

Eagles, P. F. J., & McCool, S. F., (2002). *Tourism in National Parks and Protected Areas*. CABI Publishing, Wallingford.

Eagles, P. F. J., Halpenny, E. A., McCool, S. F., & Moisey, R. N., (2000). *Tourism in National Parks and Protected Areas: Planning and Management*. CABI Publishing, Wallingford.

Eagles, P., & Hillel, O., (2008). Improving protected area finance through tourism. In: *Protected Areas in Today's World: Their Values and Benefits for the Welfare of the Planet* (pp. 77–86). Secretariat of the Convention on Biological Diversity, Montreal.

Fatanah, N. Z., Mustaffa, O., & Salleh, D., (2012). Lawad, Ye' Yo' and Tum Yap: The manifestation of forest in the lives of the Bateks in Taman Negara National Park. *Procedia - Social and Behavioral Sciences, 42*, 190–197.

Harmon, D., & Putney, A. D., (2003). *The Full Value of Parks: From Economics to the Intangible*. Rowman and Littlefield, Inc., Maryland.

Harmon, D., (2004). Intangible values of protected areas: What are they? Why do they matter? *The George Wright Forum, 21*(2), 9–22.

Juffe-Bignoli, D., Bhatt, S., Park, S., Eassom, A., Belle, E. M. S., Murti, R., et al., (2014). *Asia Protected Planet 2014*. UNEP-WCMC, Cambridge.

Kalisch, D., (2012). Relevance of crowding effects in a coastal national park in Germany: Results from a case study on Hamburger Hallig. *Journal of Coastal Conservation, 16*(4), 531–541.

Karanth, K. K., & Nepal, S. K., (2012). Local residents perception of benefits and losses from protected areas in India and Nepal. *Environmental Management, 49*(2), 372–386.

Karanth, K. K., Curran, L. M., & Reuning-Scherer, J. D., (2006). Village size and forest disturbance in Bhadra Wildlife Sanctuary, Western Ghats, India. *Biological Conservation, 128*(2), 147–157.

Kideghesho, J. R., & Mtoni, P. E., (2008). The potentials for co-management approach in western Serengeti, Tanzania. *Tropical Conservation Science*, *1*(4), 334–358.

Lindberg, K., Enriquez, J., & Sproule, K., (1996). Ecotourism questioned: Case studies from Belize. *Annals of Tourism Research*, *23*(3), 543–562.

Lopoukine, N., (2008). Protected areas - for life's sake. In: *Protected Areas in Today's World: Their Values and Benefits for the Welfare of the Planet* (pp. 1–3). Secretariat of the Convention on Biological Diversity, Montreal.

Mukrimah, A., Moh, P. M., Motoe, M., & Lim, H. F., (2015). Ecotourism, income generation, and poverty reduction: A case of Kuala Tahan. *Journal of Tropical Resources and Sustainable Science*, *3*, 40–45.

Pavlikakis, G. E., & Tsihrintzis, V. A., (2006). Perceptions and preferences of the local population in Eastern Macedonia and Thrace National Park in Greece. *Landscape and Urban Planning*, *77*(1 & 2), 1–16.

Putney, A. D., (2003). Perspective on the values of protected areas. In: Harmon, D., & Putney, A. D., (eds.), *The Full Value of Parks: From Economics to the Intangible* (pp. 3–11). Rowman, and Littlefield Publishers, Inc., Lanham, Maryland, USA.

Rolston, III, H., (1988). *Environmental Ethics: Duties to and Values in the Natural World*. Temple University Press, Philadelphia.

Stone, M., & Wall, G., (2004). Ecotourism and community development: Case studies from Hainan, China. *Environmental Management*, *33*(1), 12–24.

Vodouhê, F. G., Coulibaly, O., Adégbidi, A., & Sinsin, B., (2010). Community perception of biodiversity conservation within protected areas in Benin. *Forest Policy and Economics*, *12*(7), 505–512.

World Wide Fund for Nature (WWF) Malaysia, (1996). *Malaysian National Ecotourism Plan: Part 4 Current Status of Ecotourism in Malaysia.*

Yunus, R. M., Karim, S. A., & Samadi, Z., (2013). Gateway to sustainable national park. *Procedia - Social and Behavioral Sciences*, *85*, 296–307.

Zaaba, Z. A., (1999). *The Identification of Criteria and Indicators for the Sustainable Management of Ecotourism in Taman Negara National Park, Malaysia: A Delphi Consensus*. West Virginia University Libraries.

Zanisah, M., Nurul, F. Z., & Mustaffa, O., (2009). The impact of tourism economy on the Batek community of Kuala Tahan, Pahang. *Jurnal E-Bangi*, *4*(1), 1–12.

Zuriatunfadzliah, S., Rosniza, A. C. R., & Habibah, A., (2009). Cultural changes of Bateq people in the situation of ecotourism in national park. *Journal of Social Sciences and Humanities*, *4*(1), 159–169.

CHAPTER 13

APPRECIATING NATURE CONSERVATION AT A MALAYSIAN FIRST WORLD HERITAGE SITE THROUGH ITS RECREATIONAL AND TOURISM ACTIVITIES

HONG CHING GOH[1,2] and
WAN NUR SYAZANA WAN MOHAMAD ARIFFIN[3]

[1]Water Engineering and Spatial-Governance (WESERGE) Research Centre, University of Malaya, 50603 Kuala Lumpur, Malaysia

[2]Department of Urban and Regional Planning, Faculty of Built Environment, University of Malaya, 50603 Kuala Lumpur, Malaysia

[3]Center for Sustainable Urban Planning and Real Estate, Faculty of Built Environment, University of Malaya, 50603, Kuala Lumpur, Malaysia

13.1 INTRODUCTION

One of the key objectives of establishing a national park is for ecological conservation. In order to ensure a successful long-term conservation objective, the effort must be able to raise the public's appreciation which can be communicated through educational experience embedded in the park's recreational and tourism activities. The world's first national park, Yellowstone National Park, for instance, stated its mission as below (National Park Service, 2000):

"Preserved within Yellowstone National Park are Old Faithful and the majority of the world's geysers and hot springs. An outstanding mountain wild land with clean water and air, Yellowstone is home to the grizzly bear, wolf, and free-ranging herds of bison and elk. Centuries-old sites and historic buildings that reflect the unique heritage of America's first national park are also protected. Yellowstone National Park serves as a model and inspiration for national parks throughout the world. The National Park Service preserves unimpaired these and other natural and cultural resources and values for the enjoyment, education, and inspiration of this and future generations."

The communication of conservation is particularly important for national parks that have been inscribed as World Heritage Sites by the United Nations Educational, Scientific, and Cultural Organization (UNESCO) for their distinctive cultural or physical significance (UNESCO World Heritage Center, 2016). Today, there are a total of 1031 properties included in the World Heritage List, of which 197 are Natural Heritage Sites where Malaysia hosts two sites namely Gunung Mulu National Park and Kinabalu Park. Kinabalu Park is the first World Heritage Site in Malaysia. This chapter reveals the conservation program conducted in Kinabalu Park and the communication of its conservation program through the environmental education activities offered in the park.

13.2   COMMUNICATING CONSERVATION VALUES THROUGH RECREATIONAL AND TOURISM ACTIVITIES

In national parks, the most significant value of parks is undoubtedly its conservation values, which is actually also the main justification for park establishment. The park conservation values refer to the quantifiable values through monetary expression such as identifying the "ecosystem services" provided by national parks (Harmon, 2004). Nonetheless, these values are also communicated through the recreational and tourism activities in the parks thus creating better awareness and appreciation among the general public about the importance of parks. In a broader context, Harmon, and Putney (2003) compiled the full range of values of protected areas by categorizing them into tangible and intangible groups. Tangible values consist of conservation values and economic values (such as tourism business) that are also associated with the revenue mainly generated by tourism activities in the protected area (Harmon and Putney, 2003). In the same article, the intangible values have been highlighted through specific discussion in respective chapters. Subsequently, the International Union for Conservation of Nature and Natural Resources (IUCN)

World Commission on Protected Areas (WCPA) task force summarized these values into eleven categories, namely recreational, therapeutic, spiritual, cultural, identity, existence, artistic, aesthetic, educational, scientific research and monitoring, and peace values. In this chapter, the authors focus only on recreational, educational, and scientific research and monitoring values. While the authors recognize the interrelation of all the values, it is important for one to identify the stronger link between these values based on the empirical context - a national park.

According to the IUCN category system for protected areas, the national park is classified as Type II protected area managed mainly for ecosystem protection and recreation. The main objective of the national park, as stated in Dudley (2008) is "to protect natural biodiversity along with its underlying ecological structure and supporting environmental processes, and to promote education and recreation." Specifically, "Category II protected areas are large natural or near natural areas set aside to protect large-scale ecological processes, along with the complement of species and ecosystems characteristic of the area, which also provide a foundation for environmentally and culturally compatible spiritual, scientific, educational, recreational, and visitor opportunities" (Dudley, 2008). The emphasis on recreational and educational components distinguishes a national park from other categories of protected areas in the list (Dudley, 2008).

Recreational activities are fundamental to the establishment of parks worldwide and remain important until today. Visitors appreciate parks for the qualities "that interact with humans to restore, refresh or create a new through stimulation and exercise of the mind, body, and soul" (Harmon, 2004). These qualities are known as the recreational value. Due to the magnitude of visits to national parks, the focus of studies has been related to the recreational impacts and the subsequent management approach to these impacts (Pickering and Hill, 2007; D'Antonio et al., 2013; Ruschkowski et al., 2013; Manning, 2014; Balmford et al., 2015).

Meanwhile, the educational value is recognized as one of the most important values of a protected area today communicating the essence of scientific research results to the society at large. The educational values "enlighten the careful observer with respect to humanity's relationships with the natural environment, and by extension, humanity's relationships with one another, thereby creating respect and understanding" (Harmon, 2004). In the article titled "Convergence between science and environmental education," Wals et al. (2014) highlighted the importance of bridging the science education, environmental education and citizen science and the convergence linkage that lies among them. Particularly, the scholars were in the opinion that creating "a mature symbiotic relationship" between environmental education and science education is necessary as a result of the increasingly complex nature of the environmental issues the society is facing today. This conclusion supports the broader view of Kopnina (2012) that "education of ecologically minded future generations could help their resolution."

In protected areas, educational values are embedded in various forms of interpretative programs such as guiding, artifact exhibitions, posters, and video presentations which are normally organized by the park authority to introduce the conservation efforts to the visitors. Knapp (1998) differentiated environmental interpretation from environmental education in that interpretation is a form of environmental education commonly exercised in protected areas, an informal, fun way and essential aid to achieve the behavior change goal of environmental education. Meanwhile, Goh and Rosilawati (2014) highlighted the rising expectation of educational experience in the guiding services in the protected area, which is advocated in the same direction in a more recent study by Garcia-Rodridguez and Fernandez-Escalante (2016) based on a Spanish case study.

Scientific research and monitoring values are "those that contribute to the function of natural areas as refugees, benchmarks, and baselines that provide scientists and interested individuals with relatively natural sites less influenced by human-induced change or conversion" (Putney, 2003). In protected areas, the research can be categorized into two groups; firstly, the research conducted in order to support management and to provide solution to the conservation of the protected areas; secondly, research that aims for a better understanding of the natural and human-mediated processes which may not have any applied conservation in mind (Stab and Henle, 2009). Nonetheless, both involve the educational values, as the research itself is a process of learning (Moore, 1993). Jepson et al. (2015) further linked the research and monitoring values with the educational values by stating that the research and monitoring activities provide "inspirational and educative form of re-creative experience" to the park visitors. Hence, the nature conservation objective of the park provides a solid reason for scientific research to be translated into popular science through interpretation programs in order to connect the general public with their natural heritage thus enhancing public commitment. Moreover, it is also important to recognize that monitoring is vital to understand the changes of nature over the time and space. Specifically, Davis et al. (2003) refer the key role of ecological monitoring as an indicator to appraise the effectiveness and appropriateness of the on-going conservation activities and also to serve as an early warning system to any unfavorable change that emerges over time. Meanwhile, Davis et al. (2003) acknowledged monitoring as an essential element because "scientific knowledge and park stewardship depends on continuing improvements of knowledge and understanding of parks from scientific interactions of monitoring and experimentation (management actions) that frame, test, and falsify myriad hypotheses." Interestingly, according to Jepson et al. (2015), these scientific values of protected areas also brings in financial revenue, e.g., in the case of the government of Tanzania through travel by researchers and payment of a fee associated with working in protected areas.

13.3   KINABALU PARK

Located in the state of Sabah on Borneo Island, Kinabalu Park is the first World Heritage Site of Malaysia. It was inscribed into the UNESCO World Heritage List for its outstanding universal values by meeting two natural selection criteria, i.e., criterion (ix) and (x). Criterion ix concerns outstanding examples representing significant on-going ecological and biological processes in the evolution and development of terrestrial, freshwater, coastal, and marine ecosystems and communities of plants and animals, while Criterion ix concerns the most important and significant natural habitats for *in situ* conservation of biological diversity, including those containing threatened species of outstanding universal value from the point of view of science or conservation (UNESCO WHC, 2013). According to the IUCN management category system, Kinabalu Park is a type II protected area. The park was established in 1964 following the gazettement of Sabah National Parks Ordinance in 1962. Covering an area of 75,370 ha, the park is accessible via road within a two-hour drive (or approximately 90km) from the state capital city - Kota Kinabalu City Centre. The park headquarter has an elevation of 1,560 m above sea level and hosts the highest peak between the Himalayas and New Guinea, Mount Kinabalu which soars up to an elevation of 4,095.2 m (or 13,435 ft) above sea level. The geological formation of Mount Kinabalu is said to have begun about 40 million years ago. According to Tain Choi in Wong and Phillipps (1996), the collision process between tectonic plates under the sea millions of years ago resulted in tremendous heat that melted the rocks which later became solidified and turned into Kinabalu batholith. It was during this collision process that the magma (melted rocks) and the old rocks were brought up along the fractures of the collided tectonic plates. Nonetheless, it was around 10 million years ago that the Crocker-Trus Madi area had been pushed up onto the surface of Earth and the plate movements have since then slowed down. The uplifted rocks were later shaped and sculpted through erosion caused by rainwater and gravity. This resulted in the rocky mass summit of Mount Kinabalu today which is said to be still growing at the annual rate of 5mm.

While the park is young geologically, it hosts abundant species of flora and fauna due to its huge altitudinal range. Kinabalu Park is the Centre of Plant Biodiversity for Southeast Asia. It is also one of the world's thirteen biodiversity hotspots and has also been designated as one of the primary centers of plant diversity in the world. It is estimated that there are at least 5,000 to 6,000 vascular plants exist in the park which comprises of 200 families and 1,000 genera and made up about 2.5% of flora on Earth (Goh, 2008). Among the prominent species are the pitcher plants and Rafflesia. In terms of fauna, Kinabalu is home to the majority of Borneo's wildlife ranging from birds, reptiles, mammals, insects, and much more. There are at least 1,000 species of months, 600 out of 900 Borneo's butterflies and over 100 species of mammals found in the park.

The park has significant dry and wet seasons where the dry period is observed between February and May while the wet period is between October and January, as a result of the annual southwest and northeast monsoon. For that, the average annual rainfall in the park is relatively high where Poring station receives around 2500mm of rainwater and even higher in Park HQ (Headquarter), with an annual average of 4,000 mm. Influenced by the altitude, the average temperature in Kinabalu Park also differs between substations. Park Headquarter which is located at an elevation of 1,560 m above sea level has a daily temperature around 20°C. At Mesilau Nature Resort which is located at a higher altitude of 2,000 m, the temperature is slightly lower at 15–18°C. Poring Hot Spring which stands at an altitude of 500m above sea level has an average temperature at around 25–30°C. The temperature at Panar Laban/Laban Rata (3,344 m above sea level) is around 2–10°C and can drop below freezing point during the nights (Goh, 2008).

The park authority of Kinabalu Park is the Sabah Parks Board of Trustee. As of 2013, there is a total of 148 staff consisting of both the permanent administrative staff and those on a temporary basis, distributed in all stations within the Research and Education Division and Administrative and Enforcement Division. For the management and enforcement purpose, there are a total of seven stations in the park of which three are the main stations catering for tourism activities known as the Park HQ, Mesilau Nature Resort, and Poring Hot Spring. The other four substations, i.e., Sayap, Nalapak, Serinsim, and Monggis substations initially served as outposts along the park boundaries for monitoring purposes. Nevertheless, these substations are becoming increasingly popular among visitors.

Apart from Sabah Parks Board of Trustee, the park also comprises of several other agencies in charge of providing and managing the tourism-related activities and facilities. Sutera Sanctuary Lodge (SSL) operates the accommodations, restaurants, and souvenir shops inside the park. Koperasi Taman-Taman Sabah (KOKTAS), a cooperative organization of the Sabah Parks' staff offers the transportation services within the park and to and out of the park various stations. The Kinabalu Mountain Guide Association provides mountain guiding service. Originated from the local villages, these guides are licensed by Sabah Parks to render their guiding service to the park climbers. Porter service is also provided by the local villagers in Kinabalu Park. There are currently 193 registered mountain guides and about 118 registered porters.

13.4   TOURISM SIGNIFICANCE OF KINABALU PARK

Since it was opened to the public, Kinabalu Park has become one of the major attractions among tourists to Malaysia and particularly to Sabah. The number of tourists arriving in the park increases from year to year in all the substations. Tourism is one of the main sources of income for the park itself. Since the park was opened to the public in 1965, the number of visitors arriving in Kinabalu Park had increased annually and reached more than half a million visitors per year, and Mount Kinabalu itself is climbed by approximately 50,000 climbers annually. Kinabalu Park attracts a greater number of domestic visitors compared to foreign visitors, but the number of climbers between the two categories of visitors is relatively the same. The overall statistics on the visitors' arrival in Kinabalu Park revealed a very significant difference in the numbers of total visitors to Kinabalu Park since the park was opened until 2012. In 1965, just after the park was opened to the public, it received a total of 879 visitors, and visitor arrival began to increase over the years until it reached five figures in 1972 (11,322). Kinabalu Park received roughly around 200,000 to 300,000 visitors annually from the period of 1984 until 1999. From 2000 to 2009, around 400,000 visitors came to Kinabalu Park every year, and the number of visitors reached more than 600,000 annually starting 2010 to this day. From 2010 to 2014, the number of visitors in Kinabalu Park fluctuated between 600,000 to 700,000 visitors annually.

The number of visitors to Kinabalu Park rose gradually with a few drops in numbers at times. This proved that Kinabalu Park was becoming an increasingly popular tourist destination from year to year based on the visitors' arrival. The number of visitors in Kinabalu Park fluctuated throughout the years since it was opened to the public, but it kept on rising gradually to more than half a million visitors per year. The fluctuation of visitors could be due to a number of reasons, one of them being the status of World Heritage Site that was awarded to Kinabalu Park in the year 2000. The listing managed to attract more visitors to Kinabalu Park, and it highlighted the park as one of the main tourist attractions in Sabah and even in Borneo and Malaysia. The launch of 'Malaysia Truly Asia' in 1999 along with other promotions such as 'Think Tourism,' 'Mesra Malaysia' and 'Malaysia Welcomes the World' contributed to the rise of visitors' arrival in Kinabalu Park.

There three main stations in Kinabalu Park catering to mostly tourism and recreational activities are known as Park HQ, Poring Hot Spring, and Mesilau Nature Resort. Among the three main stations of Kinabalu Park, Poring Hot Spring received the most number of visitors every year. In Poring, the number of domestic visitors outweighed foreign visitors greatly. This was probably due to the fact that Poring is located in Ranau District, which is heavily populated with local communities and most of them regard the Hot Spring as their recreational place for regular visitation. The second most visited station in Kinabalu Park is Park HQ in Kundasang. Park HQ is considered as the main attraction in Kinabalu Park because this is where most climbers begin their climb to the mountain, and there is a wide range of activities offered here. This substation is also known for its strategic location to get a glimpse of Mount Kinabalu. The difference in a number of visitors arriving in Park HQ and Poring Hot Spring did not differ greatly. Mesilau Nature Resort received a slightly lower number of visitors compared to Poring Hot Spring and Park HQ. This was most likely due to the fact that Mesilau Nature Resort was only opened to visitors in 1998, making it one of the newest stations in Kinabalu Park that caters to visitors. The station also has lesser visitors due to its remoteness and difficult accessibility as it is located in the higher parts of Kundasang. The activities provided here by Sabah Parks are also relatively few compared to other stations. Mesilau does, however, provide attractions that other stations do not such as its mossy oak-chestnut forest and the abundance of pitcher plants especially *Nepenthes rajah* along its trails.

But due to its high number of visitors annually, there are risks of overuse and a concern that the management might not be able to cope with the high number of visitors and their demands, which can cause serious damage to the environment. However, there is an interdependent relationship between tourism and conservation in which both benefit from one another. The high numbers of protected areas existed do not guarantee the effectiveness of protection, but it is good management and governance that provide balance for both tourism and conservation in protected areas. Hence, the framework reflecting the governance adopted by the park's authorities determine the sustainability of the parks.

13.5   COMMUNICATING SCIENTIFIC FINDINGS THROUGH EDUCATIONAL PROGRAMS INTRODUCED IN RECREATIONAL AND TOURISM ACTIVITIES

Sabah Parks Board of Trustee's mission is "to preserve areas in the state of Sabah that contain outstanding natural values as a heritage for the benefits of people, now, and in the future." Five goals have been identified to achieve this mission, and those related to conservation programs are to make the parks as world-class parks, to be the center of excellence for tropical ecosystem research, nature tourism hotspots as well as the most exciting nature education programs. It is clear that the park authority intends to promote its conservation efforts by combining them with tourism and nature education programs. Initially, the park's tourism facilities and activities were both managed by Sabah Parks. After the privatization of the accommodation and eateries facilities in the park, Sabah Parks then began to focus on the 'value-chained' conservation activities including concentrating on the inventory of species, administrative, and enforcement. These findings are translated into educational programs that are then introduced to the general public through tourism activities.

There is a wide range of activities offered to visitors in Kinabalu Park. The type of activities varies among the substations. Nevertheless, all of them are nature-based activities with educational elements. Among the activities offered in Kinabalu Park HQ are the visits to the Botanical Garden, Kinabalu Natural History Gallery, Sabah Parks Exhibition Hall, video show, nature trails, mountain climbing, and Mountain Torq. The Botanical Garden at Park HQ is one of Sabah Parks 14 examples of *ex situ* conservation efforts throughout its parks in Sabah. Formerly known as Mountain Garden, it was established in 1981, and its purpose is to preserve rare and endangered species and also for education and public awareness on the importance of conservation (Goh, 2008). A guided walk is conducted in English by park's staff three times daily. The Kinabalu Natural History Gallery, on the other hand, is located at the conservation building along with the park's office. The gallery aims at telling the history of Kinabalu Park and both its natural and cultural values. The gallery is divided into different sections comprising of an introduction, the geology of Kinabalu Park, flora, and fauna, and human resources in the park. Apart from that, visitors are able to participate in jungle trekking without the assistance of guides through the nine trails available at the Park HQ. The nine trails are Liwagu trail, Silau-Silau trail, Mempening trail, Kiau View trail, Bukit Burung trail, Bundu Tuhan View trail, Pandanus trail, Bukit Ular trail, and Bukit Tupai trail (Figure 13.1a and 13.1b). However, Sabah Parks also offers a daily guided tour along one of the trails known as Silau-Silau trail.

Out of all the activities, mountain climbing is the most popular activity in the park where approximately 50,000 climbers (or 17.5% of total visitors) made an attempt to conquer the peak yearly. Climbers ascend to the summit of Mount Kinabalu from two starting points, one at Park HQ and the other at Mesilau Nature Resort. The climb normally takes two days for climbers to reach the summit and descend back down. The mountain trail stretches as far as 8.7km. Climbers must hire mountain guides from the Kinabalu Mountain Guides Association to climb the mountain.

FIGURE 13.1   (See color insert.) (a) and (b): Entrance of Pandanus Trail and Bukit Tupai Trail.

Mountain Torq was first introduced at Kinabalu Park in the year 2007 (Figure 13.2). It is known as the World's highest and Asia's first Via Ferrata providing mountaineering activities using protected mountain path consisted of suspension bridges, rungs, cables, and more (Mountain Torq, 2011). Via Ferrata activity is a more adventurous and physically challenging climb under Mountain Torq as it uses different pathways to top of the mountain than the usual climbing trail. As of this moment, Mountain Torq offers two different mountaineering activities, Low's Peak Circuit and Walk the Torq. Both are located near the summit of Mount Kinabalu. Both mountaineering activities offer visitors the opportunity to engage in rock climbing, walk on suspension bridges, and witness breath-taking sceneries into sections of the summit that are inaccessible to climbers using normal hiking trails.

Poring Hot Spring, located approximately 40km from the Park HQ is one of the substations of Kinabalu Park. The station is the most visited station in Kinabalu Park. Poring is famous for its hot sulfur baths and receives big crowds especially during the weekends and school holidays. The outdoor hot sulfur baths are free for visitors to use, and indoor sulfur baths are also available at various charges for those wanting privacy. Poring Hot Spring also houses a bunch of *ex situ* gardens such as Butterfly Farm, Lowland Tropical Garden, Orchid Conservation Center, Ethnobotanical Garden, Mini Botanical Garden, Bamboo Garden and Rafflesia Garden (Figures 13.3a–13.3c). These gardens serve the same purpose as the Botanical Garden at Park HQ. Aside from being opened to the public, researches are also conducted here in the gardens, and they are constantly monitored by the Park.

FIGURE 13.2   (See color insert.) Mountain Torq office in Kinabalu Park.

FIGURE 13.3 **(See color insert.)** (a–c): Poring Hot Spring Visitor Centre, Butterfly Farm Exhibition Gallery and Rafflesia blooming.

Another station in Kinabalu Park is Mesilau Nature Resort, opened in 1998 and located around 14km east of Park HQ at an elevation of 2,000 m above sea level (Figure 13.4a–13.4c). This station offers an alternative climbing route for climbers to the summit of the mountain using Mesilau Nature Trail. The trail is more scenic and is known to have an abundance of pitcher plants particularly one large and giant-sized *Nepenthes rajah*. One of the main attractions in Mesilau is the Nepenthes Rajah Nature Trail, a 20 minutes guided walk into the mossy forest of Mesilau. Another activity offered in Mesilau is the Guided Walk Trail, though this activity can also be done without a guide. It is a 30-minute walk starting from the Bishop Head's Hostel into the oak-chestnut forest of Mesilau. Mesilau Nature Resort has been closed due to mud flood damage since the earthquake that struck the park in June 2015.

FIGURE 13.4   (See color insert.) (a–c): Mesilau Nature Resort, Mesilau Nature Center, and map of Mesilau complex.

Not all activities are charged in Kinabalu Park. Nevertheless, there are basically two types of fees applicable to Kinabalu Park's visitors, i.e., the entrance fee and users' fee. The fees are set by the Sabah Parks and applied in Kinabalu Park. Kinabalu Park uses a two-tiered pricing system for the entrance fee as well as admission into places and activities. The users' fee in Kinabalu Park is divided into activities and facilities fees. While offering the activities with educational components to raise the awareness and appreciation of nature among the general public, charging a fee helps to ease the administrative burden of the park authority in its administrative and enforcement routine.

Climbing Mount Kinabalu requires visitors to pay for additional charges including climbing permit, insurance, accommodation, and more. It is also compulsory for climbers to hire mountain guides during their climb as a safety measure as well as to share knowledge about the park with the climbers. The rate for the mountain guides have undergone a few changes from time to time after being reviewed by the mountain guides association and Sabah Parks. The rate also depends on the number of climbers and their destinations or the trails used by the climbers. One Kinabalu mountain guide is only allowed to guide a maximum of six climbers per climb. Additional mountain guides will be added if the number of climbers exceeds six people.

13.6 CHALLENGES FACED IN COMMUNICATING CONSERVATION IMPORTANCE

In Kinabalu Park, substantial efforts have been made to combine the educational programs into its existing recreational land tourism activities. Through interesting educational programs in an informal setting of interpretation, it is more receptive among the visitors in highlighting their individual responsibility and role in supporting nature conservation. It is also proven to be a better form of communication in order to enhance the general public's level of awareness of nature conservation. In the park, fees are charged on entrance and also selected activities. This practice has helped to ease the burden of park authority in terms of its administrative and management costs. However, these efforts of promoting nature conservation through the interpretive programs introduced in the tourism activities are not without challenges.

Firstly, the lack of expertise in the interpretive program is a major challenge to the park educational programs. Without the effective interpretive program, it is quite unlikely to translate the complicated scientific findings of nature conservation into interesting learning materials in a natural setting. Secondly, the appreciation of interpretation and interpretive programs within the park authority is still low, and this is closely related to the first challenge. Many parts within the park that have the interpretive potential have not been explored while the existing, have not been fully optimized. The guides who were hired in the park were not given proper formal training prior to the actual guiding activity. In fact, the staff in the education and interpretative unit learns and improves their guiding skill through their actual guiding experience. The exhibition hall and slideshow, which have been active for decades, were no longer in operation due to the poor turnout among the visitors. The exhibition hall was equipped with information regarding Kinabalu Park and its substations, vegetation zones, and ecosystem. The hall also housed information about the various research and education programs carried out by Sabah Parks, not just in Kinabalu Park but also in other parks as well (Figure 13.5). The video show, located at the basement of the center shows viewing of a video titled "The Beacon of Diversity." Some of the feedback from the visitors (especially the repeat visitors) was about the old exhibition materials used that required replacement and should be periodically updated with the recent scientific findings. Similar feedback was received on the video slideshow which should also be replaced with a newer and updated video about the park.

FIGURE 13.5   **(See color insert.)** (a), (b) and (c): Interpretive guiding by Sabah Parks' staff, visitors at the Exhibition Hall and Mountain guiding activity in Kinabalu Park.

Thirdly, the English literacy level among existing mountain guides still requires substantial improvement, as it was consistently reflected in the climbers' feedback through the periodic questionnaire survey conducted in the park since 2005. This factor is also assumed to have thwarted the communication of park conservation values to the visitors. Fourthly, while tourism activities bring in the financial revenue that in turn contributes to the various conservation efforts in the park such as scientific research and enforcement, the heavy influx of tourists into the park may on the other hand harm the park's fragile environment. Although the "honeypot" concept has been introduced in the park by attracting the visitors to only three main stations, i.e., Kinabalu Park HQ, Poring Hot Spring and Mesilau Nature Resort, this risk must be taken very cautiously by the park authority in long run in managing tourism activities without compromising its nature conservation objective.

Lastly, the monitoring of the effectiveness of various conservation efforts in Kinabalu Park is still not obvious, or nearly absent. Monitoring is crucial as it serves as an indicator to any arising risk of changes to the ecosystem as well as the justification for any continuous or change to the management strategies introduced to the park management.

13.7   CONCLUSION

The educational program in protected areas, especially in world heritage sites is a crucial element for its role in influencing the public perception of nature conservation and to enhance the appreciation among the general public about nature. Nevertheless, protected areas in developing countries, in particular, are facing the challenges in terms of the lack of financial support thus having to generate their income through charging a fee on the tourism activities. Without careful management, it may lead to another issue of human disturbance to the sensitive ecosystem in the park. Another challenge is related to the lack of appreciation of interpretation in protected areas which must be overcome in order to sustainably promote nature conservation through the translation of research findings into interesting and easily understood language for general appreciation and to update the general public about the on-going nature conservation efforts. Furthermore, due to the lack of expertise and financial capacity, the educational programs cannot be sustained like in the case of the exhibition hall and slideshow program in Kinabalu Park. This is also applied to the scientific research programs of which many have not yet involved in the monitoring component.

ACKNOWLEDEGMENTS

Authors would like to acknowledge Ministry of Higher Education for the fund provided under Fundamental Research Grant Scheme (FRGS FP047–2013A) as well as the technical assistance and supports given by Sabah Parks' staff during the course of research work.

KEYWORDS

- **biodiversity**
- **conservation**
- **educational values**
- **environmental education**
- ***ex situ* conservation**
- **IUCN**
- **Kinabalu Park**
- **Malaysia**
- **monitoring values**
- **protected areas**
- **recreational activities**
- **recreational values**
- **sustainable tourism**
- **tourism**
- **UNESCO**
- **Yellowstone national park**

REFERENCES

Balmford, A., Green, J. M. H., Anderson, M., Beresford, J., Huang, C., Naidoo, R., & Manica, A., (2015). Walk on the wild side: Estimating the global magnitude of visits to protected areas. *PLOS Biology, 13*(2), e1002074.

Choi, D. L. T., (1996). Geology of Kinabalu. In: Wong, K. M., & Phillipps, A., (eds.), *Kinabalu Summit of Borneo* (pp. 19–29). The Sabah Society, Kota Kinabalu.

D' Antonio, A., Monz, C., Newman, P., Lawson, S., & Taff, D., (2013). Enhancing the utility of visitor impact assessment in parks and protected areas: A combined social-ecological approach. *Journal of Environmental Management, 124,* 72–81.

Davis, G. E., Graber, D. M., & Acker, S. A., (2003). National parks as scientific benchmark standards for the biosphere, or, how are you going to tell how it used to be, when there's nothing left to see? In: Harmon, D., & Putney, A. D., (eds.), *The Full Value of Parks: From Economics to the Intangible* (pp. 129–140). Rowman and Littlefield Publishers, Inc., Maryland.

Day, J., Dudley, N., Hockings, M., Holmes, G., Laffoley, D., Stolton, S., & Wells, S., (2012). Guidelines for applying the IUCN protected area management categories to marine protected areas. *IUCN, Gland,* 1–36.

Dudley, N., (2008). *Guidelines for Applying Protected Area Management Categories.* IUCN, Gland (pp. 1-86).

García-Rodríguez, M., & Fernández-Escalante, E., (2016). Geo-climbing and environmental education: The value of La Pedriza Granite Massif in the Sierra de Guadarrama National Park, Spain. *Geoheritage,* 1–11.

Goh, H. C., & Rosilawati, Z., (2014). Conservation education in Kinabalu Park, Malaysia: An analysis of visitors satisfaction. *Journal of Tropical Forest Science, 26*(2), 208–217.

Goh, H. C., (2008). Sustainable tourism and the influence of privatization in protected area management: A case of Kinabalu Park, Malaysia. *Ecology and Development Series 57,* 1–179.

Harmon, D., & Putney, A. D., (2003). *The Full Value of Parks: From Economics to the Intangible* (pp. 1–360). Rowman and Littefield Publishers, Inc., Maryland.

Harmon, D., (2004). Intangible values of protected areas: What are they? Why do they matter? *The George Wright Forum, 21,* 9–22.

Jepson, P., Caldecott, B., Milligan, H., & Chen, D., (2015). *Project for Protected Area Resilience: A Framework for Protected Areas Asset Management* (pp. 1–178). Smith School of Enterprise and the Environment, University of Oxford, UK.

Knapp, D., (2005). Environmental education and environmental interpretation: The relationships. In: Hungerford, H. R., Bluhm, W. J., Volk, T. L., & Ramsey, J. M., (eds.), *Essential Readings in Environmental Education* (pp. 293–300). Stipes Publishing, Illinois.

Kopnina, H., (2012). Education for sustainable development (ESD): The turn away from "environment" in environmental education? *Environmental Education Research, 18,* 699–717.

Manning, R. E., (2014). Research to guide management of outdoor recreation and tourism in parks and protected areas. *Koedoe, 56,* 1–7.

Moore, J. A., (1993). Science as a Way of Knowing (pp. 1–530). Harvard University Press, Cambridge (Massachusetts).

National Park Service, (2000). Strategic Plan for Yellowstone National Park (FY 2001–2005). National Park Service, Wyoming.

Pickering, C. M., & Hill, W., (2007). Impacts of recreation and tourism on plant biodiversity and vegetation in protected areas in Australia. *Journal of Environmental Management, 85,* 791–800.

Putney, A. D., (2003). Introduction: Perspectives on the values of protected areas. In: Harmon, D., & Putney, A. D., (eds.), *The Full Value of Parks: From Economics to the Intangible* (pp. 3–12). Rowman, and Littlefield Publishers, Inc., Maryland.

Ruschkowski, E. V., Burns, R. C., Arnberger, A., Smaldone, D., & Meybin, J., (2013). Recreation Management in parks and protected areas: A comparative study of resource managers' perceptions in Austria, Germany, and the United States. *Journal of Park and Recreation Administration, 31*, 95–114.

Stab, S., & Henle, K., (2009). Research, management, and monitoring in protected areas. In: Gherardi, F., Corti, C., & Gualtieri, M., (eds.), *Biodiversity Conservation and Habitat Management* (Vol. 1, pp. 128–152). Eolss Publishers Co. Ltd., Oxford.

Wals, A. E. J., Brody, M., Dillon, J., & Stevenson, R. B., (2014). Convergence between science and environmental education. *Science, 344*, 583–584.

CHAPTER 14

PUBLIC AWARENESS TOWARDS ECOLOGICAL SERVICES AND CULTURAL SIGNIFICANCE OF MANGROVES IN SOUTHERN PENINSULAR MALAYSIA

HONG CHING GOH

Water Engineering and Spatial-Governance (WESERGE) Research Centre, University of Malaya, 50603 Kuala Lumpur, Malaysia

Department of Urban and Regional Planning, Faculty of Built Environment, University of Malaya, 50603 Kuala Lumpur, Malaysia

14.1 INTRODUCTION

Mangroves are forests consisting of large shrubs and small plants found in the area where land and sea meet in the subtropics and tropic regions between latitudes 25°N and 25 S, and where the temperature of the sea surface is always above 16°C (Alongi, 2002). Mangroves are important coastal resources providing a vast range of ecological services and holds cultural values (Goh, 2016). These include climate regulation, coastal defense, shoreline stabilization, maintenance of water quality, breeding, and nursery sites for fishes and invertebrates, fisheries, home ground for plant and timber products as well as spiritual and cultural values especially among the indigenous people as well as the more recent use - tourism (Tinch and Mathieu, 2011). These values are estimated at beyond US$ 1.6 billion of worth in 2010 (Duke, 2007; Barbier, 2011; Haines-Young and Potschin, 2011; Horowitz et al., 2012).

At global context, mangroves help to regulate climate through its function as a carbon sink. In the tropic regions, it is recognized as one of the most carbon-rich forests containing an average of approximately 1000 mg carbon per ha (van Bochove et al., 2014). According to Mcleod et al. (2011), global deforestation resulted in the mangroves generating 10% of the total annual carbon emission despite the fact that mangroves only represent 0.7% of tropical forest area. This particular function of mangroves is vital in supporting the global efforts to cut down the carbon emission as declared by both the developed and developing countries in the 2014 United Nations Climate Change Conference (Mcleod et al., 2011). Protecting mangroves at destination level would significantly contribute to this global commitment.

At the site-specific level, mangroves reduce the natural disaster risk which threatens the coastal cities such as the rising of seawater and the intensified storms (Spalding et al., 2014). The unique structures of the root help lessen the currents' velocity, absorb, and reduce the impacts of strong winds, floods, and tropical storms accompanied by tidal waves (Goh, 2016). At the same time, mangroves' root structures also trap soils and sediments above ground, keeping the sediments and other solids from being washed offshore thereby securing the shorelines and reducing coastal erosion (WWF, 2016; www.mangrove.org). Mangroves also serve to maintain the water quality of an area by filtering the sediments, contaminants, and nutrients apart from playing a role in biofiltration and waste processing. These functions of mangroves are very important especially in coastal areas where aquaculture activities and have slow tides (Snedekar and Brown, 1981). Although mangroves may not solve the problem of water pollution, its degradation and loss can lead to a decrease in the water quality which has become a critical issue in many urbanized and industrialized areas (Goh, 2016).

Apart from being environmentally significant, mangroves are also an important breeding and nursery grounds for marine and terrestrial species such as birds, fishes, reptiles, and mammals thus making it economically significant to the local people as well considering it is their source of livelihood and for commercial fishing (Jusoff and Haji Dahlan, 2008). It is said that mangrove forests supported at least one-third of fishes harvested in South East Asia. Economically, mangroves are also the source of woods used for construction, firewood, charcoal, and furniture (van Bochove et al., 2014). The locals also rely on mangroves for their woods in which they utilize for cooking and heating (Amjad et al., 2007).

Apart from these tangible values, there are a range of economically immeasurable intangible values provided by mangroves. The most significant ones are the spiritual and cultural values (van Bochove et al., 2014). Many indigenous communities have been associated with the mangroves for centuries. One of the most prominent examples can be seen in the Torres Straits Islanders of Australia who had managed the mangroves for more than 40,000 years (Mangrove Watch Australia, 2013). More than five decades ago, the Seletar people who are also known as the Sea Gypsies of the Johor Straits had settled down at the southern coast of Johor state of Malaysia and had since been depending on mangroves as their home and for livelihood (Nobaya et al., 2012).

Despite the general impression that mangroves tend to be smelly and muddy and often perceived as breeding grounds for mosquito (Young, 2010), tourism, and recreational activities have been recently introduced in mangroves and its potential is currently on the rise (Ratnayake, 2012). Among the activities introduced in the mangroves are bird and wildlife watching, fishing, board-walking, and boating around the mangroves area. These ecosystem services of mangroves are popular among visitors to Dongchaigang Nature Reserve in China and Laguna de Resting in Venezuela (van Bochove et al., 2014).

14.2 DESTRUCTION OF MANGROVES

Over the years and especially after the Indian Ocean earthquake and tsunami in 2004, the significance of mangroves in providing these ecological services as well as portraying the cultural significance have been vastly discussed and given further attention (Spalding et al., 2014). In fact, an international treaty known as the Ramsar Convention for conservation and sustainable utilization of the mangrove forests was signed in 1971 to recognize the vast range of values portrayed by mangroves (IUCN, 2012). Nonetheless, the rising recognition does not stop the mangroves from gradually disappearing. According to Giri et al. (2011), the coverage of global mangrove had shrunk from 181,000 km^2 to 137,760 km^2 between the period of 2002 to 2011 in which the coverage recorded in 2002 was already about one-third remaining of the total 50 years ago. Furthermore, 16% out of the 70 mangrove species have been categorized as threatened and listed under the International Union for Conservation of Nature (IUCN) Red List (Conservation International, 2010). The destruction of mangrove is closely related with development and the growth of the human population (McGranahan et al., 2006; McDonald et al., 2008). In fighting against the mangrove deforestation, various efforts have been initiated. Among the more recent effort is the "Mangrove for the Future" (MFF) launched by the IUCN in 2006. One of the MFF objectives is 'key stakeholders empowered to engage in decision-making in support of sustainable management of coastal ecosystems' with a focus on 'building awareness and capacity of civil society and private sector' (Mangrove for the Future Secretariat, 2016). It is based on this focus area that a study was initiated in southern Peninsular Malaysia. It aimed to identify the existing public awareness on the importance of mangroves before appropriate strategies can be developed thereby helping to mitigate the impacts of development on mangrove.

14.3 SOUTHERN PENINSULAR MALAYSIA

The southern Peninsular Malaysia where the state of Johor is located accommodates a total of 29,668 ha mangroves as at 2006 (Ramsar Convention Secretariat, 2014). These mangroves fall into two management categories, i.e., mangrove forest reserves (16,127 ha) and state land mangroves (13,561 ha) (Kamaruzaman and Haji Dahlan, 2008). The mangroves here represent 27.5% of the total remaining mangroves in Peninsular Malaysia. The southern Peninsular Malaysia is also home to three Ramsar sites, i.e., Sungai Pulai Forest Reserve, Pulau Kukup National Park and Tanjung Piai National Park (Goh, 2016). Apart from that, according to the JHEOA (2008), the mangroves are also considered as home to many of the coastal communities particularly the Seletar indigenous whose population is alarmingly estimated at only 1,250 people (Paul et al., 2014). Similar to the global trend, the destruction of mangroves in Malaysia is obvious. According to the Malaysian Nature Society, only 1.8% of the land is covered by mangroves, and more than 50% of these mangroves were lost between 1950 and 1985 (Khalid, 2009). Meanwhile, Gong and Ong (1990) roughly calculated that Malaysia had lost close to 30% of its mangroves prior to 1990 and the rate is expected to continue at 1% a year. Malaysia experienced an overall reduction in the mangroves of approximately 12% since 1980. Of all the states, major loss of mangroves had taken place in Johor, the southern state of Peninsular Malaysia (Alongi, 2002). Being among the fastest growing states in

Malaysia especially with the establishment of Iskandar Malaysia region which covers an area of 2,217 km^2, thrice bigger than Singapore along the southern coastline of Peninsular Malaysia (where 12.5% of Peninsular Malaysia's mangroves are found at as of 2005), the continued survival of mangroves is threatened by the aggressive economic growth (Kanniah et al., 2015).

14.4 GENERAL PERCEPTIONS ABOUT MANGROVES AMONG THE PUBLIC

A questionnaire survey took place in 2015 in order to measure the public perception of mangroves. The survey covered the state capital city of Johor, Johor Bahru and other major towns in the southern coastal areas which consist of Johor Bahru Tengah, Kulai, Pasir Gudang, Pontian, and Gelang Petah. The prepared questionnaire was bilingual. Four enumerators were hired to conduct this survey in five districts within the metropolitan. The selection of respondents was carried out randomly in public areas namely shopping malls and public parks. Two sections in the questionnaire were designed to test the respondents' existing knowledge pertaining to mangroves. The first section gathered the respondents' demographic profile including gender, education level, age group, income level, and employment. The second section emphasized on the knowledge statements about mangroves, and a total of 19 knowledge statements were highlighted under this section, reflecting the ecological services and significance of mangroves as described in the preceding section (Table 14.1).

A total of 400 completed samples were used for analysis purposes. Statistical analysis was performed on the collected sample using the software package IBM Statistical Package for the Social Sciences (SPSS) Statistics version 21. Descriptive analysis included both frequency and cross-tabulation while the inferential analysis using the Pearson Chi-Square test was further employed in order to test whether or not there were any significant differences in the relationship between selected attributes.

The analysis results suggest that almost two-thirds of the respondents were females (61%) compared to 39% who were males. Majority of the respondents were in the four younger age groups ranging between 18–25 years old (26.0%), 26–30 (20.2%), 31–40 (18.8%) and 41–50 years old (18.0%), which made up 83% of the total respondents. 5.0% of the respondents were below 18 years old, while another 12.0% were above the age of 51 (11.0% fell under the 51–60 years old age group and 1.0% were above 60 years old). Majority of the respondents lived in Johor Bahru (30.5%) and Johor Bahru Tengah (29.8%). Respondents from Kulai, Pontian, and Gelang Patah made up 10.0% each while 9.8% of the respondents were from Pasir Gudang. In terms of education level, one-third of the respondents were either university graduates or attending the program at universities while another 13.2% were taking university postgraduate program or graduated at the master or doctoral level. 23.5% of the respondents completed secondary school education while another 16.2% were college graduates. Another 8.0% of the respondents completed primary school education. The respondents consisted of 26.5% general workers and those holding clerical and supervisory positions while 30.8% were professionals, academics, and those in management and senior management positions. Retirees made up 1.2% of the total respondents. 64.5% of the respondents had visited mangroves, and an almost similar percentage of respondents (60.2%) knew about Ramsar site (Table 14.2).

As this survey was aimed to test the general perception towards mangroves among the respondents, 'correct' answer was not determined for each statement. In general, two types of questions were tested in the questionnaire, i.e., general knowledge and the attitude towards mangroves. Statements 10, 17, 18 and 20 are statements related to the attitude among the respondents towards mangroves while the remaining are statements associated to the general knowledge held by the respondents. Specifically, the statements are arranged into five groups, namely (1) general facts about mangroves, (2) mangroves' ecological services, (3) negative perception towards mangroves, (4) cultural significance of mangroves, and (5) attitudes towards mangroves. Pearson Chi-Square tests reveal that there are statistically significant differences between respondents' demographic profiles including age, education, employment, and the residence location of the respondents in terms of their responses towards the various statements (p-value <0.05). The details of the significance are shown in the respective sub-sections according to the statement categories.

TABLE 14.1   The Questionnaire

Attribute	Questions		Answers
Demographic profile	Gender		Male, female
	Age		Below 18 years old, 18–25 years old, 26–30 years old, 31–40 years old, 41–50 years old, 51–60 years old, 61 years old and above
	Where do you live?		Johor Bahru, Johor Bahru Tengah, Kulai, Pasir Gudang, Pontian, Gelang Patah, other area
	Education		Primary school, secondary school, college, university undergraduates, university postgraduates, others
	Employment		Student, home duty/ housewife, general worker, clerical, supervisory, professional, academics, management, senior management, self-employed, retired, others
Awareness/ knowledge	Q1.	Have you ever visited a mangrove forest?	Yes, no
	Q2.	Do you know what a Ramsar site is?	Yes, no
	Q3.	Mangrove forest is located in coastal areas	True, false, not sure
	Q4.	They are found in the tropical and sub-tropical countries	True, false, not sure
	Q5.	It is located in a rural area	True, false, not sure
	Q6.	It is an important ecosystem	True, false, not sure
	Q7.	It is a place where indigenous people live	True, false, not sure

TABLE 14.1 *(Continued)*

Attribute	Questions		Answers
	Q8.	Mangrove forests reflect indigenous cultural values	True, false, not sure
	Q9.	Mangrove forest is important for coastal protection against natural disaster, e.g., the tsunami	True, false, not sure
	Q10.	Mangrove forest is important to reduce coastal erosion	True, false, not sure
	Q11.	Mangrove forests are sources of timber/wood	True, false, not sure
	Q12.	In my opinion, it is not as valuable as other tropical forests	True, false, not sure
	Q13.	I do not know much about mangroves compared to other forests	True, false, not sure
	Q14.	Mangroves are overgrown, messy, and smelly areas	True, false, not sure
	Q15.	Mangroves breed mosquitoes	True, false, not sure
	Q16.	I do not know what a mangrove forest is	True, false, not sure
	Q17.	Johor has one of the largest mangrove forests in Malaysia	True, false, not sure
	Q18.	It is facing serious destruction	True, false, not sure
	Q19.	Mangrove conservation is more important than waterfront development	True, false, not sure
	Q20.	Mangroves should be replaced with development as the land is very expensive in urban areas	True, false, not sure

TABLE 14.1 *(Continued)*

Attribute	Questions		Answers
	Q21.	Tourism/recreational activities are offered in mangroves	True, false, not sure
	Q22.	Mangroves must be reserved as protected areas	True, false, not sure

TABLE 14.2 General Perception/Knowledge about Mangroves among the Respondents (%)

No.	Statement	Yes	No	Not Sure
1	Mangrove forest is located in coastal areas	78.8	4.8	16.5
2	They are found in the tropical and sub-tropical countries	73.0	0.2	26.8
3	It is located in a rural area	89.8	4.5	5.8
4	It is an important ecosystem	96.2	0.2	3.5
5	It is a place where indigenous people live	47.8	13	39.2
6	Mangrove forests reflect indigenous cultural values	55.0	5.2	39.8
7	Mangrove forest is important for coastal protection against natural disaster, e.g., the tsunami	92.2	1.2	6.5
8	Mangrove forest is important to reduce coastal erosion	96.2	0.5	3.2
9	Mangrove forests are sources of timber/wood	73.0	16.5	10.5
10	In my opinion, it is not as valuable as other tropical forests	14.5	78.8	6.8
11	I do not know much about mangroves compared to other forests	41.8	45.8	12.5
12	Mangroves are overgrown, messy, and smelly areas	29.0	54.5	16.5
13	Mangroves breed mosquitoes	27.0	53.5	19.5
14	I do not know what a mangrove forest is	23.0	65.2	11.8
15	Johor has one of the largest mangrove forests in Malaysia	61.8	3.0	35.2
16	It is facing serious destruction	81.2	5.0	13.8

TABLE 14.2 *(Continued)*

No.	Statement	Yes	No	Not Sure
17	Mangrove conservation is more important than waterfront development	82.8	4.5	12.8
18	Mangroves should be replaced with development as the land is very expensive in urban areas	7.0	84.8	8.2
19	Tourism/recreational activities are offered in mangroves	91.2	3.0	5.8
20	Mangroves must be reserved as protected areas	95.2	2.2	2.5

14.4.1   GENERAL FACTS ABOUT MANGROVES

Table 14.3 shows the respondents' response to the general facts about mangroves. The analysis findings reveal that majority of the respondents responded correctly (yes). Nonetheless, more than one-third of the respondents (35.2%) were not aware of the fact that the state of Johor hosts one of the largest mangroves Malaysia. Further to that, quite a number of respondents were unsure about the geographical location (16.5%) and the climate zone (26.8%) of mangroves. Respondents also believed that mangroves are generally located in a rural setting (89.8%). Despite these perceptions, it was generally agreed among the respondents that mangroves are facing serious destruction (81.2%). Only 45.8% of the respondents were in the opinion that their level of knowledge regarding mangroves was at par with their knowledge about other forests types, despite only two-thirds (65.2%) knew about mangroves. Table 14.4 indicates the result of the chi-square test where significant differences were observed between the respondents' knowledge on the general facts about mangroves and their residence location, age, education, and employment ($p < 0.05$). This means the respondents' perception of mangroves under this category is influenced by their demographic and geographical attributes. The residence location also influenced the respondents' perception particularly in terms of how 'urbanized' the town is. It was found that the more urbanized the areas were, the more 'not sure' the respondents became in their answers compared to less urbanized areas. As mangroves are commonly found in the suburban and rural areas of which natural landscape is more dominant than built environment, the respondents that reside in developing towns indicated better awareness towards the general facts about mangroves. Age, education, and employment also influenced the results about the general facts of mangroves. The analysis shows that respondents over the age of 40 years old had better awareness when it comes to the general facts about mangroves. Meanwhile, education influenced the level of awareness among the respondents positively for questions 2, 3 and 14. Apart from residence location, age also played a role in the respondents' perception about the rural location of mangroves (3). This is probably caused by the fact that the older generation had the opportunity to witness the evolving landscapes of an area that was once a rural setting but has now been urbanized, compared to the younger generation who did not have the chance.

TABLE 14.3   Response to General Facts about Mangroves (%)

No.	Statement	Yes	No	Not Sure
1	Mangrove forest is located in coastal areas	78.8	4.8	16.5
2	They are found in the tropical and sub-tropical countries	73.0	0.2	26.8
3	It is located in a rural area	89.8	4.5	5.8
11	I do not know much about mangroves compared to other forests	41.8	45.8	12.5
14	I do not know what a mangrove forest is	23.0	65.2	11.8
15	Johor has one of the largest mangrove forests in Malaysia	61.8	3.0	35.2
16	It is facing serious destruction	81.2	5.0	13.8

TABLE 14.4   Chi-Square Results on Demographic Attributes and General Facts about Mangroves

No.	Statement	Location and demographic attribute	p-value
1	Mangrove forest is located in coastal areas	Residence location	0.000*
		Gender	0.034*
		Age	0.017*
		Education	0.002*
		Employment	0.788
2	They are found in the tropical and sub-tropical countries	Residence location	0.000*
		Gender	0.261
		Age	0.695
		Education	0.318
		Employment	0.596

TABLE 14.4   *(Continued)*

No.	Statement	Location and demographic attribute	p-value
3	It is located in a rural area	Residence location	0.038*
		Gender	0.804
		Age	0.039*
		Education	0.262
		Employment	0.508
11	I do not know much about mangroves compared to other forests	Residence location	0.000*
		Gender	0.223
		Age	0.001*
		Education	0.003*
		Employment	0.030*
14	I do not know what a mangrove forest is	Residence location	0.000*
		Gender	0.640
		Age	0.001*
		Education	0.101
		Employment	0.005*
15	Johor has one of the largest mangrove forests in Malaysia	Residence location	0.000*
		Gender	0.224
		Age	0.002*
		Education	0.003*
		Employment	0.034*
16	It is facing serious destruction	Residence location	0.000*
		Gender	0.156
		Age	0.001*
		Education	0.078
		Employment	0.130

14.4.2   MANGROVES' ECOLOGICAL SERVICES

There were five statements that tested on the mangroves' ecological services, i.e., an important ecosystem, coastal protection against natural disaster, reduce coastal erosion, sources of timber/wood and tourism/recreational attractions. All statements scored highly (>90%) except for the role of mangroves as sources of timber/wood. Fewer respondents were aware of the role of mangroves as sources for timber/wood compared to other roles considering 16.5% of them responded incorrectly (No) while another 10.5% were unsure. Most of the respondents were aware of the role mangroves play in coastal erosion reduction due to the issue being constantly highlighted in Malaysia and has frequently appeared in newspapers and television news. The role of mangroves in coastal protection against natural disaster was relatively less known as it is noticeably less significant in Peninsular Malaysia because of its strategic geographical location surrounded by other islands thus making it well protected from the tsunami, for instance. In terms of tourism/recreational attractions, respondents were aware that both Tanjung Piai and Pulau Kukup are mangrove forests gazetted as national parks in the southern Peninsular Malaysia and are promoted as tourism/recreational destinations. The role of mangrove as sources of timber/wood was less-known among the respondents. Instead, Malaysia is widely known as one of the tropical countries exporting hardwood from the tropical rainforests such as chengal, meranti, and belian. These well-known rainforests' timber products possibly overshadow mangroves' timber production which is limited to local use and only for regional consumption.

Table 14.5 shows the results of the chi-square test indicating significant differences between the respondents' perception of mangroves' ecological services and their demographic attributes along with residence location. In general, the respondents were well aware of the importance of mangroves as an ecosystem and its function in reducing coastal erosion. Meanwhile, residence location had influenced the respondents' perceptions towards all mangroves' ecological services except for questions 4 and 8. Mangroves' role as the sources of timber/wood was less known among the respondents and their awareness level towards this role was influenced by the age, education, and employment attributes (Table 14.6).

TABLE 14.5   Response to Mangroves' Ecological Services (%)

No.	Statement	Yes	No	Not Sure
4	It is an important ecosystem	96.2	0.2	3.5
7	Mangrove forest is important for coastal protection against natural disaster, e.g., the tsunami	92.2	1.2	6.5
8	Mangrove forest is important to reduce coastal erosion	96.2	0.5	3.2
9	Mangrove forests are sources of timber/wood	73.0	16.5	10.5
19	Tourism/recreational activities are offered in mangroves	91.2	3.0	5.8

TABLE 14.6 Chi-Square Results on Demographic Attributes and Statements on Mangroves' Ecological Services

No.	Statement	Location and demographic attribute	p-value
4	It is an important ecosystem	Residence location	0.098
		Gender	0.331
		Age	0.201
		Education	0.982
		Employment	0.814
7	Mangrove forest is important for coastal protection against natural disaster, e.g., the tsunami	Residence location	0.009*
		Gender	0.275
		Age	0.406
		Education	0.665
		Employment	0.731
8	Mangrove forest is important to reduce coastal erosion	Residence location	0.106
		Gender	0.225
		Age	0.151
		Education	0.547
		Employment	0.698
9	Mangrove forests are sources of timber/wood	Residence location	0.000*
		Gender	0.332
		Age	0.000*
		Education	0.000*
		Employment	0.025*
19	Tourism/recreational activities are offered in mangroves	Residence location	0.034*
		Gender	0.196
		Age	0.559
		Education	0.900
		Employment	0.097

14.4.3 NEGATIVE PERCEPTION TOWARDS MANGROVES

There were two statements testing on the negative perception towards mangroves, i.e., overgrown, messy, and smelly areas, and breeding grounds for mosquitoes. More than half of the respondents disagreed with the two statements (54.5% and 53.5% respectively) notwithstanding that there were still more than one-fifth of the respondents who perceived mangroves as so (29.0% and 27.0%).

Similar to the analysis findings for the 'general facts about mangroves' and 'mangroves' ecological services,' residence location had a significant influence on the negative perceptions towards mangroves among the respondents (both 12 and 13) as displayed in Table 14.7 and 14.8. More respondents residing in urbanized areas including Johor Bahru and Johor Bahru Tengah as well as urban areas away from the coast had a higher negative perception towards mangroves which is possibly due to the lack of exposure to mangroves as part of the ecosystem within their living environment. Age and education level influenced the respondents' perception towards mangroves. The perception towards mangroves was less negative among respondents who fell under the middle-age groups between 30–60 years old compared to the younger age groups (<30 years old) and the senior respondents (>61 years old). Education has a mixed influence on the perception which seems to be overshadowed by the age attribute where those with post-graduate degrees seemed to have a better awareness of the mangroves, similar to those with primary and secondary school education. Relatively, those who were undergraduates and bachelor degree holders demonstrated lesser awareness of mangroves when questioned whether or not 'mangroves are overgrown, messy, and smelly areas' or whether 'mangroves breed mosquitoes.'

TABLE 14.7 Negative Perception Towards Mangroves among the Respondents (%)

No.	Statement	Yes	No	Not Sure
12	Mangroves are overgrown, messy, and smelly areas	29.0	54.5	16.5
13	Mangroves breed mosquitoes	27.0	53.5	19.5

TABLE 14.8 Chi-Square Results on Demographic Attributes and Statements on 'Negative Perception Towards Mangroves'

No.	Statement	Location and demographic attribute	p-value
12	Mangroves are overgrown, messy, and smelly areas	Residence location	0.000*
		Gender	0.146
		Age	0.022*
		Education	0.174
		Employment	0.008*
13	Mangroves breed mosquitoes	Residence location	0.000*
		Gender	0.148
		Age	0.002*
		Education	0.017*
		Employment	0.052

14.4.4 CULTURAL SIGNIFICANCE OF MANGROVES

The cultural significance of mangroves was tested through two statements: firstly, mangrove is home to indigenous people and secondly, mangrove forests reflect indigenous cultural values. Both statements indicate the highest percentage of uncertainty among the respondents. 39.2% of the respondents were unsure if mangrove is a place where indigenous people live. Similarly, 39.8% were also uncertain about the indigenous cultural values reflected by mangroves (Table 14.9).

Chi-square test results as portrayed in Table 14.10 indicate the influence of residence location and respondents' demographic attributes including education, employment, and age on their perception towards the cultural significance of mangroves. Although most of the respondents were unsure about the cultural significance of mangroves, those that recognized this aspect the most were either postgraduate students or had already completed post-graduate studies. Meanwhile, among others who were also aware of the cultural values of mangroves include those who were self-employed, worked at the supervisory level, professionals, and academics were. In terms of residence location, the results reveal that the respondents from relatively more urbanized areas (Johor Bahru and Johor Bahru Tengah) disagreed with the notion that mangroves carry cultural meanings.

TABLE 14.9   Response to the Cultural Significance of Mangroves (%)

No	Statement	Yes	No	Not Sure
5	It is a place where indigenous people live	47.8	13.0	39.2
6	Mangrove forests reflect indigenous cultural values	55	5.2	39.8

TABLE 14.10   Chi-Square Results on Demographic Attributes and Statements on 'Cultural Significance of Mangroves'

No.	Statement	Location and demographic attribute	p-value
5	It is a place where indigenous people live	Residence location	0.000*
		Gender	0.141
		Age	0.001*
		Education	0.000*
		Employment	0.004*

TABLE 14.10   *(Continued)*

No.	Statement	Location and demographic attribute	p-value
6	Mangrove forests reflect indigenous cultural values	Residence location	0.000*
		Gender	0.004*
		Age	0.215
		Education	0.037*
		Employment	0.017*

14.4.5   ATTITUDE TOWARDS MANGROVES

The attitude towards mangroves was evaluated by inquiring the respondents with a series of statements pertaining to the relative value of mangroves against other types of tropical forests, the relative importance between mangrove conservation and waterfront development, mangrove conservation versus development in general due to the high land price in urban areas, and the mechanism of mangrove conservation. The analysis findings show that the respondents strongly recognized the act of conserving mangroves through gazetting them as protected areas (95.2%). More than four-fifths of the respondents also in agreed that mangroves conservation is more important than waterfront (82.8%) or any type of land development (84.8%). Furthermore, more than three-quarters of the respondents disagreed with the statement saying that mangroves are less valuable than other tropical forest types (78.8%).

Residence location and education level indicate an influence on the respondents' attitudes towards mangroves as shown in Table 14.11. While the general attitude towards mangroves was very positive, less desirable attitude towards the protection of mangroves, in general, was observed among the respondents living in more urbanized areas (Johor Bahru and Johor Bahru Tengah) and areas away from the coast (Pontian). The chi-square test findings also suggest that the higher the level of education is, the better the respondents' awareness about the importance of mangroves became thus implying the need to protect them (Table 14.12).

TABLE 14.11 Attitude Towards Mangroves among the Respondents (%)

No.	Statement	Yes	No	Not Sure
10	In my opinion, it is not as valuable as other tropical forests	14.5	78.8	6.8
17	Mangrove conservation is more important than waterfront development	82.8	4.5	12.8
18	Mangroves should be replaced with development as the land is very expensive in urban areas	7.0	84.8	8.2
20	Mangroves must be reserved as protected areas	95.2	2.2	2.5

TABLE 14.12 Chi-Square Results on Demographic Attributes and Statements on 'Attitudes Towards Mangroves'

No.	Statement	Location and demographic attribute	p-value
10	In my opinion, it is not as valuable as other tropical forests	Residence location	0.019*
		Gender	0.973
		Age	0.373
		Education	0.000*
		Employment	0.405
17	Mangrove conservation is more important than waterfront development	Residence location	0.002*
		Gender	0.199
		Age	0.560
		Education	0.666
		Employment	0.693
18	Mangroves should be replaced with development as the land is very expensive in urban areas	Residence location	0.042*
		Gender	0.873
		Age	0.207
		Education	0.133
		Employment	0.155
20	Mangroves must be reserved as protected areas	Residence location	0.049*
		Gender	0.361
		Age	0.823
		Education	0.034*
		Employment	0.590

14.5 CONCLUSION

This chapter highlights the general awareness towards mangroves conservation among the citizens in southern Peninsular Malaysia. Overall, the findings suggest that respondents had a higher level of awareness when it comes to statements pertaining the ecological services provided by mangroves thus scoring more positively in terms of their attitudes towards mangrove protection, but a slightly lower level of awareness related to the general facts (knowledge) and cultural significance of mangroves. Further to that, only two-third out of all the respondents had visited mangroves before and knew about Ramsar site.

While the general awareness is above average, the research also showed that residence location, age, education, and employment had affected the respondents' level of awareness regarding mangroves. Fewer respondents residing in fully or more urbanized areas such as Johor Bahru and Johor Bahru Tengah were aware of the arrays of ecological services and cultural significance provided by mangroves and in general, they portrayed a higher level of negative perceptions towards mangroves. Similar results were yielded for a residence that are located away from the coast such as Pontian. These findings denote the lack of exposure to mangroves among the urbanized residents attributable to the absence of mangroves in the built environment where they reside in. Relatively, mangroves still exist in other residence locations. Furthermore, respondents from the older age groups indicate better sense of awareness on the importance of mangroves across all aspects. The older age groups were more aware and had experienced landscape changes that evolved from natural to a more urbanized setting during the urbanization process. Education and employment had also influenced the awareness level among the respondents across various aspects. Despite not uniformly, respondents with higher educational background and those in the supervisory positions, along with professionals and academics displayed better awareness related to mangroves' significance compared to other groups.

These analysis findings provide evidence on the influence of the publics' prior experience and exposure associated with the location of residence, as well as their age, education, and employment on their level of awareness towards the significance of mangroves. Subsequently, the efforts of raising the awareness should set its target on the urban population with lower strata in the society whom mostly were blue-collar employees with the relatively lower educational profile. Campaigns and activities that are designed to raise the public's awareness towards mangroves must then be tailored according to these target groups.

KEYWORDS

- cultural significance
- demographic profile
- Dongchaigang Nature Reserve
- ecological services
- ecosystem
- indigenous people
- IUCN
- mangroves
- mangroves ecological services
- mangroves ecosystem
- Pearson Chi-Square test
- public awareness
- Pulau Kukup National Park
- Southern Peninsular Malaysia
- Sungai Pulai Forest Reserve
- Tanjung Piai National Park
- tourism

REFERENCES

Alongi, D. M., (2002). Present state and future of the world's mangrove forests. *Environmental Conservation, 29*(3), 331–349.

Amjad, A. S., Kasawani, I., & Kamruzaman, J., (2007). Degradation of Indus delta mangroves in Pakistan. *International Journal of Geology, 3*(1), 27–34.

Barbier, E. B., Sally, D. H., Chris, K., Evamaria, W. K., Andrian, C. S., & Brian, R. S., (2011). The value of estuarine and coastal ecosystem services. *Ecological Monographs, 81*(2), 169–193.

Duke, N. C., Meynecke, J. O., Dittmann, S., Ellison, A. M., Berger, U., Cannicci, S., et al., (2007). A world without mangroves? *Science, 317*, 41–42.

Giri, C., Ochieng, E., Tieszen, L. L., Zhu, Z., Singh, A., Loveland, T., Masek, J., & Duke, N., (2011). Status and distribution of mangrove forests of the world using earth observation satellite data. *Global Ecology and Biogeography, 20*, 154–159.

Goh, H. C., (2016). Assessing mangrove conservation efforts in Iskandar Malaysia. In: Cruikshank, J., (ed.), *Working Paper Series* (pp. 1-33). MIT-UTM Malaysia Sustainable Cities Program.

Gong, W. K., & Ong, J. E., (1990). Plant biomass and nutrient flux in a managed mangrove forest in Malaysia. *Estuarine, Coastal, and Shelf Science, 31*, 519–530.

Haines-Young, R., & Potschin, M., (2011). *Common International Classification of Ecosystem Services* (*CICES*) (pp. 1–17). Centre for Environmental Management, University of Nottingham, UK.

Horwitz, P., Finlayson, M., & Weinstein, P., (2012). *Healthy Wetlands, Healthy People: A Review of Wetlands and Human Health Interactions, Ramsar Technical Report No. 6* (pp. 1–114). Secretariat of the Ramsar Convention on Wetlands, Gland, Switzerland, and The World Health Organization, Geneva, Switzerland.

Kamaruzaman, J., & Haji, D. J. T., (2008). Managing sustainable mangrove forests in Peninsular Malaysia. *Journal of Sustainable Development, 1*(1), 88–96.

Kanniah, K. D., Sheikhi, A., Cracknell, A. P., Goh, H. C., Tan, K. P., Ho, C. S., & Rasli, F. N., (2015). Satellite images for monitoring mangrove cover changes in a fast-growing economic region in southern Peninsular Malaysia. *Remote Sensing, 7*, 14360–14385.

Khalid, S. (2009). *Move to restore mangrove forests - Community | The Star Online.* Retrieved February 1, 2019, from https://www.thestar.com.my/news/community/2009/11/03/move-to-restore-mangrove-forests/

Lewis, M. P., Simons, G. D., & Fennig, C. D., (*2014*). *Ethnologue: Languages of Asia* (pp. 1–558). SIL International, Global Publishing.

Mcdonald, R. I., Kareiva, P., & Forman, R. T. T., (2008). The implications of current and future urbanization for global protected areas and biodiversity conservation. *Biological Conservation, 141*(6), 1695–1703.

McGranahan, G., Marcotullio, P., Bai, X., Balk, D., Braga, T., Douglas, I., et al., (2006). Urban systems. In: Hassan, R., Scholes, R., & Ash, N., (eds.), *Ecosystems and Human Well-Being: Current State and Trends* (pp. 795–826). Island Press, Washington DC.

McLeod, E., Chmura, G. L., Steven, B., Rodney, S., Mats, B., Carlos, M. D., et al., (2011). A blueprint for blue carbon: Toward an improved understanding of the role of vegetated coastal habitats in sequestering CO_2. *Frontiers in Ecology and the Environment, 9*(10), 552–560.

Nobaya, A., Asnarulkhadi, A. S., Hanina, H. H., & Marof, R., (2012). The Seletar Community (Orang Laut) of Johore and the challenges against development. *International Proceedings of Economics Development and Research, 48*(30), 139–142.

Ratnayake, P. U., (2012). A collaborative approach between tourism and coastal communities: A present-day need and opportunity for mangrove management and conservation in Sri Lanka. In: Macintosh, D. J., Mahindapala, R., & Markopoulos, M., (eds.), *Sharing Lessons on Mangrove Restoration* (pp. 63–74). Mangroves for the Future, Bangkok, and IUCN, Gland.

Snedekar, S. C., & Brown, M. S., (1981). *Water Quality and Mangrove Ecosystem Dynamics*. US Environmental Protection Agency, Washington DC, EPA/600/4–81/022.

Spalding, M., McIvor, A., Tonneijck, F., Tol, S., & Van Eijk, P., (2014). *Mangroves for Coastal Defense: Guidelines for Coastal Managers and Policy Makers* (pp. 1–42). Wetlands International and The Nature Conservancy, Arlington.

Tinch, R., & Mathieu, L. (2011). *Marine and coastal ecosystem services: Valuation Methods and their Practical Application* (pp. 1–46). UNEP-WCMC Biodiversity Series No. 33. Cambridge, UK.

Van Bochove, J., Sullivan, E., & Nakamura, T., (2014). *The Importance of Mangroves to People: A Call to Action* (pp. 1–128). United Nations Environment Programme, World Conservation Monitoring Centre, Cambridge.

INDEX

T

T - #0807 - 101024 - C472 - 234/156/21 - PB - 9781774634455 - Gloss Lamination